REVOLUTIONARIES OF THE COSMOS

Revolutionaries of the Cosmos

The Astro-Physicists

I.S. GLASS
South African Astronomical Observatory

OXFORD
UNIVERSITY PRESS

OXFORD

UNIVERSITY PRESS

Great Clarendon Street, Oxford OX2 6DP

Oxford University Press is a department of the University of Oxford.
It furthers the University's objective of excellence in research, scholarship,
and education by publishing worldwide in

Oxford New York

Auckland Cape Town Dar es Salaam Hong Kong Karachi
Kuala Lumpur Madrid Melbourne Mexico City Nairobi
New Delhi Shanghai Taipei Toronto

With offices in

Argentina Austria Brazil Chile Czech Republic France Greece
Guatemala Hungary Italy Japan Poland Portugal Singapore
South Korea Switzerland Thailand Turkey Ukraine Vietnam

Oxford is a registered trade mark of Oxford University Press
in the UK and in certain other countries

Published in the United States
by Oxford University Press Inc., New York

British Library Cataloguing in Publication Data

Data available

Library of Congress Cataloging in Publication Data

Data available

Typeset by the author
Printed in Great Britain
on acid-free paper by
Biddles Ltd., King's Lynn

ISBN 0–19–857099–6   978–0–19–857099–8

1 3 5 7 9 10 8 6 4 2

PREFACE

As a teenager I derived great pleasure and inspiration from reading the book *Men of Mathematics*, written by a friend of the cosmologist Edwin Hubble, Professor Eric Temple Bell (1937) of Caltech. In colourful language he described the lives and achievements of the great mathematicians and I later regretted that nothing similar was available in my own field—astrophysics. This book was initially conceived to fill the gap but, as the project matured, I found that I preferred to take a more serious approach than Bell and therefore chose a smaller number of subjects.

Ultimately, I selected eight 'astro-physicists', each of whom made at least one important discovery by applying the methods of physical science to astronomy. Indeed, several dramatically altered the prevailing world picture with consequences that went far beyond science and became part of the intellectual foundations of their times. Their ideas ultimately affected the outlook of every thinking person, even if the details of what they had done were not always easy to understand. Though most numerate people are aware of Galileo, Newton, Hubble, and their contributions to science, in their own times the other five were also important figures, known even to the general public.

Today 'astronomy' and 'astrophysics' are almost indistinguishable. The latter word was introduced in the nineteenth century, somewhat gratuitously, by Huggins and his contemporaries in order to to emphasise the spectroscopic revolution that they were associated with. In fact, the development of astronomy has always been stimulated by ideas drawn from physics, whether theoretical or practical. Before the word 'astrophysics' came into existence, the phrase 'Physical Astronomy' was commonly used. In 1852, it had been the subject of a famous book, the *History of Physical Astronomy*[1] by Robert Grant, professor of astronomy in Glasgow. The hyphen I have used in the sub-title of this book is there to emphasise the fact that earlier physical astronomers than Huggins and his successors have been included.

Contributions to our understanding of the universe were also, of course, made by many others besides the subjects of the present book. A good few of them have been referred to in passing but there is a limit to what can reasonably be included. Though every particular development depended on the work of many predecessors, not all of these could be mentioned. It is therefore hoped that the lives of the chosen eight will illustrate in a general way the scientific and social

---

[1]Full title: *History of Physical Astronomy from the Earliest Ages to the Middle of the Nineteenth Century. Comprehending a Detailed Account of the Establishment of the Theory of Gravitation by Newton, and its Development by his Successors; with an Exposition of the Progress of Research on All the Other Subjects of Celestial Physics.*

attitudes of the times they lived in. English speakers may predominate but not a few significant figures could have been chosen from other cultural milieux.

The outline of each life is based on what seems to me to be the best available biography and I acknowledge here my debt to the works I have chosen. I have tried to paint each individual as a complete character, an ordinary fallible being who could sometimes be capable of irrational or even unpleasant behaviour. Material has also been gathered from research articles and books. Each person has been the subject of a considerable number of studies by astronomers and scientific historians: Galileo and Newton, in particular, have given rise to a substantial literature. For Galileo I have mainly followed Stillman Drake (1978): *Galileo at Work, His Scientific Biography*; in Newton's case, Richard Westfall's (1980) *Never at Rest*. The most comprehensive biography of William Herschel (and his indefatigable sister Caroline) is the *Herschel Chronicle* by Constance Lubbock (1933), one of his descendants. Huggins's memory and that of his wife are served, through various quirks of fate, by a very short biography written by C.E. Mills and C.F. Brooke (1936). Another valuable source has been the PhD thesis of Barbara J. Becker (1993) *Eclecticism, Opportunism, and the Evolution of a New Research Agenda: William and Margaret Huggins and the Origins of Astrophysics*. Hale is the subject of a biography by Helen Wright (1994), who was beholden to his family. Eddington's biographer, Alice Vibert Douglas (1956), was asked by his sister to complete a work which had been started by his best friend, C.J.A. Trimble. Shapley (1969) wrote an autobiography *Through Rugged Ways to the Stars* which, although useful and revealing, does not do him much justice. Hubble has been the subject of two full-length biographies, viz A.S. Sharov and I.D. Novikov, (1989) *Edwin Hubble, The Discoverer of the Big Bang Universe*, and Gale E. Christianson (1995) *Edwin Hubble, Mariner of the Nebulae*. I have followed the second of the two, which is the much the more comprehensive. Material taken from these books has been acknowledged and original references have been carried over.

I would like to express my gratitude to the following individuals for their help and advice at various times and in various ways during the realisation of this project:

Barbara Becker (University of California, Irvine), Mary Brück (Penicuik, Scotland), John Butler (Armagh Observatory), Shireen Davis (SAAO), Ian Elliott (Dublin), Richard French (Wellesley College), the late David Evans (University of Texas), Owen Gingerich (Harvard), David Glass (Dublin), Elizabeth Griffin (Dominion Astrophysical Observatory), Vincent Icke (Leiden), Ethleen Lastovica (SAAO). Dan Lewis (Huntington Museum), Jeff McClintock (Smithsonian), Maria McEachern (Harvard), Don Osterbrock (Lick Observatory), Piero Salinari (Arcetri), Auke Slotegraaf (Stellenbosch), Peter Spargo (Cape Town), Frances Stoy (Cape Town), John Strom (Carnegie Institution), Siep Talma (Pretoria), Cliff Turk (Cape Town).

I have also benefited from using libraries and archives at the following institutions: Cambridge University, European Southern Observatory (Munich), Harvard

College Observatory and Harvard University Archives (Cambridge, MA), Huntington Library (San Marino, CA), Institut d'Astrophysique (Paris), Liber Liber (www.liberliber.it), National Astronomical Observatory of Japan (Mitaka), Observatoire de Paris, University of Cape Town, South African Astronomical Observatory (Cape Town), Royal Society (London), Trinity College (Cambridge), Trinity College (Dublin), Wellesley College (Wellesley, MA).

I also wish to thank those who have supplied illustrations and permissions to use copyright material. For quotations of significant length I have attempted to obtain copyright clearance. Should I have inadvertently gone beyond 'reasonable use' I will be more than happy to correct any future edition.

References

Becker, Barbara J., 1993. *Eclecticism, Opportunism, and the Evolution of a New Research Agenda: William and Margaret Huggins and the Origins of Astrophysics*, Ph.D. Thesis, 2 Vols., Johns Hopkins University, Baltimore, MD.

Bell, E.T., 1937. *Men of Mathematics*, Simon and Schuster, New York, reprinted 1953, Penguin Books, London, 2 vols.

Christianson, Gale E., 1995. *Edwin Hubble, Mariner of the Nebulae*, University of Chicago Press, Chicago, IL.

Douglas, A. Vibert, 1956. *The Life of Arthur Stanley Eddington*, Thomas Nelson and Sons Ltd, London.

Drake, Stillman, 1978. *Galileo at Work, His Scientific Biography*, University of Chicago Press, Chicago, IL.

Grant, Robert, 1852. *History of Physical Astronomy*, Henry G. Bohn, London.

Lubbock, Constance A., 1933. *The Herschel Chronicle: The Life-story of William Herschel and His Sister Caroline Herschel*, Cambridge University Press, Cambridge.

Mills, C.E. and Brooke, C.F., 1936. *A Sketch of the Life of Sir William Huggins K.C.B., O.M.*, Privately printed, London.

Shapley, Harlow, 1969. *Ad Astra per Aspera: Through Rugged Ways to the Stars*, Charles Scribners' Sons, New York.

Sharov, Alexander S., and Novikov, Ivan D., 1993. *Edwin Hubble, The Discoverer of the Big Bang Universe*, tr. Kisin, V., Cambridge University Press, Cambridge.

Westfall, Richard S., 1980. *Never at Rest. A Biography of Isaac Newton*, Cambridge University Press, Cambridge.

Wright, Helen, 1994. *A Biography of George Ellery Hale: Explorer of the Universe*, American Institute of Physics, New York.

CONTENTS

1

INTRODUCTION: TALENT AND OPPORTUNITY

1.1 The astro-physicists

Galileo, Newton, Herschel, Huggins, Hale, Eddington, Shapley, Hubble: these eight astronomers were responsible for dramatic changes to the world-pictures they inherited. They showed that celestial objects are made of materials familiar to us on earth and that they obey the same physical and chemical laws. They displaced first the earth, then the sun, and finally the galaxy from being the centre of the universe. Yet the inspiration for their insights was drawn from ideas already in circulation. They could each, like Isaac Newton, have repeated the statement of Bernard of Chartres, who lived in the twelfth century, 'We are like dwarfs on the shoulders of giants, able to see further than they ... because we are carried high and raised up by their giant stature'. For the most part their advances can be attributed to the application of new ideas drawn from the simultaneously developing field of physics.

To the scientist, the modern era succeeded the medieval around 1600. The transition is strongly associated with the life and work of Galileo who broke away from the medieval worship of Aristotelian views by trusting his own observations and basing new conclusions upon them. Not only did he make progress in the field of motion, but he used the newly discovered telescope to show that the physical nature of the heavenly bodies is similar to that of the earth. His observation that Jupiter has moons demolished a favourite tenet of the Aristotelians, that the earth was the only centre of rotation in the universe. Further, his finding of the moon-like phases of the planet Venus demonstrated conclusively that it is in orbit around the sun and thus that Copernicus's previously debatable model was almost certainly correct. Kepler, a contemporary of Galileo, discovered the laws followed by planetary orbits. While this was an essential step in the process that led to Newton's theory of gravitation, his way of thinking was a hybrid of the mystical and the scientific.

Though in the early decades of the seventeenth century there were several who elaborated the finds of Kepler and Galileo, it was the masterly synthesis by Newton that laid the foundations of mechanics and universal gravitation. It was also he who began the study of the composite nature of light and developed the first reflecting telescope, even though he was not an observational astronomer as such. His picture of a rational universe, governed by fixed laws, was the stimulus for the first telescopic surveys of the sky undertaken by William Herschel in the second half of the eighteenth century. Herschel himself built large telescopes with unprecedented light-collecting power but his real contribution lay in his systematic examination of the sky. While incidentally this led to the discovery of the first planet not known to the ancients, his quantitative investigation of the Milky Way and his discovery of stars that were in orbit around one another were even more important. The latter showed that gravity was not confined to our own solar system but controlled the orbits of bodies far away in the depths of space.

In the nineteenth century, the application of laboratory spectroscopic methods to astronomical observations opened a new vista. It was now possible to find out the chemical compositions of celestial objects. The absorption lines of the sun's spectrum were first mapped by Fraunhofer in 1814. In the 1860s, following the demonstration by Kirchhoff and Bunsen that each line can be attributed to a particular chemical element, Secchi, Rutherfurd, and Huggins began to apply the new technique to the stars. Huggins, using one of the first spectrographs to be attached to a telescope, solved a long-standing puzzle by showing that many nebulous (cloud-like) objects are formed of gas rather than of closely packed stars. With improved equipment, the movement of stars along the line of sight could soon afterwards be measured.

Hale, starting in the last decade of the same century, believed that the way forward in astronomy was through a new kind of observatory that combined innovative instrumentation with giant telescopes capable of reaching deep into space or of studying bright objects in great detail. He was adept at persuading wealthy benefactors to support the construction of new observatories and instruments. Among these were the Yerkes Observatory of the University of Chicago and the Mount Wilson Observatory in California. The first contains the largest refracting telescope ever built and the second a reflector which was the world's largest for three decades. He himself invented or co-invented the spectroheliograph for investigating the physics of the sun's surface and advanced our knowledge of its magnetic field.

The twentieth century is universally acknowledged to have been a golden age for discoveries in physics, starting with Einstein's Special Relativity and Planck's Quantum Theory. Astronomy was no laggard either: Shapley was soon to discover that the centre of the Milky Way lies far from the Solar System and was able to form the first idea of our galaxy's size. When Einstein's General Theory of Relativity emerged during the First World War, Eddington verified it by observing the gravitational bending of starlight during an eclipse of the sun. Almost simultaneously, he was making his name by applying physical methods to understanding the interiors of the stars.

Still falling within the modern period, but 300 years after its beginning, was the last of the figures in this book, Edwin Hubble. He proved that the galaxies lie beyond the Milky Way and determined their distances. From his lifelong study of the redshift-distance relation he is regarded as the founder of modern observational cosmology.

Since about the time of Hubble's death in 1953, the number of astronomers has increased overwhelmingly and current progress is less the result of a few individuals than the interlinked activities of many people. Radio and space research have opened new vistas to astrophysics, leading to the discovery of many new phenomena, of strange objects, and of better ways to investigate the nature of space-time. The number of large observatories and the quantity of papers published annually have increased so much that an overall biographically orientated view of present-day astronomy would be excessively detailed. Though it is possible to name the dominant figures of recent years, some time will have to pass before a coherent history of the period can be written.

1.2   Setting the scene

As the Middle Ages came to a close, the number of those recognisable as scientists was small. The rare examples that existed were isolated and had to be self-reliant. A good education and a lifestyle that encouraged curiosity were available to very few. Such privileged people rarely came from land-bound or labouring families, whose lives were filled with long hours of work and who neither could read books nor had any encouragement to change their status. Only the most fortunate could travel abroad to study. Whereas today one receives a never-ending stream of new results from colleagues worldwide, three or four hundred years ago information diffused slowly. International contacts at a personal level were limited; travelling was expensive, uncomfortable, and dangerous.

Galileo, the first of our group, lived towards the end of the Italian Renaissance and was lucky enough to be the son of an intellectually inclined father; an amateur scientist interested in musical instruments who, unusually for the time, had carried out practical experiments on how string length and tension affected the pitch of a note. Newton, on the other hand, came from a family of working farmers, but one that could boast of some well-educated relations. Thanks to them, his potential was recognised and he was sent to study in Cambridge. That university was then at a low ebb but there were able scholars about who recognised and encouraged his talent. Herschel grew up on the fringe of a small German court in a musical family with an exceptional love for philosophical discussion and was largely self-taught. Huggins came from a 'dissenting' family, i.e. one that did not belong to the Church of England, then the state religion. Because of legal discrimination against them, particularly in regard to higher education and state employment, dissenters were highly conscious of the need to be well-educated and self-reliant. As a consequence, they were well-represented among the ranks of the engineers and scientists.

Though many other persons in history have possessed similar opportunities, security, physical ability, and energy to the people in this book, they often lacked the 'divine spark'. The successful seem to have had a heightened degree of 'physical intuition', the ability to perceive what is relevant and what is not when approaching a new question. They were able to recognise and tackle the kind of problems likely to lead to new developments and to seize opportunities as they arose. In every age there were others, often very able people, who became lost in the bye-ways of science or whiled away their lives polishing results already established. Thus, the ability to concentrate on new and fruitful ideas must be recognised as one of the characteristics that distinguishes the great scientist.

The teachers of the sixteenth and early seventeenth centuries believed that Aristotle and the other philosophers of classical times had known everything worth knowing. They saw no need to conduct experiments or to make observations themselves and claimed that all phenomena could be explained in terms of existing ideas. The magnitude of Galileo's task in overcoming the prejudices implanted by his education was thus immeasurably greater than what his followers had to face. Nothing characterised the times that followed more than a sudden willingness to acquire all sorts of practical knowledge, even if sometimes in a random and unsystematic way. Going beyond the mere accumulation of data, Galileo and other pioneers of the 'scientific method' not only gathered information but used it to build up rules that could predict new results.

1.3   The talented individual

It is salutary to note that many important scientific advances have followed technological ones. Kepler's laws could not have been discovered without the increase in precision of Tycho Brahe's instruments compared to those available to the ancients and the medieval Arabs. In this book we will come across several examples of scientists who were just one step ahead of their contemporaries. Though Galileo's application of the telescope was essential to his important discoveries, rivals were beginning to use the instrument for astronomy at the same time as him, but without his understanding and degree of success. Thus one cannot help enquiring whether a particular discovery was critically dependent on the individual who made it or if it was just waiting for the first suitably trained person to appear. On the other hand, the degree of opposition that often arose towards a new fact shows that contrarian thinking and a determination to proceed in spite of opposition from the established members of the profession were also essential.

Another significant personal trait is that many of the innovators in this book combined practical ability with their mathematical competence. The fact that they had a feel for materials and were not shy to use their own hands enabled them to design, build, and quickly modify the apparatus that they needed in their investigations. Galileo's experiments, in which he timed weights sliding down inclined planes, though they sound simple enough, would have required considerable improvisational and experimental skill—for example, split-second timing was no easy matter before the existence of stop-watches. Quite probably he made the lenses for his telescopes himself. Newton, following his construction of the first reflecting telescope, wrote enthusiastically about the techniques he used, from the casting of the metal to its grinding and polishing. Herschel made mirrors so large that he had to work on an almost industrial scale. He thought nothing of spending long hours at his lathe or his polishing machines. Such practical energy and a willingness to work hard were strong characteristics of these pioneers. The effort they invested in technical innovation was not just the pursuit of a hobby. It led to the development of unique equipment unavailable to rivals and almost inevitably to new discoveries.

1.4 Motivation

An important motivation for many scientists seems to be, and in the past to have been, the acquisition of a kind of immortality through their achievements. This was manifested by the obsessive concern for priority or publicity shown by almost all those in this book. How it expressed itself varied from person to person. Galileo was keen that his ideas should reach a wide audience and he often wrote in Italian for the benefit of 'intelligent laymen'. Newton, on the other hand, was not interested in making his most famous work, the *Principia*, easily understood by those he viewed as 'little Smatterers in mathematicks'. Instead he wrote an impressive treatise that required a serious effort to read, not to mention a knowledge of Latin, then the international language of the academic world. Later in life he notoriously argued his priority in mathematical discoveries above the rival claims of Leibniz. Herschel, uniquely among the group represented here, was admired for his modesty. He was satisfied to publish in scientific journals, but those who followed him wrote popular books and gave frequent public lectures to place their discoveries before a wide public. Some, like Hubble, who at one time went so far as to employ a publicity agent, carried self-glorification to an extreme.

Most undoubtedly had good opinions of themselves. For example, Galileo's self-image was revealed in this attack on a fellow scientist who used the pseudonym of Sarsi:

> I believe that [good philosophers] soar like eagles rather than fly about in flocks like starlings. It is true that, because of their rarity, they are little seen and less heard, while those that fly like starlings fill the sky with shrieking and noise, making a mess wherever they land ... Signor Sarsi, the rabble of fools who know nothing is infinite. Those who know very very little of philosophy are numerous. Few indeed are they who know even some small part of it well, and there is only One who knows all.[1]

Were the eight inspired by religious beliefs? None were dogmatic. Galileo seems to have been a fairly conventional Catholic, though willing to question the authority of a church which, during his lifetime, was becoming intellectually rigid as a consequence of the Counter-Reformation. Newton, a deeply religious person, made extensive researches into the works of the early Christian theologians and became an Arian—a Christian who denied the divinity of Christ. Of necessity he was secretive about his opinions, opening his heart only to friends with similar views. He generally avoided religious matters in his scientific works, but did describe his concept of a God, present throughout space, in the *General Scholium* to the second edition of his *Principia* (see Section 3.22). Herschel was strongly influenced by the latitudinarian views of Locke and other philosophers and seems

---

[1] Galileo 1623; Il Saggiatore (The Assayer), tr. ISG. See also Drake (1957).

to have been a deist—one who sees God in all nature. Huggins drifted away from his non-conformist Congregational background and was later described by his wife as 'Christian unattached'. Hale seems to have become an atheist; the same can be said of Hubble. Shapley showed a deep interest in religion but was far from being a conventional believer. Eddington was a sincere Quaker (member of the Society of Friends) as well as something of a mystic, but he was careful not to mix his religion with his science. The following quotation from *Science and the Unseen World*, a lecture series that he delivered at Swarthmore College, a Quaker foundation, shows that he kept science and faith in separate compartments of his mind:

> In the case of our human friends we take their existence for granted, not caring whether it is proven or not. Our relationship is such that we could read philosophical arguments designed to prove the non-existence of each other, and perhaps even be convinced by them—and then laugh together over so odd a conclusion. I think it is something of the same kind of security we should seek in our relationship with God. The most flawless proof of the existence of God is no substitute for it; and if we have that relationship the most convincing disproof is turned harmlessly aside. If I may say it with reverence, the soul and God laugh together over so odd a conclusion. (Eddington 1929, p. 43)

1.5   The need for support

Needless to say, a life devoted to research requires some form of financial support. Sponsorship for science in the past was even more *ad hoc* than it is now. Galileo relied on political patronage. He was paid by the Republican government of Venice and later by the Grand Dukes of Florence. In his younger years he operated a scientific instrument-making business to supplement his then rather meagre income. Newton during his productive years was a Fellow of Trinity College Cambridge. He later held a University Professorship. William Herschel started out as a professional musician and only following his discovery of Uranus was he able to pursue astronomy full-time, thanks to a pension provided by an appreciative King. Later in life, he became comparatively rich by making and selling telescopes but marriage to a wealthy widow was also a factor. William Huggins was a shopkeeper who sold an inherited business and became a *rentier*, though he was never as rich as some of the other nineteenth-century British amateurs. George Ellery Hale was the pampered son of a rich family with a private income besides what he received from his employment ... only the most recent of our group had the kind of 'career open to talent' that we are familiar with today.

1.6   Lifetimes in science

Though one thinks of life in the days before modern medical technology as having been short and uncertain, most of our eight lived to good ages. They tended to remain productive even when old; in this respect they form a contrast to mathematicians who are usually at their best before middle age. Galileo was about 35 years old when he made his dynamical discoveries and about 45 when he used the telescope to find the moons of Jupiter. His last major discovery, of the libration of the moon (see Section 2.19), was made when he was 73! Newton, more typical of a mathematician, was at his most original as a young man in the plague years 1665–1666 'For in those days I was in the prime of my age for invention & minded Mathematics & Philosophy more then than at any time since[2]'. He was then around 23. He ceased to be original when he was in his early fifties, though he was regarded as mathematically competent until his late seventies. William Herschel made one of his major discoveries—infrared radiation—at about 62 years of age. It is often felt today that many of those who remain in research as they get older become stale and may even have a negative effect on their fields by discouraging new ideas and blocking the advancement of the young. Others turn towards administration, sometimes with conspicuous success. Examples of each type occur among the eight in this book. While the 'third age' of a scientific life is usually of less significance in terms of original discovery than the other two, it is nevertheless a part of the individual's complete history and is thus worthy of being described.

References

Chandrasekhar, S., 1995. *Newton's Principia for the Common Reader*, Clarendon press, Oxford.

Drake, S., *Discoveries and Opinions of Galileo*, Doubleday, New York. This book contains *inter alia* a translation of *Il Saggiatore* (1623) [*The Assayer*].

Eddington, A.S., 1929. *Science and the Unseen World*, Allen & Unwin, London.

---

[2]Memorandum in the Portsmouth Collection (Cambridge University Library), quoted by Chandrasekhar (1995, pp. 1–2).

2

GALILEO: SEEING AND BELIEVING

Philosophy is written in the greatest book, one that stands open before
our eyes (I speak of the universe). But it cannot be comprehended with-
out first understanding the language and knowing the characters in which
it is written. That language is mathematics, and the characters are tri-
angles, circles, and other geometric figures. Without these, it is humanly
impossible to understand the words; without these, one wanders vainly
about in a dark labyrinth.[1]

Galileo Galilei was born in Pisa, the second city of the independent Italian Grand
Duchy of Tuscany, on or around 15 February 1564—a few days before the death
of his countryman, the artist Michaelangelo. The capital of the Duchy, Florence,
was just over a hundred kilometres away and had been the leading city of the
Italian Renaissance, though its most brilliant period was now past. Its dialect had
become the standard language of Italian literature and it remained the cultural
capital of a politically fragmented Italy. Social divisions in Tuscany, as elsewhere
in those times, were strong, though the leaders of society came from commercial rather than aristocratic families. At the top were the Medici, whose wealth
originally derived from banking, and who had dominated the state for most of
the previous 130 years. During Galileo's childhood the head of the family was
given the title 'Grand Duke of Tuscany', by Pope Pius V. There was a relatively
well-developed middle class of lawyers, businessmen, doctors, and academics but
most of the population of about one million people were peasants, attached to
the land, spending their lives tending grapes and olives and hardly witnessing
any change from one generation to another. Economically, the Grand Duchy's
income depended on the cloth trade. Many citizens worked in the numerous small
weaving and dyeing workshops. Shopkeepers, stall-holders, and artisans made a
modest living on the fringes. A person's status was immediately obvious from
his clothing. Homes were sparsely furnished, with perhaps hard chairs, a table, a
linen cupboard, and some beds. There were many clergy, monks, and nuns of the
Catholic church, who depended on the rest of the population for their support.
The Church was by far the largest organisation of the times, dominating life in
every way. Apart from taking care of the souls of the population, it provided for
various social needs, such as looking after the poor and the sick. It offered most
of the school-level education that was available. Far from being a monolithic
body, it was divided into many different religious orders, which could be bitter
rivals.

[1]Galileo, *Il Saggiatore*, 1623, tr. ISG. See also Drake (1957).

2.1 Early years

Galileo's parents were Vincenzio Galilei and Giulia Ammannati. Although his birth occurred in Pisa, the Galilei family had once been among the leaders of Florentine society: their ancestors included members of the governing body of the pre-Medicean Republic as well as professors and lawyers. Galileo's father, though far from wealthy, was cultivated and possessed many interests. As a good 'renaissance man', besides making his living in the cloth trade, he was an accomplished lutenist who also wrote on the history of music. He belonged to a group called the Camerata, active in the artistic life of the city. His treatise, published in 1581, *Dialogue Concerning Ancient and Modern Music*, dealt with the ancient Greek emphasis on voice in music and pointed in the direction followed by opera, which was invented in the following generation (Brown 1976). He is known to have conducted experiments to find out how the musical note produced by a vibrating string depended on its length and tension and it may be that that the young Galileo helped him in this work and received inspiration from it. It is interesting that he showed little reverence for established viewpoints, like his son. In his *Dialogue*, he wrote:

> It appears to me that they who in proof of anything rely simply on the weight of authority, without adducing any argument in support of it, act very absurdly. I, on the contrary, wish to be allowed to raise questions freely and to answer without any adulation [of authorities], as becomes those who are truly in search of the truth. (Allan-Olney 1870)

Galileo was first sent to a local school to learn Greek and Latin. Around the age of twelve he went to the Camaldolese (related to Benedictine) monastery of Vallombrosa, near Florence, for a literary education. At fifteen he was tempted to become a monk, but his father acted quickly to save him from the clutches of the Church, using the excuse that he was suffering from an eye disease to fetch him home.

Vincenzio planned that Galileo should, like him, become a cloth dealer, since this offered the possibility of earning real wealth and recouping the family fortunes. But his son was already showing signs of precocity in other directions. As a child he had constructed toy machines, as Newton was to do in his generation, and he had acquired from his father a taste for playing the lute, a recreation which he enjoyed throughout his long life. In addition, he was a passable artist and painter, an occupation he once claimed he might have followed if he had been allowed his own choice. In short, he was a suitable candidate for further education. His enlightened father eventually decided that he was more likely to be successful by becoming a medical doctor, even though the period of study meant a financial burden that the family could hardly afford.

2.2 Studies in Pisa

Thus in 1581, at the age of 17, Galileo was sent to study medicine at the University of Pisa. The course included Natural Philosophy, which we now call 'science'.

At that time, the universities were intellectually subservient to the Church. They had become fixated on the Aristotelian philosophical viewpoint and could tolerate no other. The Aristotelians, or 'peripatetics[2]', as they were often called, believed among other things that the earth was the centre of the universe and that the heavens in contrast were unchangeable. Aristotle had concluded that all terrestrial substances were made of earth, air, fire, and water; the heavens were made of something perfect but unspecified and their contents moved only in perfect circles, basically around the earth, which itself stood still! The use of new observations and experiments, followed by logical deduction, had not yet made a serious impact on mainstream scientific thought, though progress was being achieved in practical areas such as medicine and various technological disciplines. The conservative academic world believed that the wisdom of the ancient Greek and Roman philosophers could not be matched by living persons.

The 'standard' Aristotelian viewpoint was nevertheless steadily being undermined through the expansion and cheapening of the printed word, from its beginning in Germany during the previous century. New ideas began to percolate through Europe on a timescale that would have been inconceivable to the ancients. The Italians were among the leading publishers. In particular the Venetian firm of Aldus Manutius, specialists in editions of the Greek and Roman classics, was one of the most famous.

Galileo was an outspoken student and did not hesitate to question the views of his professors. He soon acquired a reputation as an inquisitive young man who was not content with parroting the standard teachings. Probably he was protected from serious chastisement by his status as a quasi-nobleman. His first contribution to physics was made about this time, when he recognised the value of the pendulum as a precise timekeeper, supposedly by using his own pulse to time a swinging lamp in the cathedral of Pisa, while his mind was wandering from the subject of some tedious sermon! He noticed that the time taken by one swing of a pendulum is essentially independent of the arc of the swing.

The real scientific awakening for Galileo came when he met Ostilio Ricci, a mathematician who was acting as a tutor to the court pages of the Grand Duke of Tuscany. He happened to hear Ricci presenting some lessons on geometry and was attracted by the beautiful logic of Euclid, the ancient Greek mathematician. Mathematically talented youngsters have often been inspired by Euclid's derivation of complicated and surprising geometrical results from simple axioms and Galileo, now aged 18, was no exception. Mathematics seized his attention and took over from medicine: to his father's dismay he dropped out of the University in Pisa in 1585 without taking a degree and left to join his family in Florence, the usual residence of the Tuscan court, to be able to spend more time with Ricci. At this time, his notes show that he still accepted the conventional Aristotelian viewpoint in physics.

---

[2]From the Greek word for walkers; a reference to Aristotle's habit of walking about as he taught.

2.3   Florence and Siena

In the next four years much of Galileo's time was spent in giving private mathematical tuition in Florence and Siena. His scientific investigations included an improvement to the design of the balance, designed to increase its precision. He also began to sort out the confusion that existed concerning the behaviour of falling bodies and the effects caused by the resistance of the media through which they were moving. It is believed that at this time he first became interested in the connection between mathematics and the measurement of quantities such as the mass and dimensions of bodies. During this period he was also invited to contribute to the Academy of Florence a talk on the geometry and sizes of the places described in Dante's *Inferno*! This should not really surprise us: Galileo was throughout his life fond of Italian literature.

In another of his early investigations, he repeated the work of the Greek physicist Archimedes. The latter is famous for having realised while in the bath how an alloy could be distinguished from pure gold—a discovery so exciting that he is reputed to have jumped out and ran naked into the street shouting 'Eureka, Eureka' (I have found it, I have found it)! The report of his procedure, as handed down by the Roman writer Vitruvius, did not make sense to Galileo and he set about finding out what the real method must have been. He devised a simple form of lever balance that could be used to determine the mass and volume of an irregular lump of metal using Archimedes' principle, that the apparent loss of weight when an object is weighed in air and then in water is equal to the mass of water displaced. By dividing the mass of the body by the mass of water displaced, he obtained the density of the metal and could tell whether it was pure or an alloy. This clear result he wrote up as a pamphlet which he circulated among his friends—the way that new scientific results were often announced in those days. In 1587 he visited Rome, the capital of a state which was governed directly by the Church, where he made contacts in the Roman College (Collegio Romano), the intellectual headquarters of the Jesuit order, which had been founded in 1534 as part of the Catholic reaction to the spread of Protestantism. Among those he met was Clavius (Christopher Clau), their leading mathematician, who had been behind the replacement in 1582 of the inadequate Julian Calendar by the 'Gregorian' one, named after Pope Gregory XIII. His early unpublished writings show that he learned much from the professors of the Roman College. One able mathematician who saw his work around this time was the Marquis Guidobaldi del Monte, a scientifically minded courtier who became his patron.

2.4   Professor at Pisa 1589–1592

By the age of 25, Galileo felt he was a worthy mathematician and he made several attempts to obtain a university professorship, applying for various posts that became vacant in Bologna, Padua, and Pisa, but without success. Nevertheless, his circle of influential friends was growing and, after two fruitless years, with the help of Guidobaldi and his brother, Cardinal del Monte, he at last in July 1589 obtained a three-year appointment as a professor in Pisa, albeit with a minimal salary of 60 florins per year.

Vicenzio, his father, died in 1591, leaving him as head of the family, with the duty of looking after his apparently rather troublesome mother, his irresponsible brother Michelangelo, and his sisters Virginia and Livia, for the latter of whom he had to provide a dowry of almost his total annual salary. She was to prove a strain on his finances for many years.

During this (Pisan) period of his life Galileo started to develop his ideas on mechanics, though he did not publish them until his old age, by which time they had been clarified and improved very much. He believed strongly in the importance of experimentation so that any individual could show for himself the truth of statements concerning physics. It is generally doubted that he actually dropped spheres of different sizes from the Leaning Tower of Pisa to show that they took the same time to fall, irrespective of mass. Aristotle's claim was that the heavier masses should have fallen more quickly. Galileo's notes show that he did, in fact, make many experiments with inclined planes of wood, timing the movement of metal balls as they rolled down. His equivalent of a stopwatch involved catching a constant stream of water in a container over the interval concerned and weighing it afterwards.

Galileo was something of an *enfant terrible* in Pisa and once was even fined for not wearing proper academic dress. In 1590, he wrote a satirical poem *Against the wearing of gowns (Contro il portar la toga)*, in which he suggested that clothing was the source of the world's evils and that humans should go about naked.

After the expiry of his Pisan contract, his friends recommended him for the chair of mathematics at the more prestigious University of Padua, in the Venetian republic. By this time, his reputation had grown to such an extent that he secured the appointment with relative ease.

2.5   Padua 1592–1610

As an old man, when reminiscing, Galileo considered his period in Padua to have been the happiest part of his life. His starting salary was 180 florins per year. It was there that he did his most important work on the behaviour of falling bodies and discovered other principles of kinematics (the laws followed by moving bodies without reference to causes). The cosmopolitan and intellectually free city of Venice, in whose territory Padua fell, was within easy reach, only 40 km away.

Scientifically, one of his most important innovations was to distinguish clearly between constant speed and acceleration, or rate of change of speed. Though this seems almost instinctive to the present generation, who are used to driving cars and readily understand coasting at constant speed or accelerating to a higher one, in Galileo's time the difference was not at all clear. As mentioned, it was believed that the weight or mass of a body affected the speed at which it fell, as also did its density and the density of the medium through which it was falling. The Aristotelians had no concept of inertia, the ability of a body to keep moving by itself, without the aid of some force to push it along. They thought that an arrow had to be pushed by the air which surrounded it. They also could not understand why, if Copernicus was right, and the earth was really speeding through space, objects should not fall off its surface. Among Galileo's achievements during his long life were new approaches to all these phenomena. Instead of searching for the *cause* of the behaviour of moving objects, he concentrated on formulating a mathematical description of their *actual* behaviour—what would nowadays be called a 'kinematical model'. The appropriateness of such a model could be judged by what it was able to predict. In this, he was the originator of the 'scientific method' in use at the present time, having broken away from the earlier heavy emphasis on logical deduction from initial assumptions. Part of his genius was a sort of 'physical intuition', an ability to see which properties were capable of being modelled and an ability to tell when a discrepant measurement was simply the result of experimental error. His realisation that a falling body accelerates uniformly with time, and that the distance it has dropped varies with the square of the time, occurred around 1603. This insight was central to the new mechanics.

By 1597 he is known to have been interested in Copernican astronomy. Before Copernicus, and indeed for more than sixty years after his death in 1543, the Aristotelian view that the earth is the centre of the universe prevailed. The movements of the 'perfect' heavenly bodies were described in terms of circles or transparent ('crystalline') spheres riding upon other circles or spheres (the notorious epicycles). For a precise description, a very complicated series of such epicycles had to be invoked.

Copernicus, a church official in Toruń, Poland, had seen that a very considerable simplification could be made if the sun was taken to be the centre of the planetary system. For example, the periods of revolution of the planets would then increase in accordance with their distances from the centre, which was not the case in the Aristotelian scheme. Unfortunately, it was difficult to produce evidence that his model was the correct one. He did not, of course, know that the actual planetary paths were ellipses and he had to continue with the use of epicycles to explain their non-circularity. Copernicus's treatise *De Revolutionibus ... or Concerning the Revolutions of the Heavenly Bodies* was published only as he lay dying in 1543. The preparations for its publication were undertaken by Andreas Osiander, a Lutheran minister, who feared an adverse reaction from the conventionally minded. He wrote an introduction, against the will of Copernicus, in which the new world-picture was passed off as a mere hypothesis that might be of some help in making calculations. Partly as a result, the impact of the work at the time was not what it should have been.

In 1596, Johannnes Kepler published the first of his books on planetary orbits, his *Mysterium Cosmographicum*, based on the precise observations of the Danish astronomer Tycho Brahe, which again demonstrated the simplicity of a sun centred planetary system.[3] He sent a copy of his work to Galileo, who thanked him and mentioned that he too was a Copernican:

> I have preferred not to publish, intimidated by the fortune of our teacher Copernicus, who though he will be of immortal fame to some, is yet by an infinite number (for such is the multitude of fools) laughed at and rejected.[4]

In fact, it is evident that he already believed that the tides could only be explained on Copernican principles. Rather strangely, he never seems to have digested Kepler's later discovery (1609) of elliptical orbits. It is true, however, that Kepler's style was very verbose and that many of his results were deeply buried in irrelevant material.

In 1599, Galileo had a salary increase to 320 florins and he acquired a mistress, Marina Gamba, who lived in a separate establishment and by whom he had three children, Virginia (1600), Livia (1601), and Vicenzio (1606). This relationship lasted until 1610, when he left to return to Florence. Such behaviour was quite acceptable in the easygoing moral atmosphere of the time. It is said that some university official even proposed raising his pay on the grounds that with a mistress his expenses would be greater! 'Si non è vero, è ben trovato'.[5] His life in Padua followed a regular pattern. He lectured daily at 3 p.m. and had numerous private pupils, including many foreigners, who often boarded with him. He loved company and enjoyed a glass of wine, which he once referred to as 'light held together by moisture'. In his household also lived a workman who was employed to manufacture for sale his 'military compass', a convenient and simple sector-like instrument with graduated scales that had applications in surveying and gunnery. Many were made, and besides the income derived from sales, he gave lessons in their use and sold a booklet containing instructions for them.

He found agreeable companions among the intellectually minded and enjoyed discussing new discoveries with these friends. Unfortunately, on a certain occasion in the summer of 1603, he and two acquaintances were taking a siesta in a cool room at a villa outside Padua when, supposedly as a result of exposure to a chilly draft, all three became seriously ill. One of the two subsequently died and the other lost his hearing. For the remainder of his life Galileo himself suffered from lengthy bouts of a sort of rheumatism or arthritis, especially when the weather changed. He was often laid up for months at a time and had to work from his bed.

The printed form of his booklet on the Military Compass came out only in 1606, and was dedicated to Cosimo de' Medici, the son of the reigning Grand Duke and his eventual successor, who had been taking lessons from Galileo. The following year, a certain colleague, Baldessar Capra, claimed that it was he who

---

[3] Tycho himself favoured a hybrid model in which the sun went around the earth but Mercury and Venus went around the sun.

[4] Galileo to Kepler, 4 August 1597, tr. Drake (1978, p. 41).

[5] Even if its not true, its a good story!

had made the invention, plagiarising both the compass and the book. Galileo had him taken before the governors of the University. Following an examination, it became clear that Capra did not understand what he claimed to have written, and he was duly expelled. The unsold copies of his book were confiscated.

Much of Galileo's time before the arrival of the telescope was spent in continued investigations of mechanics, strength of materials, hydrostatics, and magnetism. However, this steady stream of researching was interrupted in October 1604, when a supernova or stellar explosion occurred, creating a 'new' star of incredible brightness that stayed in the same position with respect to the 'fixed' stars (non-planets) until it faded many months later. A similar outburst had occurred in 1572 when he was a child. These two events were bad news for the Aristotelians: the heavens could no longer be regarded as perfectly unchangeable. Supernovae are, in fact, extremely rare: the two that exploded in Galileo's lifetime are the last known to have occurred in our own galaxy, the Milky Way.

Galileo enjoyed a good fight and had no hesitation in engaging his Aristotelian rivals in acrimonious debates. Given his aristocratic self-confidence and remembering his outspoken youth, it is not surprising that as a mature scientist he did not suffer fools gladly and could be excessively harsh towards those who opposed his ideas, which, we now know with the benefit of hindsight, could occasionally be wrong. He delighted in antagonising Aristotelian academic philosophers whenever possible. He took full advantage of the occurrence of the supernova to criticise them, giving popular lectures and composing a supposedly anonymous pamphlet in which two peasants made fun of the philosophers in their crude dialect.

2.6   The telescope

The invention of the telescope, or at least the first public knowledge of the instrument, occurred in the Netherlands in 1608, when Hans Lipperhey, a spectacle maker of Middleburg, attempted to obtain a patent for it. Galileo heard the news on a short visit to Venice in 1609. He set about experimenting with spectacle lenses, and soon succeeded in making an instrument for himself. His approach seems to have been quite primitive, since the theory behind lenses was developed only later, even though the technology of their manufacture had been known for over a century. The original 'Galilean' telescope was a simple device with a weak convex lens at the outer end and a strong concave one at the eye end (Fig 2.1).

Galileo, or possibly a nearby spectacle maker, probably made his lenses by cutting out round blanks from the clearest pieces of glass available and then grinding and polishing them. Spectacle makers had little need for real precision, but to make a good telescope lens the surfaces have to be true to shape. Tests carried out in recent times on five of his lenses that have survived show that they are of surprisingly high optical quality (Greco et al. 1992). The performance of his telescopes would mainly have been limited by their chromatic aberration— the inability to focus all colours to the same place. Two of the lenses have flat (plano) sides while the other three are curved on both. Grinding a lens surface

FIG. 2.1. Optics of a Galilean telescope, shown schematically.

to a satisfactory spherical shape is not very difficult because of the nature of the grinding process itself, but true flatness is very hard to attain. It is not surprising to learn that Galileo had to select among his production for the best-performing lenses when doing his most critical astronomical work.

Within a month he took an 8-power telescope to Venice and demonstrated it to members of the government of the Republic from the highest convenient location, the Campanile in Piazza San Marco. By its means they could spot an approaching ship two hours or more before it was apparent to the naked eye. It was no wonder that even the most elderly senators were prepared to climb the many flights of stairs to experience the new instrument. Conscious of the military significance of his invention and anxious to turn it to his advantage, he decided to make a present of a telescope to the Doge himself. He did not have to wait long for the expected result: while waiting to be admitted

> there appeared presently the Procurator Priuli, who is one of the gover-
> nors of the University. Coming out of the College, he took my hand and
> told me how that body ... would at once order the honorable governors
> that, if I were content, they should renew my appointment for life and
> with a salary of one thousand florins per year ... to run immediately.[6]

He continued to work on improvements to the instrument and by November 1609 had achieved a magnification of twenty times. By placing a mask over the convex lens to reduce its usable diameter he improved its definition somewhat, probably because the shape of its outer part was imperfect.

2.7 The nature of the moon

Galileo's greatest period of astronomical discovery then began with a systematic study of the moon, which he observed in all its phases over a month. He soon became aware that its surface was not perfectly smooth and spherical, as previously believed by the Aristotelians, but showed mountains, valleys, and plains. It was therefore very similar to the earth, and probably consisted of similar materials. He calculated that the mountains had to be higher than the highest known earthly ones. This discovery, that at least one heavenly body was made

[6]Galileo to Landucci (brother-in-law), August 1609, tr. Drake (1978, p. 141).

of familiar materials, rather than a mysterious Aristotelian 'fifth element', was a giant leap forward and led to the suggestion that they probably obeyed the same physical laws as terrestrial substances.

By 7 January 1610, he had also observed Jupiter and discovered its four main moons. This was a true revelation, because it showed that moons or little planets could revolve around a centre other than the earth, and it dealt another blow to the Aristotelian concept that the earth was the centre of all heavenly motion.

He lost no time in informing Cosimo de' Medici, his former summer student, who had just succeeded to the Grand Dukedom of Tuscany as Cosimo II. He had already shown the Duke how the moon appeared through the telescope:

> But, most marvelous of all, I have discovered four new planets and observed their proper and particular motions, differing among themselves and from the motions of the other stars; and these new planets move about another very large star [i.e. Jupiter] just as Venus, Mercury and perhaps[7] the other known planets move about the sun.[8]

The Grand Duke and his brothers were appropriately 'astonished by this new proof of his almost supernatural intelligence', as he reported.

He found that far more stars could be seen with the aid of the telescope than by eye. Another important discovery was that they had diameters too small to measure. Before his work, the bright stars were thought to have angular sizes that could be determined, making them seem to be quite close on a cosmic scale.

These discoveries had to be published without delay, because others were also known to be using telescopes for astronomical studies. In almost indecent haste he composed a treatise, to be called *Sidereus Nuncius*, meaning 'Starry Message' or 'Starry Messenger', which he decided to dedicate to Cosimo de' Medici (Fig 2.2). On the title page he described himself proudly as 'Patrician of Florence and Public Mathematician of the University of Padua'. The printing was complete by 19 March and the first available copy was sent to the Duke. The four newly discovered moons, or Jovian planets, were named by him the 'Medicean Stars'. Some 550 copies were printed and distributed far and wide with the help of the Florentine diplomatic service.

Within weeks, Galileo's work was known throughout Europe and he had become famous. However, the Aristotelian community refused to believe him, some even declining offers to look through the telescope to see for themselves.

2.8 Return to Florence

Unfortunately for his freedom in later years, Galileo felt homesick in Padua and longed to return to Florence, with its lively spirit, even though he was able to make frequent visits there during his holiday periods. The Grand Duke was

---

[7]This is a reference to Brahe's hybrid model, already mentioned, in which only the inner planets revolved around the sun.

[8]Galileo to Belisario Vinta, Secretary to Cosimo II de' Medici, 30 January 1610; tr. ISG. From Galileo, *Opera*, **10**, www.liberliber.it. See also van Helden (1989).

FIG. 2.2. Title page of *Sidereus Nuncius*, the book in which Galileo announced the discovery of Jupiter's moons. He describes himself as a 'Patrician of Florence' as well as 'Public Mathematician of the University of Padua'.

petitioned for an appointment, preferably without teaching duties, that might enable him to carry on with his fundamental researches. It was particularly requested that the title of such a position should not only be that of mathematician, but also of 'philosopher'. Following the successful result, he became 'Chief Mathematician of the University of Pisa and Philosopher to the Grand Duke of Tuscany' with an annual salary of 1000 florins but no actual obligation to reside in Pisa. His salary was one of the highest in the Duchy at the time, according to Biagioli (1990), who speculates that the Grand Duke was highly taken with the symbolism of Jupiter and its four moons as heavenly emblems of his family—the giant planet represented Cosimo I, the first of the Grand Dukes, and the moons or 'Medician stars' Cosimo II himself and his brothers.

The move away from Venice upset not only his friends, but also the rulers of the Republic, who felt he had shown a lack of proper gratitude, considering that he had just obtained life tenure of his university position at Padua and that his salary had recently been doubled. He stated in his defence:

To obtain from a republic, however splendid and generous, a stipend without public duties attached, is impossible. To get a useful income one has to satisfy the whole populace and not one particular individual. So long as I am capable of lecturing and serving, nobody can exempt me from this obligation while allowing me to be paid: and, in sum, I can only expect these benefits from an absolute ruler. (Galileo to 'S. Vesp.', February 1609, tr. ISG)

His best Venetian friend, the nobleman Giovanfranscesco Sagredo, who he immortalised as one of the protagonists in his later 'Dialogues', wrote to him to express 'inconsolable unhappiness' at his departure. He pointed out, very presciently as it later emerged, that Galileo was leaving the certainty of a well-governed republic for the unpredictable caprices of a prince who could change his mind or even die at any moment. Furthermore, the Jesuits, who were the 'thought police' of those religiously troubled times, were very influential in Florence, whereas Venice had seen fit to expel them some time before when they had attempted to interfere with the university. Sagredo had even been threatened by another of his friends that he would not speak to him again if he continued his friendship with the disloyal Galileo. Overall, he was clearly concerned that Galileo had made a serious mistake.

Galileo took up residence in Florence in September 1610. He left Marina in Padua and tried to have his two daughters placed in a convent. However, they were too young and had to remain with him until 1613 when, with the help of a friendly cardinal, he managed to get them into the convent of the Poor Clares, known as San Matteo, at Arcetri, nearby. Virginia became Sister Marie Celeste and Livia became Sister Arcangiola. Conditions in the convent were rather grim, as indicated by the letters from the grown up Sister Marie Celeste to Galileo. They describe her bad health, probably due to poor living conditions and meagre food. The Poor Clares were not allowed to own property and were completely dependent on charity.

Although Galileo's action seems cruel to us, it was not unusual at the time to place middle- and upper-class girls in a convent, as there was no future for them other than marriage. In Galileo's case, his daughters would have needed large dowries to be marriageable, especially in view of their illegitimate births. With all his other responsibilities towards his family, he probably did not have the wherewithal for them to consider matrimony.

2.9 Mature scientist

Settled now in Florence and with his mistress and daughters off his hands, Galileo could dedicate more of his time to science. Carrying on with his telescopic observations, he found that Saturn always seemed to be accompanied by two smaller round planets, one on each side. It was only after several decades had passed that the Dutch astronomer Christiaan Huygens revealed this appearance to be the degraded view in a low-resolution telescope of the planet's now well-known ring system. Galileo also saw for the first time the moon-like phases of Venus, which showed conclusively that it was in orbit around the sun. Surprisingly, to the

FIG. 2.3. Engraving by N. Schiavoni of a portrait of Galileo by Domenico Tin-
toretto painted around 1610–1615 (Abbot, 1937).

naked eye, Venus looks brightest when at crescent phase and faintest when full.
It had required the increased resolution of Galileo's telescopes to make out the
changes in its shape.[9] A former pupil, the Benedictine monk Benedetto Castelli,
thought that this discovery would be sufficient to convince the most obstinate of
the validity of the Copernican viewpoint, but Galileo was not at all optimistic:

> Did [he] not know that the demonstrations produced earlier were sufficient
> to convince those who can reason and want to know the truth, but that
> to convince those who are obstinate and care only for the empty applause
> of the stupid mob the witness of the stars come down to Earth to speak
> for themselves would not suffice? (tr. Sharratt 1994, p. 90)

In March 1611 he again visited Rome. His discoveries had been investigated
and confirmed by Jesuit scientists of the Collegio Romano (Roman College). He
was received with great honour by leading churchmen and the nobility; he had
a friendly interview with the Pope and was fêted at the Roman College.

---

[9] Phases would also have occurred if Brahe's hybrid model had been right; Galileo's observations therefore did not rule it out.

Only Cardinal Robert Bellarmine, a man of great personal holiness, famed for his definition of Roman Catholic doctrine in his three-volume work of 1581, *Disputationes de Controversiis Christianae Fidei Adversus hujus Temporis Haereticos*,[10] and made a saint in 1930, was suspicious of his views, making discreet enquiries as to his orthodoxy. As Bellarmine was head of the Roman College and also a leading member of the Roman Inquisition, the tribunal which decided on matters of religious orthodoxy, this was potentially a serious matter. He had been a member in 1600, when it condemned the thoroughly heretical philosopher Giordano Bruno to death for advocating among other things that there could be other inhabited worlds besides our own. He was a scholastic theologian of the most uncompromising kind and had spent his whole life in combatting heresies (Westfall 1989), a fundamentalist for whom truth lay in the literal words of the Bible. Bruno, feeling to the end that he was right and the Inquisition had been unjust, told the Cardinals of the Inquisition 'You must feel more fear in pronouncing this sentence than I do upon hearing it'. Drake (1980) believed that these frightening words may have given Bellarmine pause later when he had to deal with Galileo's case.

Some years before, a new academy, the Accademia dei Lincei (Academy of the Lynx-eyed; Fig 2.4) had been founded by a Roman nobleman, the Marquis Federigo Cesi, and some of his friends. Several academies existed for various intellectual purposes in Italy about this time, and were a very important development in that they provided a route for the rapid dissemination of new ideas. They were the precursors of the French Academy and the Royal Society of London. The Lincei elected Galileo as its sixth member and provided him with intellectual companionship as well as an avenue for publishing his discoveries.

On his return to Florence, Galileo published, at the Grand Duke's request, a *Discourse on Floating Bodies* (Galileo 1612) which showed in the Italian language and in the clearest and most easily verifiable manner, that many of Aristotle's assertions were completely incorrect. He cleverly anticipated and demolished the counter-arguments of the academic philosophers and their supporters among the clergy in a way that even lay people could understand. This did not, of course, increase his popularity in university circles.

2.10   Sunspots

Simultaneously, other astronomers had been using the telescope to explore the sun. Several of them had independently discovered the existence of sunspots (another unwelcome sign of imperfection in the heavens!). One, a Jesuit priest called Christopher Scheiner, of Ingolstadt, published some letters anonymously, under the name of Apelles, through a merchant friend Mark Welser of Augsburg.[11] After Welser published Scheiner's letters, Galileo replied with his own *Letters on*

---

[10] *Disputations concerning Controversies between the Christian Faith and the Heretics of our Times.*

[11] Appelles was a painter at the time of Alexander the Great who hid behind a screen in order to hear what people were really saying about one of his works.

FIG. 2.4. The emblem of the Accademia dei Lincei of Rome, which Galileo joined in 1611.

Sunspots, later published by the Accademia dei Lincei (in 1613). Apart from his claims to priority, these contained many other observations of interest, including the sizes, durations, and latitude distribution of sunspots, the rotation of the sun, the phases of Venus, and the supposed triple nature of Saturn. It incidentally mentions a view that comes very close to Newton's law of inertia (that a body continues in a state of rest or motion in a straight line unless acted on by an external force). He also showed that he was not against Aristotelian philosophy as such, but only the then current over-literal interpretation of it:

> Now, to harvest some fruit from the unexpected marvels that have stayed concealed until this age of ours, it will be well if in the future we turned an ear to those wise philosophers who think differently from Aristotle about celestial material. The self-same Aristotle would not have distanced himself so far from their point of view if his knowledge had included our present sensory evidence, since he not only admitted manifest experience was among the most powerful ways for forming conclusions about the problems of natural philosophy, but even considered it the most important. So while he argued the immutability of the heavens from the fact that no alteration had been seen in them at any time, it is credible that if his senses had shown him what is now clear to us, he would have followed the [contrary] opinion to which these wonderful discoveries have now led us. (From Galileo, *Second letter on Sunspots*, tr. ISG)

and later:

> The only people who remain in opposition [to my work] are some strict defenders of every minute point made by the Peripatetics [followers of Aristotle], those who, so far as I can see, were educated and nourished from infancy on the opinion that philosophising is and can be nothing other than making a big survey of the texts of Aristotle ... They want never to have to raise their eyes from those pages—as if the great book

of the universe had been written to be read by nobody but Aristotle, and his eyes alone had been destined to see for all those who followed him. (From Galileo, *Third letter on Sunspots*, tr. ISG)

This work was completed a few miles to the west of Florence at the Villa del Selve in December 1612. Its owner was Filippo Salviati, another close friend, a member of one of the Florentine banking families, whose name Galileo later used for the spokesman for his own views in his two famous books, written in the form of dialogues.

2.11 Beginnings of clerical opposition

At about this time Galileo's views and those of Copernicus were attacked in conversation as hostile to scripture by a conservative Dominican friar called Lorini, a Florentine noble. He showed himself rather ignorant of the true issues and afterwards made an apology, but nevertheless, it was a warning sign. His later actions show that the apology may not have been sincere.

In 1613, Castelli, perhaps Galileo's most successful pupil, became Professor of Mathematics at Pisa. He was informed by the overseer of the university that he was on no account to mention the motion of the earth and other topics contrary to Aristotle's teaching in his lectures. However, he was soon afterwards drawn into a conversation while at dinner in the Ducal court, during which he had to explain Galileo's discoveries. To his consternation, the Grand Duchess Christina, mother of the reigning Duke Cosimo II, together with a member of the philosophy faculty, began to question him about the scriptural orthodoxy of Galileo's views. For the moment, he succeeded in defending his teacher's position with the help of sympathetic courtiers present.

The real beginning of Galileo's troubles came when another Dominican priest, Thomas Caccini, denounced Galileo, Copernicus, and mathematicians in general as diabolical, enemies of the Christian religion, and a danger to the state in a sermon delivered on 21 December 1614. He facetiously commenced with the biblical words 'Men of Galilee, why do you stand looking into heaven?' (Acts I, v. 11). Caccini, in the words of the Galilean scholar Stillman Drake (1957), was the true villain of the piece. His attack was deliberate and preconceived. He was ambitious, wrongheaded, bigoted, and willing to make unsupportable statements under oath. The preacher-general of the Dominicans apologised to Galileo, but the damage had been done. His Roman friend Cesi of the Lincei warned him to be careful in how he responded to the attack. Cardinal Bellarmine was now on the warpath and his anti-Copernican views were well-known.

Lorini, the previous Dominican opponent of Galileo, now became more convinced than ever that he was a heretic and a danger, especially having seen a letter to Castelli, in which he discussed his views on the proper relationship between religion and science. He went so far as to write to the Inquisition in Rome, also known as the Holy Office, requesting them to investigate the views of Galileo and his followers. Their initial response was to ask Castelli to send them the original of this letter.

At about the same time, towards the end of 1615, a book explicitly supporting Copernicus was published by a Carmelite priest called Foscarini. Cardinal Bellarmine read this work and did not like it. He more-or-less insisted that Copernican theory should be treated as purely hypothetical, as a mere calculating tool. He made the point that the orthodox Catholic interpreters of the Bible were quite explicit about the fact that the sun moves and the earth stands still; to claim any other interpretation was prohibited by the Council of Trent, which had defined acceptable Catholic belief in the middle of the previous century. The hypothetical approach was in effect an offer to compromise, but it was not one that appealed to Galileo.

2.12   Science and scripture

Galileo decided to visit Rome, arriving in December 1615, in order to promote his views, in spite of warnings that it would have been more prudent to keep a low profile. He knew that, unlike the academic philosophers, many of the clergy were open-minded or sympathetic to him and felt that he could save the Church much embarrassment if obvious truths could be made to prevail. To explain his philosophical position and his genuine concern for the reputation of the Church in detail, he had written during that year his *Letter to Madame Christina of Lorraine, Grand Duchess of Tuscany, Concerning the Use of Biblical Quotations in Matters of Science* (see Drake 1957). He showed that his accusers were denying simple facts and were trying to refute them by quoting passages from the Bible which they clearly had not understood. At the same time, he pointed out that some of his enemies were purely malicious and were trying to do damage to his reputation. He explicitly stated:

> They know I hold concerning the constitution of the universe that the sun, without changing its place, stays situated in the center of the revolution of the celestial orbs, while the earth, rotating on its own axis, revolves about the sun. (From Galileo, *Letter to the Grand Duchess Christina*, tr. ISG)

He went on that to deny facts that anybody can verify for himself would make the Church look foolish to non-believers. The translators and interpreters of the Bible had obviously tried to make it understandable to ordinary uneducated people to whom it *seemed* that the earth stands still while the heavens move; the texts should be examined more closely to see if other interpretations were possible. Throughout, he supported his views with quotations from the early Fathers of the Church and classical sources, explaining that even they should not always be taken literally.

As expected, Galileo received considerable support from the more progressive churchmen, to such an extent that even Cardinal Bellarmine seems to have wanted to avoid a real confrontation. However, the controversy became annoying to the reigning Pope, Paul V, a notorious anti-intellectual, who asked the Inquisition in February 1616 for a ruling on the matter of the movement of the earth and the stability of the sun. As might have been predicted, the reply had to be that Copernicus's views were probably contrary to the Bible. In consequence, the Pope told the Cardinal to inform Galileo that he could no longer hold or defend such views and that, if he refused, he was to be threatened with imprisonment. In the event, there was no choice but to agree. Even now, on a personal level, he was treated in a fairly friendly manner by both Pope and Cardinal. To counter rumours that he was in bad grace with the Church, Bellarmine issued a certificate to show that he had neither been forced to abjure nor to perform any penance: he only had been told that Copernican views were contrary to Holy Scripture and could not be defended or held. Galileo interpreted the outcome as a mild rebuke which would allow him to treat Copernicanism at least as a 'useful hypothesis'. However, the unfortunate Foscarini's book was placed on the Church's official *Index of Prohibited Books* and Copernicus's work was forbidden reading 'until corrected' to make its assertions more hypothetical. Merely 'forbidding', rather than absolutely condemning, Copernicus was the suggestion of two liberal cardinals, one of whom was Maffeo Barberini, later Pope Urban VIII. The corrections to Copernicus's book were announced in 1620, but were disregarded outside Italy (Gingerich 1982).

At Galileo's trial before the Inquisition seventeen years later, a document was produced which appeared to show that he had, during the interview with Bellarmine demanded by the Pope, promised to obey a much harsher injunction: that he was not to hold, teach, or defend Copernicanism in any way, verbally or in writing. During the trial Galileo claimed that he had no recollection of any such promise, and indeed there is evidence that it may have been a forgery or a document prepared by the more extreme faction but never made use of. Unfortunately, many of those who could have shed light on the matter were no longer alive by then.

2.13   Towards the Ptolemaic–Copernican 'Dialogue'

Following his return from Rome to Florence, Galileo worked for a while on a proposal for improving navigation. No precise clock existed at the time, let alone one capable of being used as a ship's chronometer. Since a knowledge of the time was essential to determining a ship's longitude before the advent of satellite-based navigational aids, he proposed that observations of the positions of Jupiter's moons could be used, together with tables calculated by himself, for this purpose. Although he negotiated with the Spanish court, he was unsuccessful in 'selling' his idea.

In 1617 he moved to a house at Bellosguardo, in the hills outside the city. Towards the end of the year he fell seriously ill and had to take to his bed. When he recovered, in May 1618, he made a pilgrimage in to Loreto, a manifestation of what seems to have been a genuine, though rarely expressed, religious belief. His son, Vicenzio, was legitimized by the Grand Duke in 1620, an important improvement to his status, especially if he hoped to get married to a middle- or upper-class girl.

When he was fully settled in the countryside he began to prepare a systematic review of the theorems and discoveries that he had made as a young man, his pupils Mario Guiducci and Niccoló Arrighetti assisting him, since he was ill much of the time. Three comets appeared in 1618 and Guiducci published an account of two lectures he had given about them—*A Discourse on Comets*—based on Galileo's ideas. The work of a Jesuit, Father Horatio Grassi, Professor of Mathematics at the Roman College, was heavily criticized. This had two undesired results: an increased enmity towards Galileo on the part of many Jesuits and a book under the pseudonym of Sarsi, but written in fact by Grassi, *The Astronomical and Philosophical Balance,* containing a bitter attack. The 'Balance' was to weigh his opponents' arguments. Galileo replied in kind in 1623 with a heavily sarcastic book called *The Assayer* (*Il Saggiatore*) (see Drake 1957), again using the weighing metaphor, but with the implication that *his* balance was a much more accurate instrument than Sarsi's. In it he made an important statement about scientific method:

> It seems to me Sarsi firmly believes that in philosophising one must bolster oneself upon the opinions of some famous author, as if our intellect, when not married to the discourse of another person, must remain completely sterile and unfruitful. Possibly he thinks that philosophy is a book based on somebody's imagination, like the *Illiad* or *Orlando Furioso*, works in which the least important thing is whether what is written is true. Well, Sarsi, it is not like that. Philosophy is written in the greatest book, one that stands open before our eyes (I speak of the universe). But it cannot be comprehended without first understanding its language and knowing the characters in which it is written. That language is mathematics, and its characters are triangles, circles, and other geometric figures. Without these, it is humanly impossible to understand the words; without these, one wanders vainly about in a dark labyrinth. (From Galileo, *Il Saggiatore*, 1623, tr. ISG)

The *Assayer* was to be published by the Academy del Lincei in Rome. However, just before its printing was completed, Cardinal Maffeo Barberini, one of the liberal cardinals in 1616, and a former member of the Lincei, was elected Pope as Urban VIII.[12] The book was hastily dedicated to him in the expectation that a newer and more rational age had dawned. The Pope had *The Assayer* read to him at table and was pleased by it. He was soon visited by Galileo. The latter was promised a pension for his son and given some valuable presents. Galileo came away feeling that he was well appreciated and could freely start to write a book which would be the definitive comparison of the Copernican and Ptolemaic systems, even if he had to avoid an explicit decision in favour of one or the other.

---

[12]This was the same Pope who is thought to have ordered the melting down of the bronze beams of the ancient Pantheon to make the baldacchino or altar canopy of St Peter's Basilica. Afterwards it was said 'Quod non fecerunt barberi, fecerunt Barberini'—What the barbarians did not do, the Barberini did.

From this time until 1634 we catch occasional glimpses of his private life through letters written to him by his daughter Marie Celeste (Virginia) from her convent. These are composed in an overly respectful, often obsequious, style. Although we have none of his side of the correspondence, it is evident that he wrote at times on a daily basis, though his visits were none too frequent. She frequently mentions sending simple gifts such as a few pieces of fruit or marzipan, cinnamon water, and, on one occasion, a rare rose out of season. He in turn sends wine, a lamb, pieces of cloth, or gifts of money. She makes and bleaches cuffs and collars for Galileo and her brother Vicenzio. She asks for his help in getting a better class of confessor for the convent than the loose-living, grasping, and ignorant priests they were having to put up with. In one letter she tells her father not to risk his health by working too much in his garden—he should deprive himself of this pleasure as a Lenten sacrifice!

The preparation of the future *Dialogue Concerning the Two Chief World Systems—Ptolemaic and Copernican* (Galileo 1632) was under way by 1625. One of the central problems that Galileo faced was that of 'indivisibles'. A famous Greek paradox asked how a body could move at all: for, before it could have got a certain distance, it had to have gone half that distance; before that it had to have gone quarter; and so on; so, how did it ever get started? Similarly, a body starting from rest and acquiring a certain speed had to have passed through all intermediate speeds. This question he was able to address to some extent, but its real resolution came in the hands of others later in the century, when the mathematical concepts of continuity and infinite series were understood better. It is of central importance to the differential calculus developed later by Leibniz and Newton.

Around 1627, Galileo's difficult brother Michelangelo decided he would like to return to Florence from Munich, where he was a musician and had been safely out of the way. He sent only his wife at first, to be Galileo's housekeeper, but his whole family soon followed and an unsatisfactory domestic situation arose. Fortunately, this experiment came to an end in the following year and the family returned to Munich. It was arranged that Michelangelo's son, another Vicenzio, was given the pension intended for Galileo's own son and packed off to Rome to learn music. Galileo's former student Castelli took an interest in him, helping him with finding a place to stay. The boy turned out to be a wastrel and, what was even worse, he was contemptuous of religion: 'he could not see why he should join others in worshipping a painted wall' (Sharratt 1994). His landlord threatened to denounce him to the Inquisition: he was likely to be burned alive like Giordano Bruno. Galileo himself would be harmed by association. The solution, following a proposal by Castelli, was to remove him altogether from Rome.

The other Vicenzio, Galileo's own son, who Sister Marie Celeste called a calm and wise young man, married in 1629 a girl of a good Florentine family.

As late as February 1630, Galileo was held in high enough esteem that the Pope gave him two prebends or official church positions carrying small salaries. He was admitted to holy orders by the Bishop of Florence the following year.

FIG. 2.5. Galileo in 1624, sketch by Ottavio Leoni in the Louvre (RMN ©;
Photo: Michèle Bellot).

However, he was not expected to wear a clerical outfit or actually perform any
serious duties in connection with these offices.

In May and June 1630, Galileo visited Rome, staying with the helpful and
friendly Florentine Ambassador, Francesco Niccolini, with the intention of seeing
the *Dialogue* through the press. He was well received in most circles and it
appeared at first that the book was going to encounter little opposition. It was
to have been published by the Linceans. Unfortunately, however, Prince Cesi,
their leading spirit, died intestate just after Galileo left Rome, and the Academy
fell apart. To make matters worse, in September a serious outbreak of bubonic
plague occurred in many Italian cities. Florence lost about one in eight of its
inhabitants during the first year of the outbreak and it continued at a less severe

level for two years more. Because of quarantine regulations, travel was forbidden.

2.14 Publication

As a result, it was decided to publish the *Dialogue* locally in Florence and the Florentine censor was asked to examine the manuscript. Galileo believed that his (mistaken) theory of the tides, which supposedly arise from the composite Copernican rotation and revolution of the earth, was a conclusive proof of Copernicanism. He was told not to mention the tides in the title, and was reminded that all discussions of Copernicanism were to be on a hypothetical level. The Pope's own arguments concerning divine omnipotence were to be included in the ending of the book.

In his introduction 'To the Discerning Reader', after mentioning that he had had the approval of the Roman censors for his supposedly hypothetical view of Copernicanism, he wrote, taking a dig once again at the Aristotelians:

> I have taken the Copernican side in the discourse, proceeding as with a pure mathematical hypothesis and striving by every artifice to represent it as superior to supposing the earth motionless—not indeed, absolutely, but as against the arguments of some professed Peripatetics. These men indeed deserve not even that name, for they do not walk about; they are content to adore the shadows, philosophising not with due circumspection but merely from having memorised a few ill-considered principles. (Galileo 1632)

The 'dialogue' in *Dialogue Concerning the Two Chief World Systems* was between two characters with the names of old (but now dead) friends of Galileo and an unreformed but rather dense Aristotelian called Simplicio, named after an early commentator on Aristotle who had actually existed. Galileo's own views were represented by Salviati, and were supported by Sagredo, the host, an intelligent gentleman of Venice, who had however to have many matters explained to him. Written in Italian instead of the Latin then customary for scholarly works, the *Dialogue* was intended to be read by interested laymen. This 'dramatic' method of presentation appealed to Galileo but in a caustic review of his other dialogue, the *Discourse*, some years later, René Descartes, the French mathematician and philosopher, complained 'his fashion of writing in dialogues, where he introduces three persons who do nothing but exalt each of his inventions in turn, greatly assists in [over]pricing his merchandise [by making the book much longer!]' (Drake 1978, p. 388). In fact, it is a rambling and slow-moving work, rather irritating to the modern scientific reader, who is used to precise descriptions of experimental results expressed in unemotional language.

The contents are divided into four 'days', on the first of which most of Galileo's previously described ideas are put forward in more detail and contrasted with rigidified Aristotelian viewpoints. The literal followers of Aristotle are mocked, as believers in the words of an out-dated philosopher who are unable to see the completely obvious. The second day contains a description of what is called today 'Galilean Relativity'—the idea that carrying out an experiment in, for example, a smoothly moving ship's cabin does not affect the outcome.[13] Similarly, since everything on the surface of the earth is moving smoothly as the earth turns, this movement is not noticeable. Dropping an object from the mast of a moving ship, one sees that it falls at the foot of the mast and is not affected by the ship's motion. The third day is devoted to the detailed case for Copernicus and the fourth deals with Galileo's erroneous theory of tides, ending, as required, with the Pope's argument, which amounted to saying that God was all-powerful and could cause the tides to operate just how he liked!

The printing of an edition of a thousand copies was started in June 1631. On completion, it was an instant success and was sold out within a few months.

2.15   Private life

The letters from Sister Marie Celeste (Galileo's daughter) during these years contain more detail than her earlier ones. Although they are full of stories of dying nuns, small gifts and requests for money and good wine, the personal life of Galileo himself is more frequently reflected. A practical man, he helped, for example, by mending a clock for the convent and he glazed a window for his daughter's room. Vicenzio, her brother, and his wife and firstborn, little Galileo, lived with his father or nearby and visited her at times. She again castigates her father often for spending too much time in his garden, which she was sure was the cause of his bad health, and was also concerned that he might be working too hard on his *Dialogue*:

> [6 April 1630] 'And I would not want you, while seeking to immortalise your fame, to cut short your life; a life held in such reverence and treasured so preciously by your children, and by me in particular.' (tr. Sobel 2000)

It is clear that their relationship had developed over the years, becoming one of loving and caring for each other.

In March 1631, Galileo's brother Michelangelo died, leaving him to find support for the young family. He evidently attempted to put his niece La Virginia in the same convent as Marie Celeste, but was rebuffed, partly on the grounds of the poverty of the institution and partly because of rules regarding the admission of relatives. Michelangelo's son, Vicenzio, expected an allowance from him also.

Marie Celeste kept a close eye at this time on the property market, hoping to persuade her father to move nearer to her convent. She mentioned various available houses, until in September 1631 he settled into a villa called 'Il Gioiello' (The Jewel), in Arcetri (Fig 2.6). This house was to be his residence for most of his remaining years. It still stands today.

When he was away from home, especially during the troublesome period of his trial in 1633, his daughter took over the administration of the villa, in spite of being confined within the walls of her convent. She acted by instructing Galileo's servants (a housekeeper and a serving boy) and friends as to what to do. One

---

[13]Einstein's 'Special Relativity' deals with observations of systems moving with very high velocities, comparable to the speed of light, relative to an observer.

FIG. 2.6. Galileo in the courtyard of Il Gioiello, where he lived from 1631 onwards, mostly under house arrest. From a print dated 1833 by G.B. Silvestri. (Arcetri Astrophysical Observatory)

learns about the quite extensive domestic activities there: the growing of fruits such as lemons, oranges, plums, pears, and pomegranates, vegetables such as beans, the maintenance of a dovecote, the caring for a hungry mule which was getting refractory through lack of a rider. In addition, the barrels of wine had to be watched carefully: their construction of wood meant that their contents could easily be exposed to the air and go sour. These concerns and many financial transactions filled her correspondence.

2.16 Trial—A 'vehement suspicion of heresy'

The printing of the *Dialogue* was completed in February 1632 and one of the first copies was presented to the Grand Duke. It took a few months to reach Rome because of the quarantine measures connected with the continuing plague. By August, Galileo's friends were reporting difficulties with the Holy Office, which had been incited to take action by various enemies, such as the Jesuit Scheiner, by then the leading expert on sunspots. The Tuscan ambassador had an interview with the Pope, who was very angry, indicating that the *Dialogue* was likely to be prohibited. The reason for his anger seems to have been that Galileo had not told him of the questionable 1616 injunction at the time of his asking for permission to publish. Further, his own arguments against Copernicus had been put in the *Dialogue* into the mouth of the character Simplicio. The apparent contempt implied by the use of this name, although that of a historical person, he took as an insult. The Pope convened a special commission which concluded that the matter was serious enough for the Roman Inquisition to deal with. Thus, early in October, Galileo received an order to present himself before that body.

At this point, Galileo could probably have escaped to Venice or even to some more liberal country outside Italy (de Santillana 1955), but he seems partly to have lacked the will to do so and partly to have hoped to conquer his enemies.

By pleading illness and old age, the trip to Rome was delayed until January 1633. He had heard a rumour that 'the Jesuit Fathers have impressed the most important persons "in Rome" with the idea that this my book "the Dialogo" is execrable and more dangerous to the Holy Church than the writings of Luther and Calvin' (Pedersen 1985). At nearly seventy years old and in poor health, the journey of nearly 300 km from Florence to Rome on horseback in the depth of winter was likely to be too much for him. Eventually, having made his will, he travelled by litter—a type of bed carried by porters or animals—taking twenty-three days. On reaching Rome, he was allowed, as a favour to the Grand Duke, to stay in the Florentine Embassy, the delightful Villa Medici, with the ever-friendly Niccolinis, rather than in a prison cell. He was to be away from home for almost a year. The Ambassador's wife Caterina was particularly fond of Marie Celeste and kept her informed of events as they developed.

On 12 April he had to move to quarters in the offices of the Inquisition, but was given a pleasant apartment and allowed to keep his own servant. The issue to be discussed was not that of the validity of Copernicanism. Rather, it was whether Galileo had deliberately disobeyed the order contained in the disputed document found in the record of his previous brush with the Inquisition. Galileo could show Bellarmine's certificate of good standing; the Inquisitors could not produce the original of the document which had allegedly been disobeyed; Cardinal Bellarmine had died in the meantime as had other potential witnesses. In principle, he had won his case, but the heavy-handed authorities had to do something about him if they were not to appear foolish. The Inquisition threatened an interrogation concerning the Copernican views he had clearly expressed. However, the Pope's nephew, Cardinal Francesco Barberini, a discrete supporter of Galileo, worked out a 'plea bargain' whereby Galileo would publicly confess his error and thus allow the Inquisition to show some mercy towards him. This he duly did, claiming to have been led astray by vain ambition, ignorance, and inadvertence. At this point he was evidently jubilant at having settled the matter so easily, as we know from his daughter's congratulatory letters.

Unfortunately the Pope was vindictive. He was not prepared to lose face by letting Galileo appear to have won his case. Furthermore, he was under pressure from the Hapsburg dynasty in Spain and Austria, who accused him of being insufficiently supportive of the Catholic side in the ongoing war against the King of Sweden, Gustavus Adolphus. Thus, he needed to demonstrate his orthodoxy as a Catholic (Finocchiaro 1989). The ongoing trial was a perfect opportunity. He decided that Galileo's intention in publishing should be examined and, if it proved satisfactory, the punishment should be imprisonment at the pleasure of the Holy Office and confiscation of the *Dialogue*. Otherwise, worse should follow.

The examination as to intent was made on 21 June 1633. Galileo stated, and confirmed on pain of torture, that he was there to obey, and had not held Copernican opinions since the injunction of 1616. The following day he was given a sentence of life imprisonment, read to him at a ceremony in the Great Hall of the Dominican monastery of Santa Maria Sopra Minerva, where the Inquisition had its headquarters. The cardinals of the Inquisition were present; three of them including Barberini however refused to sign the sentence. Galileo was made to kneel before the assembly and abjure. His crime was declared to be a 'vehement suspicion of heresy'. This technical term, which sounds strange, if not indeed comical, to the modern ear, meant that he was somewhere between being 'slightly suspected' of heresy and a full-blown heretic (Finocchiaro 1989).

According to de Santillana (1955), Galileo treated his abjuration as a 'form of words' which, though forcing him to keep quiet about Copernicanism in public, had been extracted by force: he felt no moral obligation to keep silent in private. Any refusal to abjure would probably have led to complete imprisonment or even a sentence of death. He was to feel to the end of his life that his conviction had been unjust.

There was no way that such an important work as the *Dialogue* could have been suppressed effectively. Arrangements were soon made to have a Latin translation published outside Italy (1637) so that it could be understood by scholars in other countries. An English translation was published only in 1661. A modern English translation, with a forward by Einstein, was made by Stillman Drake in 1953 (see Galileo 1632).

2.17 Aftermath of the trial

When she heard of his sentence, Marie Celeste wrote to comfort him:

> [2 July 1633] My dearest lord father, now is the time to avail yourself more than ever of that prudence which the Lord God has granted you, bearing these blows with that strength of spirit which your religion, your profession, and your age require. And since you, by virtue of your vast experience, can lay claim to full cognisance of the fallacy and instability of everything in this miserable world, you must not make too much of these storms, but rather take hope that they will soon subside and transform themselves from troubles into as many satisfactions. (tr. Sobel 2000)

Cardinal Barberini arranged that the place of imprisonment should at first be the Florentine embassy in Rome. In June, he was placed under the friendly custody of Archbishop Piccolomini in Siena, who helped him to regain his composure after the trauma he had recently faced. At the start of his stay it was reported that he could not sleep and was behaving as though demented. But the loving care of Piccolomini had its effect and he soon became his usual self. One scandalised priest reported that Galileo was being treated more as an honoured guest than as a heretic. By the end of the year he was allowed to return to his own home in Arcetri, but he remained under house arrest for the rest of his life. This meant in effect that he was severely, but not completely, restricted as to who he might see and what he might teach or publish.

Galileo's famous words of defiance 'Eppur si muove' (but it does move) were almost certainly not uttered at the time of his abjuration. The tradition that he did say them at some point dates from as early as 1643 (or perhaps 1645), when they were included in a Spanish painting showing his confrontation with the Inquisition (Drake 1978, p. 357).

The philosophical world outside Italy was generally shocked by his treatment. The suppression of his views was only successful in those Catholic countries where most of the intellectuals were under the control of the Church, such as Spain and Italy. France and the Protestant countries contained by this time many independent thinkers who carried on his work. Galileo's persecution by the Church has been the subject of numerous books by both those who detested its actions and those who have tried to apologise for them, almost as if his problems with the Church were more important than his scientific achievements. The Catholic church made such an issue of the matter that it had great difficulty in coming to terms with its mistake. In recent years it has largely adopted Galileo's ideas about the proper relationship between science and scripture as expressed in his *Letters to the Grand Duchess*. Galileo's own faith seem to have been that of a liberal Catholic. His position was similar to that of Erasmus, who preferred to see the unity of Christendom retained, though strenuously wishing for the reform of abuses.

Alas, only a few months after his return to Arcetri, his daughter Marie Celeste was stricken with dysentery, from which she died on 2 April 1634. Coming to terms with this tragedy, so soon after his own troubles with the Inquisition, took several months. At the same time, his health problems were becoming more severe. He wrote to a friend:

> the state of my health is most troubling. The hernia has become larger than before, and my pulse is irregular because of palpitations of the heart; I am immensely sad and melancholic, have absolutely no appetite and cannot stand myself; and most of all I seem all the time to hear my beloved daughter calling me ... in addition to all this a perpetual sleeplessness makes me afraid.[14]

2.18   Discourses: 'Two New Sciences'

Galileo's last work, *Discourses and Mathematical Demonstrations Concerning Two New Sciences* (Galileo 1638) is concerned with two topics—the strength of materials and mechanics. The latter contained his collected thoughts on acceleration, which is perhaps his most important contribution to physics. He did not dare to publish it himself. Since there was no objection to the circulation of handwritten copies it was decided that one of these should be taken outside Italy

---

[14]Galileo to Boccherini, 27 April 1634, tr. Pedersen 1985.

by François de Noailles, the French Ambassador to Rome and a former pupil. It would then be published, ostensibly at his request, in some more liberal country. After some difficulty, it was eventually printed by Louis Elzevir in Leiden. The title page was dated 1638. In the dedication to the Count of Noailles, Galileo artfully expresses surprise that he should have decided to print it!

In 1636 he sent the States-General of the Netherlands details of a method for determining the longitude of a vessel at sea again using his idea that the satellites of Jupiter could serve as a clock. They were sufficiently impressed that they offered him a gift of a gold chain worth 500 florins, even though the method was hardly practicable as it involved the use of a telescope on a moving ship. However, he was advised in his own interest not to accept this present.

2.19   Last years

In July 1637 he lost the sight of his right eye. In spite of this, he reported a major astronomical discovery in November, the libration of the moon.[15] By the end of the year, his sight had gone completely. He never actually saw the printed *Discourses*.

Only once was Galileo allowed to go to Florence, for medical treatment, though artists seem to have had little difficulty in getting access to him: the Flemming Justus Sustermans, who had settled in Florence, became a friend and painted him at least twice (Fig 2.7). Portraits by two others exist from this period also. Foreign Protestants may have found it easier to visit him than good Catholics. One of these was the English philosopher, Thomas Hobbes. In 1638, the poet John Milton seems not to have encountered any problems. In his famous *Areopagitica* of 1644, a protest against censorship of the press in England, he wrote:

> I could recount what I have seen and heard in other Countries, where this kind of inquisition tyrannizes; when I have sat among their lerned men, for that honor I had, and bin counted happy to be born in such a place of *Philosophic* freedom as they suppos'd England was, while themselvs did nothing but bemoan the servil condition into which lerning amongst them was brought; that this was it which had dampt the glory of Italian wits, that nothing had bin there writt'n now these many years but flattery and fustian. There it was that I found and visited the famous *Galileo* grown old, a prisner to the Inquisition, for thinking in Astronomy otherwise than the Fransciscan and Dominican licencers thought. (Milton 1898; originally published 1644)

Although he found mental activity more difficult, and even admitted he could hardly follow some of his earlier proofs, he continued to work and correspond with other scientists until his final illness. His last pupil, Vincenzio Viviani, a promising sixteen-year old Florentine, who had come to the attention of Cosimo

---

[15]Libration is a 'nodding' type of motion which makes the moon show slightly different parts of its surface to us, rather than always presenting exactly the same face, as had previously been believed.

FIG. 2.7. Galileo's appearance around 1635. Lithograph dated 1897 based on a portrait painted by Justus Sustermans (SAAO).

II's successor, Grand Duke Ferdinand II de' Medici, arrived to live with him in Arcetri in October 1639 and acted as his secretary. Viviani later became a significant mathematician in his own right and edited the first collected edition of Galileo's works. He was a lifelong admirer and devoted much of his time and fortune to keeping alive his teacher's memory. In describing Galileo in his old age he wrote:

> His conversation was full of wit and conceits [opinions], rich in grave wisdom and penetrating sentences. His subjects were not only the exact and speculative sciences but also music, letters and poetry. He had a wonderfully retentive memory and knew most of Virgil, Ovid, Horace, and Seneca[16]; among the Tuscans, Petrarch almost whole, the rhymes of Berni and all of the poems of Ariosto, who was his favourite author (tr. de Santillana 1955).

---

[16]Latin poets and writers.

In 1640, Galileo made some almost paradoxical remarks in a letter to an unreconstructed Aristotelian, a former colleague at the University of Padua, Fortunio Liceti:

> I am impugned as an impugner of the Peripatetic doctrine, whereas I claim (and surely believe) that I observe more religiously the Peripatetic or I should rather say Aristotelian teachings than do many who wrongfully put me down as averse from good Peripatetic philosophy; and since one of the teachings given to us admirably by Aristotle in his *Dialectics* is that of reasoning well, arguing well, and deducing necessary conclusions from the premises . . . I think I may rightfully deem myself a better Peripatetic, and [consider] that I more dextrously use that doctrine than anyone who makes use of it clumsily.[17]

Liceti was obviously amused at this apparent turn-about and replied:

> That you profess not to contradict the Aristotelian doctrine is most welcome to me, just as (to speak frankly) it is news to me, I seeming to have gathered the contrary from your writings; but it may be that on this particular I was mistaken, along with others of the same opinion.[18]

In a later letter to the same Liceti, Galileo showed that he was as acerbic as ever:

> and if philosophy were what is contained in Aristotle's books, you would in my opinion be the greatest philosopher in the world, so well does it seem to me that you have ready at hand every passage he wrote.[19]

His former pupils kept in touch. Galileo showed great personal affection for them. For example, also in 1640, he wrote to Benedetto Castelli

> Bereft of my powers by my great age and even more by my unfortunate blindness and the failure of my memory and other senses I spend my fruitless days which are so long because of my continuous inactivity and yet so brief compared with all the months and years which have passed; and I am left of no other comfort than the memory of the sweetnesses of former friendships of which so few are left although one more undeserved than all the others remains, that of corresponding in love with Your Reverence.[20]

With difficulty Castelli obtained permission to visit him in March 1641, bringing with him a book by his own pupil Evangelista Torricelli, after whom the vacuum at the top of a mercury barometer is named. Torricelli took up residence with him in October, but Galileo's health took a turn for the worse the following month, with manifestations of severe kidney trouble.

[17]Galileo to Liceti, 25 August 1640, tr. Drake (1978, pp. 407–408).
[18]ibid, p. 408.
[19]Galileo to Liceti, January 1641, tr. Drake (1978, p. 412).
[20]Galileo to Castelli, 16 April 1640, quoted by Pedersen (1985).

In the same year, Galileo was able to discuss with his son Vicenzio his ideas for a pendulum-clock escapement. Vicenzio made a drawing of it and had a model made, though it was not completed before his father died. It does not seem to have become known to others, though it is very similar to practical mechanisms made later.

Some idea of how he reconciled his injunction 'not to hold, defend or teach [Copernicanism] in any way' with his private views comes from a letter he wrote on 29 March 1641 to the Florentine ambassador in Venice. He says that

> the falsity of the Copernican system ought not to be doubted in any way, and most of all not by us Catholics who have the undeniable authority of Holy Scripture, interpreted by the best theologians.

He then disingenuously added

> if the observations and conjectures of Copernicus are insufficient, those of Ptolemy, Aristotle and their followers are in my view even more false.[21]

In other words, even if the Copernican theory may not have all the answers it is nevertheless better than the others known to him!

He died on 8 January 1642 in his 78th year and was buried quietly with Florence's other heroes in Santa Croce, but in the Novice's Chapel, away from the main basilica. Although the Grand Duke Ferdinando II de' Medici wanted to erect a fitting monument opposite that of Michelangelo in this Florentine Pantheon, the Pope and the Inquisition let it be known that such recognition was not appropriate for someone who was still doing penance for a serious offence. Any inscription or funeral oration was to say nothing that would harm the Inquisition. On the advice of the Florentine ambassador to Rome, the Grand Duke obeyed. The fine monument that is seen today in Santa Croce was erected in terms of the will of his last pupil Viviani, who died in 1703. It was completed in 1737.

As perhaps befits a scientific 'saint', the middle finger of his right hand was removed at the time of his remains being transferred to his second tomb and preserved in a glass egg-shaped reliquary. It can be seen to this day in the Science Museum in Florence, together with some of his telescopes and other instruments.

References

Abbot, C.G., 1937. *Annual Report of the Smithsonian Institution for 1936*, U.S. Government Printing Office, Washington, D.C.

[Allan-Olney, M.], 1870. *The Private Life of Galileo*, Macmillan, London (published anonymously).

Biagioli, M., 1990. Galileo the Emblem Maker, *Isis*, **81**, 230–258.

Brown, H.M., 1976. *Music in the Renaissance*, Prentice-Hall, Englewood Cliffs, NJ.

---

[21]Galileo to Rinnucini, 29 March 1641, quoted by Pedersen (1985).

de Santillana, G., 1955. *The Crime of Galileo*, University of Chicago Press, Chicago, IL.

Drake, S., 1957. *Discoveries and Opinions of Galileo*, Doubleday and Co. Inc., New York. This book contains translations of *The Starry Messenger* (1610), *Letters on Sunspots* (1613), *Letter to the Grand Duchess* (1615), and an abridged version of *The Assayer, (Il Saggiatore) (1623)*.

Drake, S., 1980. *Galileo*, Oxford University Press, Oxford.

Drake, S., 1978. *Galileo at Work: His Scientific Biography*, University of Chicago Press, Chicago, IL.

Finocchiaro, M.A., 1989. *The Galileo Affair, A Documentary History*, University of California Press, Berkeley, CA.

Galileo, G., 1612. *Cause, Experiment & Science, A Galilean dialogue incorporating a new English translation of 'Bodies That Stay Atop Water, Or Move In It'*, tr. Drake, S., 1981, University of Chicago Press, Chicago, IL.

Galilei, G., 1632. *Dialogue Concerning the Two Chief World Systems, Ptolemaic & Copernican*, tr. S. Drake, 1953, University of California Press, Berkeley, CA.

Galilei, G., 1638. *Dialogues Concerning Two New Sciences*, (usually known as the *'Discourses'*), tr. by Henry Crew and Alfonso de Salvio, 1954, Dover, New York.

Gingerich, O., 1982. The Galileo Affair, *Scientific American*, **247**, August, 118–127.

Greco, V., Molesini, G., and Quercioli, F., 1992. Optical tests of Galileo's lenses, *Nature*, **358**, 101.

Milton, J., 1898. *Areopagitica*, Clarendon Press, Oxford (Originally published 1644, London).

Pedersen, O., 1985. Galileo's Religion, in *The Galileo Affair: a Meeting of Faith and Science, Proceedings of the Cracow Conference 24 to 27 May 1984*, eds. Coyne, G.V., Heller M., and Życiński, J., Specola Vaticana, Vatican City.

Sharratt, M., 1994. *Galileo, Decisive Innovator*, Cambridge University Press, Cambridge.

Sobel, D., 2000. *Galileo's Daughter*, Fourth Estate, London.

van Helden, A, (tr.), 1989. *Sidereus Nuncius or The Sidereal Messenger*, by Galileo Galilei (1610), Univ. of Chicago Press, Chicago, IL.

Westfall, R.S., 1989. The Trial of Galileo: Bellarmine, Galileo, and the Clash of Two Worlds, *J. Hist. Astr.*, **20**, 1–23.

3

ISAAC NEWTON: RATIONALISING THE UNIVERSE

And the same year [1666] I began to think of gravity extending to ye orb of the Moon . . . I deduced that the forces wch keep the Planets in their Orbs must [be] reciprocally as the squares of their distances from the centers about wch they revolve: and thereby compared the force requisite to keep the Moon in her Orb with the force of gravity at the surface of the earth, & found them answer pretty nearly. All this was in the two plague years of 1665–1666. For in those days I was in the prime of my age for invention & minded Mathematicks and Philosophy more than at any time since.[1]

The year that saw the death of Galileo (1642) also saw the birth of Newton, although this coincidence relies on the fact that England was still using the Julian calendar. According to the Gregorian calendar, which England did not adopt until 1752 because it was promulgated by a Pope (in 1582), his date of birth would have been 4 January 1643. His father, another Isaac, though illiterate, was a farmer of sheep and cattle and a member of a family which had increased in prosperity from generation to generation. He was 'Lord of the Manor' of Woolsthorpe (Fig 3.1) in Lincolnshire, meaning that as owner of that particular property he was entitled to act as judge in certain law cases. By the date of his death he owned 270 acres, a comfortably large farm by the standards of the time. Isaac was born three months after his father's death, on Christmas day. He was a premature baby and not expected to live. 'They could put him into a quart pot & [he was] so weakly that he was forced to have a bolster all round his neck to keep it on his shoulders'.[2]

His mother, Hannah, however, had some well-educated relations. Her brother William Ayscough was a clergyman and a graduate of Cambridge University. Three years after the death of Isaac's father, Hannah got married again; this time to a wealthy widower, the Rev Barnabas Smith, who thus became Newton's stepfather. She left her young son at Woolsthorpe to be brought up by her own mother and moved into her husband's rectory at North Witham, a village 2 km away. She only returned to her son when he was ten years old, following the death of Barnabas. Among a secret list of his sins that he compiled nine years later, the young Isaac listed 'Threatening my father and mother Smith to burne them and the house over them'.[3] It is thought by some that this early

[1] Memorandum in the Portsmouth Collection (Cambridge University Library), quoted by Westfall (1980, p. 143).

[2] Conduitt, Keynes ms (King's College, Cambridge) 130.10, quoted by Westfall (1980, p. 49).

[3] From a notebook in the Fitzwilliam Museum, Cambridge; quoted by Westfall (1963).

FIG. 3.1. Woolsthorp Manor, where Newton grew up. It appeared much the same in a sketch dated 1721 and reproduced by Stukeley (1936). According to the latter's *Memoirs of Sir Isaac Newton's Life*, it had become 'much decay'd' and was re-built by Newton's stepfather, Rev Barnabas Smith, soon after Newton's birth (from Mitchell 1856).

deprivation contributed to his later neurotic personality and soured his relations with women.

Isaac was sent at the age of 12 years to the nearby town of Grantham, to attend the grammar school there. His youth coincided with the 'Commonwealth', or period of puritan rule of England under Oliver Cromwell. The eastern counties were the centre of this anti-monarchical movement. The seriousness of the age in religious matters had its effect on Isaac, who remained a puritan in outlook throughout his life.

Much of what we know of Newton's early years comes from the stories of old people who remembered him, collected by William Stukeley, his junior by about 45 years. Stukeley was a student in Cambridge just before Newton's final resignation from the University, by which time he was already a very famous man. In 1718 Stukeley, then a medical doctor in London, had become a fellow of the Royal Society. Because he took part in its administration he came to know Newton fairly well. In 1726, for reasons of health, he had to retire and decided to move to Grantham where he made a special effort to collect information about Newton's youth. No doubt, fame and the passage of time caused some exaggeration, but the fact that Stukeley was an educated man probably enabled him to winnow out the less likely 'facts'.

Newton was fortunate to have had as a teacher in Grantham a certain Henry

Stokes who was interested in mathematics and did not confine himself entirely to the traditional grammar-school curriculum based on Latin, which was all too often the sole subject taught. Being too far from home to travel to school each day, he stayed with friends of his mother's, an apothecary and his wife, surnamed Clark. His later interest in chemistry may well have been stimulated by watching Clark at work. Those who remembered him from the days before he became famous said that he enjoyed playing with girls much more than with boys. He could also be tough: a story was told that an older boy kicked him in the stomach on his way to school. A fight developed and 'tho Isaac was not so lusty as his antagonist he had so much more spirit & resolution that he beat him till he declared he would fight no more' and rubbed his nose in the dust.

While boarding at Grantham, Isaac occupied his spare time with making wooden models, using tools on which he spent all his pocket money. His output included a toy windmill, a mouse-operated treadmill and a cart that he could propel by a crank as he sat in it. On one occasion he attached a lantern to a kite and frightened the neighbours by flying it at night. He was known as a builder of sundials and water-clocks. Some of the devices he made were copied from a sort of practical manual called *The Mysteries of Nature and Art* by one John Bate; others were original. He taught himself to draw. At home, his room contained wooden shelving made by himself out of old boxes. His model-building occupied most of his attention, so that every now and then his studies suffered. However, a little serious work invariably restored him to his position at the top of the class.

Mr Clark, Isaac's landlord, had a stepdaughter, a Miss Storey, who was a few years younger than Newton. When interviewed in old age by Stukeley, she described Isaac as:

> always a sober, silent, thinking lad, never was known scarce to play with the boys abroad [outside], but would rather chuse to be at home, even among the girls, and would frequently make for them little tables, cup-boards and other utensils, for her, and her play fellows, to set thir babys and trinkets on. (Stukeley 1936, pp. 45–46)

She also mentioned that Newton was interested in gathering herbs. Stukeley went on:

> Sir Isaac and she being thus brought up together, it is said that Sir Isaac entertaind a passion for her when they grew up; nor dos she deny it. 'Tis certain he always had a great kindness for her. He visited her whenever in the country, in both her husbands days, and gave her, at a time when it was useful to her, a sum of money. She is a woman but of a middle stature, of a brisk eye, and without difficulty we may discern that she has been very handsom. (Stukeley 1936, p. 46)

In 1659, nearing the age of 17, his mother recalled him from school to Wool-sthorpe to learn farming. For such work he showed no interest or aptitude, earn-ing the contempt of the farm labourers by his neglect of the tasks at hand, slipping away to read books or make mechanical devices. After some months, and only with difficulty, his uncle William and his teacher Stokes persuaded Hannah to send her son back to school for another year in order to prepare him for university entrance. So strong was Stokes' feeling that he offered to subsidise Isaac's school fees and allow him to board in his own home.

3.1 Trinity College, Cambridge

In June of the following year (1661) Newton duly set off for Cambridge, a three day journey from his home. He was examined as to fitness to enter Trinity College, of which his uncle William Ayscough had once been a member. He was not without a friend there: the brother of Mrs Clark, wife of the apothecary with whom he had previously boarded, was Humphrey Babington, an influential fellow of the college, who must have previously met the young Isaac.

Rather surprisingly, Newton entered as a 'subsizar', a term used for poor students who paid no fees but were expected to act as servants to normal students (called pensioners), fellow-commoners (rich students who could dine with the fellows) or even the fellows themselves. Since his family was reasonably well off, it is strange that he was not a pensioner himself. His mother may simply have been unwilling to pay for him.

The Cambridge that Isaac Newton found was by no means the intellectually competitive institution that it is today. It was very largely a seminary for potential clergy of the Anglican church, and it attracted mainly those who hoped to make careers in that field. The academic staff were compelled to stay single. After some time they tended to resign in order to get married and take up 'livings' as clergymen in particular parishes falling under college control. This had the effect of keeping the average age of staff members quite young. Appointments to the academic staff were very often made by royal directive, merit being a minor consideration. Political appointments to government and academic posts were quite normal in the seventeenth and eighteenth centuries and were the usual method of rewarding the politically faithful until the great period of administrative reform around 1830. Headships of colleges were regarded as particular plums. A recent biographer of Newton, Richard Westfall (1980), has painted a bleak, though perhaps exaggerated, picture of an institution where scholarship was not important and even ordinary teaching duties were frequently neglected. Yet it managed to harbour a number of active intellectuals.

Newton was a serious and solitary student as well as being a year or so older than the average new student of the time. He was religious, a puritan at a time when many others were taking advantage of the easy-going atmosphere that had become prevalent as the Cromwellian period ended and the monarchy of Charles II became established. At nineteen and a half his conscience troubled him and he confessed in writing the sins he had committed up to this time, but concealed in a shorthand cipher. Some of them were (Westfall 1963):

> Making a [pen] on thy day.
> Denying that I made it.
> Squirting water on they day.

Making pies on a Sunday night.
Swimming in a [tub] on Thy day.
Putting a pin in John Keys hat on Thy day to prick him.
Threatening my father and mother Smith to burne them and the house
over them.
Wishing death and hoping it to some.
Striking many.
Having unclean thoughts words and dreamese.
Stealing cherry cobs from Edward Storer.
Punching my sister.
Calling Dorothy Rose a jade.
Striving to cheate with a brass halfe crowne.

These 'sins' have been taken to be early signs of the petulance which he occasionally showed as an older man, but on the whole do not indicate anything more abnormal than a rather sensitive conscience. 'Wishing death' sounds like a manifestation of teenage depression.

His first room-mate he found 'disorderly', but one day while walking dejectedly about he met another student called John Wickins who had experienced the same problem with his chamberfellow. Finding they had similar habits, they decided to room together. They continued to share for more than twenty years, until Wickins resigned to take up a living.

The official curriculum remained medieval in structure and heavily emphasised Aristotelian logic and ethics. Classical authors, both Latin and Greek, were studied extensively. Aristotle's philosophy was also taught, including cosmology. It is clear from his surviving notes that Newton was already reading modern material privately, such as some of Galileo's works. He also read widely in history and biblical chronology.

3.2 Intellectual awakening

The beginning of Newton's independent questioning of the world about him is found among his notes of around 1664 on the Aristotelian texts that he had to study. From two pages of *Quaestiones quaedam Philosophicae* (Questions concerning Philosophy) that he wrote in his notebook, it is clear that he had read the works of the French mathematician and philosopher Descartes, as well as other scientific or philosophical writings by his near contemporaries Pierre Gassendi, Thomas Hobbes, and Robert Boyle. In contrast to the end of the previous century, when Galileo was young, the number of those one might call scientists had increased considerably and progress was beginning to occur over a wider front.

René Descartes (1596–1650), whose lifetime straddled those of Galileo and Newton, was a 'philosopher' who approached geometry from the point of view of algebra, one of the most significant steps in the whole history of mathematics, and one which influenced Newton very strongly. He also had less reliable ideas concerning optics and had a complex and ultimately untenable theory concerning the orbits of the planets, which he believed to be controlled by invisible 'vortices' in space. Pierre Gassendi (1592–1655) is remembered for making some progress in the field of Galilean mechanics; Hobbes (1588–1679), an Englishman, was a leading philosopher and the Anglo-Irishman Boyle (1627–1691) was an experimenter now commonly remembered for his work on pressurised gases and his book *The Sceptical Chemyst*.

The questions Newton asked showed that he was seeking a logical consistency in nature. They concerned subjects such as the daily variations of the tides and the extent of the Moon's influence. These topics were important to his later investigations.

Within Cambridge itself there were at least a few people aware of current developments in mathematics, such as Isaac Barrow, the first person to hold the Lucasian professorship of that subject. Henry More, another former Grantham schoolboy, was a significant intellectual force at the University, a Platonist philosopher, interested in the work of Descartes and other contemporaries. Although there is no evidence of contact between the student Newton and these senior university figures, it seems likely that their views may have exerted some influence on him. Curiously, Newton's first acquaintance with geometry came through the algebraic or [Des]Cartesian approach and only later did he come to know the logically appealing work of the ancient Greek mathematician, Euclid. Euclid's proofs of geometrical theorems were based on almost mechanical operations, such as laying triangles on top of each other to prove their equality and the construction of figures using instruments such as rulers and compasses.

Very soon Isaac plunged into independent study. He would get so involved with the investigation of a problem that he would forget meals and sometimes stay up all night. This was the beginning of his real life. Stukeley, although an idolater when it came to discussing Newton, wrote with an admiration that was completely justified:

> It seems to me likely enough that Sir Isaac's early use and expertness at his mechanical tools, and his faculty of drawing and designing, were of service to him, in his experimental way of philosophy; and prepar'd for him a solid foundation to exercise his strong reasoning facultys upon; his sagacious discernment of causes and effects, his most penetrating investigation of methods to come at his intended purpose, his profound judgment, his invincible constancy and perseverance in finding out his solutions and demonstrations, and in his experiments; his vast strength of mind in protracting his reasonings, his chain of deductions; his indefatigable attachment to calculations; his incomparable skill in algebraic and the like methods of notation; all these united in one man, and that in an extraordinary degree, were the architects that raisd a building upon the experimental foundation, which must stand coeval with material creation (Stukeley 1936, pp. 54–55).

Newton's status as an undergraduate improved in April 1664 when he became a 'Scholar' of Trinity College after a competitive examination. This would have set him among the intellectual elite of the student body and brought him to the attention of the College's senior members. He now ceased to be a sizar and

was entitled to support from the college for a further four years. By 1665 he had
completed his courses and had graduated as a Bachelor of Arts. That summer, the
bubonic plague broke out in England and the university shut down for most of the
following two years. Newton returned to Woolsthorpe for its duration. Already
during the previous year he had embarked on his career of experimentation and
mathematical deduction. It was during the plague years that he experienced
many of his greatest insights.

3.3   Discoveries during the plague years

The problem of dealing with continuous movement and change which had so
much bothered Galileo and his contemporaries had been investigated with some
success in the meantime by Descartes and the English mathematician John Wallis. Newton read of these developments, which he quickly understood, and started
making innovations of his own. In some ways, he visualised curves as being generated mechanically, by a point moving according to some mathematical law.
Because he had not been introduced to geometry in the customary way, i.e.
through the simple (or, at least, non-algebraic) methods of Euclid, he automatically thought of lines and curves in terms of the equations which described them.
In constructing tangents to curves and working out the areas enclosed by them
he gradually formulated the basic ideas of differential calculus (called by him
'fluxions') and its inverse, integral calculus (called by him 'quadrature'). During
the two years from 1664 to 1666 he developed most of his basic mathematical
ideas and, though he wrote memoranda about them, he remained completely
secretive and failed to publicise them in any way. Probably only Isaac Barrow
might have heard of and understood the great strides he was making.

His 'laws of motion' began to be formulated during and to some extent following his mathematical discoveries, starting in 1665. Quite soon he had clarified
the vague notions of Descartes, already an improvement on those of Galileo,
concerning forces acting on bodies and the fact that they would stay still or
remain in a state of uniform motion unless acted upon by such forces. The old
Aristotelian notion that there had to be a 'cause' behind continuous motion did
not enter his picture at all. His concept of 'force' was the key to progress in
mechanics. He thought about circular motion and worked out the force (now
called centrifugal) in a string attached to a whirling mass. It was not long till
he compared the centrifugal force due to the motion of the moon with that of
gravity at the earth's surface. By using Kepler's third law, that the cube of the
mean radius of a planet's orbit is proportional to the square of its period, he
deduced the inverse square law of gravitation. This work he formalised in an
unpublished memorandum called *The Lawes of Motion*.

Did his insight into gravity really depend on the legendary falling apple? One
of his early biographers, John Conduitt, the husband of his niece, included the
following story:

> In the year 1666 [after a few months back in Cambridge] he retired again
> ...to his mother in Lincolnshire & whilst he was musing in a garden it

came into his thought that the power of gravity (wch brought an apple from the tree to the ground) was not limited to a certain distance from the earth but that this power must extend much farther than was usually thought. Why not as high as the moon said he to himself & if so that must influence her motion & perhaps retain her in orbit, whereupon he fell a calculating what would be the effect of that supposition . . . [4]

Newton related this story to him in the year of his death. It is probably a romanticised version of the gradual process by which the real discovery was made. However, Voltaire heard it also from Catherine Barton, Newton's niece, who lived with her uncle in his old age.

In a memorandum dated about 1714 he made the statement concerning this period of his life already quoted at the start of this chapter:

And the same year [1666] I began to think of gravity extending to the orb of the Moon and having found out how to estimate the force with which [a] globe revolving within a sphere presses the surface of a sphere, from Kepler's Rule of the periodical times of the planets being in a sesquialternate proportion of their distances from the centers of their Orbs I deduced that the forces which keep the Planets in their Orbs must [be] reciprocally as the squares of their distances from the centers about which they revolve: and thereby compared the force requisite to keep the Moon in her Orb with the force of gravity at the surface of the earth, and found them answer pretty nearly. All this was in the two plague years of 1665 and 1666, for in those days I was in the prime of my age for invention, and minded Mathematicks and Philosophy more than at any time since. [5]

The intense effort needed twenty years later to prepare a logical structure to his work suggests that at this early time he had not taken much care over rigorous development and presentation, but had excitedly followed up each idea as it occurred to him. All this was to be laid aside and might have been forgotten were it not for the pressure applied by Halley and other friends much later on. Had he died during the next two decades relatively little would have been known of his genius.

It is also probable that in 1666 Newton turned, as he remarks in his first published paper (Newton 1672a), to the investigation of light and colour. These topics were very much in the minds of his contemporaries such as Descartes and Boyle, and featured in their writings. Using sunlight falling on a prism he made a spectrum appear on a wall. However, his main optical experiments seem to have been made several years later. 'Amidst these thoughts I was forced from Cambridge by the intervening plague, and it was more than two years before I proceeded further'.

[4]Conduitt, Keynes ms 130.4, pp. 10–12, quoted by Westfall (1980, p. 154).

[5]Memorandum in the Portsmouth Collection (Cambridge University Library), quoted by Westfall (1980, p. 143).

3.4 Beginnings of recognition

In April 1667, following the abatement of the plague, Newton returned to Cambridge. He then had to face the next hurdle in his academic career: the fellowship examinations. At the examination of 1667 there were nine places to be filled and some of these were likely to be awarded to political favourites nominated by the King, who could be refused only with peril. The Master and eight Senior Fellows had the ultimate say in the remaining cases. The fellowship examination was an oral one that took place over four days in the college chapel. Newton duly became a fellow—an indication that his abilities must have been recognised by the senior members of the college. He could now look forward to years of comfortable tenure without much need to justify his existence and could only be dismissed for crime, heresy, or marriage. He had a small but adequate salary, a set of rooms, and the right to dine in College ('Commons'). In the following year he paid a fee to become a Master of Arts, a formality then as now, and increased his income to about £60 per year.

Recollections of Newton's behaviour at this time were assembled by Stukeley and by Newton's distant relation, Humphrey Newton, who was his secretary for five years in the 1680s and shared his rooms. His absent-mindedness when thinking about a problem became legendary—forgetting to eat, wearing the wrong clothes to College events, forgetting to comb his hair, etc. 'He would with great acuteness answer a Question, but would very seldom start one',[6] wrote Humphrey. He supposedly laughed only once, when he lent a copy of Euclid to a friend who asked what was the use of studying it. 'Upon which Sir Isaac was very merry'.[7] He was often described as unsociable, but it is clear that he did entertain friends, play cards, and drink wine from time to time. One friend, a chemist called Vigani, he later fell out with because he 'told a loose story about a nun'.[8] This is an example of his extreme prudishness, which appeared on other occasions also.

By 1669, Newton's isolation was brought to an end. John Collins, an accountant who was the librarian of the Royal Society in London and had a strong interest in mathematical developments, passed on to Isaac Barrow a new book that gave a series for calculating logarithms. The latter replied that a friend of his had written some papers of a more general nature on the same subject. He sent Collins a paper of Newton's entitled *De analysi per aequationes numero terminorum infinitas* or 'On analysis by equations with infinite numbers of terms'. This was, in effect, his first publication. New work then frequently passed from hand to hand in manuscript form. Even so, Newton was apprehensive as to how it would be received. Collins described or circulated it among a wide circle of friends and acquaintances both inside and outside England.

---

[6]Humphrey Newton, Keynes ms 135, quoted by Westfall (1980, p. 192).

[7]Stukeley (1936, p. 57).

[8]Keynes ms 130.6, Book 2. Quoted by Westfall (1980, p. 192).

3.5 Lucasian Professor

In October 1669, Isaac Barrow resigned the Lucasian chair of mathematics, whose first holder he had been, in order to concentrate on his ecclesiastical career. He evidently persuaded the executors of Henry Lucas's will, who still controlled the appointment, that Newton was the most fitting person to succeed him. Almost certainly, he was able to appreciate Newton's ability, even if he was far from being his scientific equal. The post of Professor, to which Newton was duly appointed, meant a further increase in salary, to £100 per annum, as well as greater security and status. The holder was supposed to deliver lectures each week during term and to deposit written copies of them each year in the university library. These conditions Newton complied with only occasionally. Other conditions included the avoidance of serious crimes, including heresy.

Newton's first lectures concerned optics, which he was researching intensively at the time. He showed that 'white light' could be split up into the primary colours. If the already separated beams of coloured light were passed through another prism they did not generate more colours. Further, the separated colours could be recombined to again make white light by passing them through a similar prism. He considered that light consists of 'corpuscles', while his contemporary, who at this stage he barely knew of, the great Dutch scientist and mathematician Christiaan Huygens, considered it to be a wave phenomenon. This was something that Newton never accepted even though certain of his own discoveries could best be explained on this basis. It is remarkable that, almost at the same time, Huygens was working on many of the same issues that Newton was, but without ultimately the same degree of success. In fact, the present quantum theory of light regards it as having both wave and 'corpuscular' aspects.

His lectures on this theme were written up and deposited as required, but were not published until much later, in 1728, for their curiosity value and to establish his priority, well after his definitive *Opticks* which itself had to wait until 1704. His original researches in optics ended during 1670. In the written version of his lectures he made a prophesy that has *not* withstood the test of time:

> The late invention of Telescopes has so exercised most of the Geometers, that they seem to have left nothing unattempted in Opticks, no room for further improvements ... (Newton 1728)

In November 1669, Newton undertook a visit to London, during which he met Collins for the first time. Collins drew him out on topics such as algebra and the formula for calculating compound interest and was anxious to publish his results. However, Newton grew apprehensive and, while agreeing to what was asked, tried to remain anonymous 'For I see not what there is desirable in publick esteeme, were I able to acquire and maintain it. It would perhaps increase my acquaintance, ye thing[9] wch I cheifly study to decline'.[10] This rather misanthropic

---

[9]y^e is an abbreviation of 'the', which is how it is meant to be pronounced.

[10]Newton to Collins, 18 February 1670, quoted by Westfall (1980, p. 224).

attitude was typical for more than a decade: he could not resist his drive to
develop his mathematics and science, but at the same time could not bring
himself to publish. As a result, he was to lose much of the credit for the discovery
of the calculus to the German mathematician and philosopher Gottfried Wilhelm
Leibniz, who was stimulated by many of the same contemporary discoveries that
Newton was. He may also have been inspired in part by Newton's manuscript *De
analysi*, which was shown to him by Collins, as well as the work of the Scottish
mathematician James Gregory on infinite series.

3.6   The first reflecting telescope

Early in 1669, Newton constructed the first reflecting telescope, which was 6
in. (15 cm) in length and had a magnification of about 40 times (Fig 3.2). In
1671, he made a somewhat larger model with a magnification of 150 times. John
Conduitt, the husband of Newton's niece, recorded in 1726:

> I asked him where he had it made, he said he made it himself, & when I
> asked him where he got his tools said he made them himself & laughing
> added if I had staid for other people to make my tools & things for me,
> I had never made anything of it ... [11]

He published his methods in detail a few years later, (Newton 1672b), and
much later again, in his *Opticks* (1704). The process he had developed for making
mirrors anticipated William Herschel and others and is indeed that familiar to
amateur telescope makers at the present time. The blank to be ground to a
spherical shape was made of a copper–tin alloy similar to what was later called
speculum metal. Whereas today hand-grinding is usually done by grinding two
circular pieces of glass of the same diameter together so that one becomes concave
and the other convex, Newton first worked two large pieces of copper against each
other to make a large pair of tools. He then ground the actual mirror to shape
against the convex tool to make it concave. For polishing, he covered the tool
with pitch and used 'putty powder' (tin oxide) as a polishing agent (today one
uses cerium oxide or rouge). He had first to learn how to treat the tin oxide to
extract the fine particles that would polish without scratching. His final product
would have been a polished surface that was probably within micrometres of a
truly spherical one.

Unlike the simple refractors of the time, the reflector gave sharp images without coloured fringes because mirrors reflect all wavelengths of light at equal angles
to focus in the same place. Simple glass lenses, on the other hand, focus each
colour at a slightly different place. For this reason, and because it works equally
well for ultraviolet and infrared rays, the reflector is the basis of the modern
telescope. The combination of a spherical (or, ideally, slightly paraboloidal) mirror with a flat 45° mirror near the eyepiece is still referred to as a 'Newtonian'
telescope. Collins and others outside Cambridge heard about the exciting new

---

[11]Conduitt, Keynes ms 130.10, quoted by Westfall (1980, p. 233).

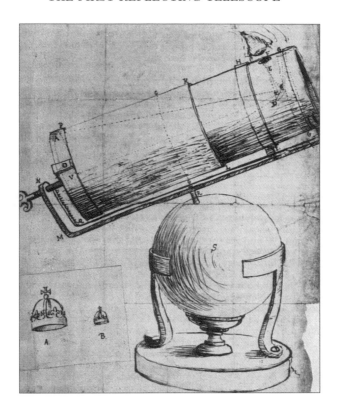

FIG. 3.2. Reflecting telescope made by Newton and presented to the Royal Society (see Turnbull 1959 for details). The light was collected by the main mirror A, of about $6\frac{1}{3}$ in. (16 cm) focal length at the left end of the tube and passed via a flat mirror DE to the eyepiece F of about $\frac{1}{6}$ in. (4.2 mm) focal length (see also Fig 3.3). The magnification was about 37 times. Focusing was achieved by turning the screw at the end to shorten or lengthen the tube. The large and small crowns were images of a distant weathercock as seen through this telescope and a refractor 25 in. (64 cm) long. (Illustration used by permission of the Royal Society.)

instrument around the end of 1671 and the Royal Society in London requested to see it.

The Royal Society had been founded in 1662 and, although many of its members were dilettantes, it was the meeting place of the real scientists of the time such as Boyle, Robert Hooke of Hooke's law, and the astronomers John Flamsteed and Edmond Halley. Continental scientific visitors frequently attended its sessions.

In response to their request, Newton presented his second telescope to the Society. It created a sensation, particularly impressing the Fellows by its short

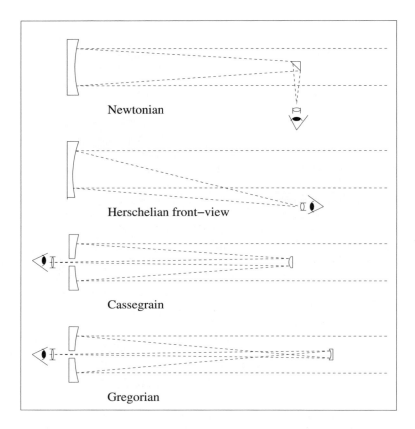

FIG. 3.3. The optics of various reflecting telescopes. At the top is Newton's design, with a paraboloidal and a flat mirror. Herschel invented the 'front view'
to avoid the substantial losses associated with a second reflection; however,
its image quality is poor except in very long instruments. The Cassegrain
combines a paraboloidal primary with a hyperboloidal secondary. The design of James Gregory, a friend of Newton, is similar, but with an ellipsoidal
secondary. Most modern telescopes are Cassegrains or variants thereof.

overall length compared with a refracting telescope of equivalent power. He was
immediately proposed for membership and the news of his invention was sent to
Huygens in Holland to ensure his priority. Newton was pleased at the recognition
he received. He was almost immediately elected a Fellow of the Society (11
January 1672).

3.7   'The oddest, if not the most considerable, detection'

One of Newton's first actions as a Fellow was to submit a paper on his theory
of colours, based on his own experiments. On January 18 he offered it to the

Society as

> an accompt of a Philosophicall discovery wch induced mee to the making
> of the said Telescope, & wch I doubt not but will prove much more grate-
> full than the communication of that instrument, being in my Judgment
> the oddest, if not the most considerable detection, wch hath hitherto been
> made in the operations of Nature'.[12]

In it he described how he had investigated why a simple lens, used in a telescope like that of Galileo, gave a rather blurry image. It simply could not focus the different colours of light to the same position! This paper was the first public announcement of the discoveries he had made during the previous few years; in particular, that white light is a mixture of colours. Further, he included the details of his 'experimentum crucis', that a particular colour of light emerging from a prism can be passed through another similar prism where it will be bent by the same angle. He also showed that light dispersed into colours by a prism can be re-assembled by passing it through an identical prism. This revolutionary paper appeared in the Society's *Philosphical Transactions* for 19 February 1672 (Newton 1672a) and circulated widely, to Paris and elsewhere. It was to establish his reputation as a serious 'natural philosopher'.

At first his work was received with doubt in some quarters, but the fact that the experiments he had described could easily be verified soon led to its acceptance. Unfortunately, he encountered some trouble from people who still held to the old Aristotelian approach, emphasizing theory without troubling to accept the truth before their eyes. Their continual objections, expressed in print, began to irritate Newton exceedingly. He withdrew into his shell and very nearly refused to communicate any more discoveries on the grounds that he was attracting controversy that he could well do without. Hooke, in addition, felt that Newton was stealing his ideas and his frequently voiced complaints were part of the reason for Newton's silence. Although Hooke was not a mathematician, he had a very broad scientific outlook, possessing good powers of intuition, and was an excellent experimenter. At one point Newton added fuel to the fire by describing Hooke's theory of light as 'not onely *insufficient*, but in some respects *unintelligible*'.[13]

Newton constantly emphasised that he was describing experimentally veri-fiable phenomena rather than theories and that he was not trying to explain colours by some mechanical model. The criticisms of Huygens also annoyed him greatly, to the extent that he burst out to Henry Oldenburg, the secretary of the Royal Society, in 1673

> S^r I desire that you will procure that I may be put out from being any
> longer fellow of y^e R. Society. For though I honour that body, yet since I

---

[12]Newton to Royal Society, 18 January 1672, quoted in Westfall (1980, p. 237).

[13]Newton to Collins 11 June 1672, quoted by Westfall (1980, p. 247).

FIG. 3.4. Late seventeenth century engraving showing the part of Trinity College between the gatehouse and chapel, where Newton lived from 1673. The garden belonged to his rooms and was accessed from the first floor by the central verandah structure. His alchemical laboratory was in the small suspended building on the right, next to the chapel. (Loggan: Cantabrigia Illustrata 1688, Photo: Trinity College Library, Cambridge).

see I shall neither profit from them, nor (by reason of this distance) can partake of the advantage of their Assemblies, I desire to withdraw.[14]

However, he went on to publish his ideas on light and thin films as *An hypothesis explaining the properties of light*. In a brief moment of reconciliation with Hooke in 1676, he had included in his praise:

[14]Newton to Oldenburg, 8 March 1673, quoted by Westfall (1980, pp. 249–250).

> What Des–Cartes did was a good step. You have added much several ways
> ... If I have seen further it is by standing on ye shoulders of Giants.[15]

This famous sentiment is by no means original to Newton. Its earliest attribution is to Bernard of Chartres who died around 1130!

From 1674 Newton had suffered criticism from an English Jesuit in Liege who went under the Latin name of Linus. In 1676, his anger over the obtuseness of his critic came to a head. He wrote to Oldenburg:

> I see I have made myself a slave to Philosophy, but if I get free of Mr Linus's business I will resolutely bid adew to it eternally, excepting what I do for my privat satisfaction or leave to come out after me. For I see a man must either resolve to put out nothing new or to become a slave to defend it.[16]

3.8 Leibniz and the early papers of Newton

From about 1673, Leibniz had been interested in British mathematical affairs and had corresponded with Collins, who told him a great deal about current developments. As an original thinker, Leibniz stands on the same plane as Newton, though the emphasis of his work was often in other directions such as philosophy (as opposed to 'natural philosophy') to which he also made many original contributions.

By about 1676 Newton was indirectly corresponding with him, dropping obscure hints about his mathematical advances but not really stating his more advanced results in an open, readily intelligible, fashion. He wrote two important letters, later called the *Epistola Prior* and the *Epistola Posterior*. What Leibniz learned from these 'letters', perhaps better described as tracts, led him further towards the independent development of the calculus. In fact, his mathematical notation was superior in clarity to Newton's and became the standard one, though national loyalty led to the adoption in England of the latter's often cumbersome 'dots' to indicate differentials. Later, their relatively independent discoveries led to bitter disputes concerning priority. As seen from the vantage point of the present time, Newton's loss of priority must to a great extent be blamed on his own secretiveness. In the eyes of scientists today, not to have published a result is tantamount to not having discovered it!

The *Epistolae*, essentially private letters, and the earlier *De analysi* were, according to Newton and his supporters in later years, the evidence of his priority, though they did not explicitly reveal his results in the calculus. These were only published in 1685, and then only partially, by the Oxford mathematician John Wallis, essentially as a patriotic gesture.

---

[15]Newton to Hooke, 5 February 1676, quoted by Westfall (1980, p. 274).

[16]Newton to Oldenburg, 18 November 1676, quoted by Westfall (1980, pp. 275–276).

3.9   Newton as heretic

Newton was little heard of outside Cambridge for the next few years. He was now financially secure and had settled into college life. He felt affluent enough to make a significant contribution towards the construction of the magnificent new library then under construction - the Wren Library of today. What seems to have been occupying him more seriously than mathematics, optics, and mechanics during this period of comparative withdrawal were chemical and religious interests. Chemistry in Newton's time was still linked to alchemy and was a subject wrapped in secrecy and mysticism. It had not yet undergone the transformation to a rational science as physics and astronomy had done. Newton left innumerable manuscripts concerning his extensive reading in the literature of alchemy as well as his experiments and speculations on chemical matters. These papers, unlike his other studies, are only now being examined in the light of modern scholarship. He owned a large library of alchemical books and strove to make sense of their contents, often by conducting experiments which he recorded with precision in chemical symbols of his own devising. He corresponded with Robert Boyle who, as a chemist, shared some of his interests. His chemical work contributed to some of his ideas concerning the interactions between particles of matter and had already been of service in the development of alloys for mirror-making. Later on, it must have proved useful when he was assaying gold as Master of the Mint.

He also studied Christian theology in great depth and became thoroughly conversant with the extensive writings of the early Christian theologians—the 'Fathers of the Church'. He became convinced that at the Council of Nicaea (325) the true Christian faith had been hijacked by Athanasius and that Arius had been the true follower of Christ. As a result, he denied the importance of the Trinity and regarded the equality of Christ with God as blasphemous: God was supreme. The Bible and the early Christian literature were meticulously searched for evidence supporting his point of view and he wrote copious notes on the subject. Monasticism he regarded as particularly obnoxious and Trinitarianism as equally bad. Thus he gradually became a true heretic, holding views far beyond the bounds of what was then acceptable. Had his ideas become known he would have been expelled from his post and treated as an outcast, as indeed happened to Whiston, his successor. After Newton died, Whiston wrote that his Arianism 'was occasionally known to those few who were intimate with him all along; from whom, notwithstanding his prodigiously fearful, cautious and suspicious Temper[ament], he could not always conceal so important a Discovery'.[17] His puritanical background also ensured that he had no sympathy for the careerists of the church, whether Roman Catholic or Protestant.

To retain his appointments he should in the ordinary course of events have become a minister of the Church of England, something that his conscience would almost certainly not have permitted him. In spite of opposition from

---

[17]Whiston, quoted by Westfall (1980, p. 650).

Barrow, by then Master of Trinity College, to letting a previous applicant off this requirement, he seems in 1675 to have been persuaded to help Newton. With support from influential persons at court, including a confidant of King Charles whose son was a fellow of the College, the condition was eventually relaxed. As a result of Newton's successful petition, the Lucasian professorship received this right in perpetuity. It is probable that Barrow and others knew something of his scruples but recognised his ability and were anxious to keep him as an asset to the College. As will also become evident from later events, Newton was not always unworldly and could be quite adept at obtaining favours from courtiers and politicians.

In 1679, his mother became seriously ill and he went home to nurse her in her final days, or at least to take care of her affairs. It was said that he lovingly changed the dressings applied to her blisters. She died within the year and Newton spent several months afterwards in Woolsthorpe finalising her estate and collecting the debts owed to it. He even brought a successful action in the Chancery court against one Dorothy Elston who had encroached on his land and was calling herself 'Lady of the Manor of Woolsthorp'.

In 1683, Wickins, his roommate of twenty years, decided to resign his fellowship to get married. He was replaced by a young student from Newton's old school in Grantham, Humphrey Newton, who stayed for five years and acted as his secretary. Westfall (1980) conjectures that Wickins and he may have parted on bad terms: Newton appears cool towards him in a few letters that survive from his later years.

In his private writings on the evils of monasticism at around this time he dwelt on monks' problems concerning sexual temptation, remarking that

> The way to be chast is not to contend & struggle with unchast thoughts but to decline them keep the mind imployed about other things: for he that's always thinking of chastity will be always thinking of weomen & every contest wth unchast thoughts will leave such impressions on the mind as shall make those thoughts apt to return more frequently.[18]

Studies of the Bible occupied his attention and perhaps were his main enthusiasm. His philosopher colleague Henry More mentioned in a letter to a friend:

> he came to my chamber where he seem'd to me not onely to approve my Exposition [of the Apocalypse] as coherent and perspicuous throughout from the beginning to the end, but (by the manner of his countenance, which is ordinarily melancholy and thoughtfull, but then mighty lightsome and chearfull, and by the free profession of what satisfaction he took therein) to be in a maner transported.[19]

He (Newton) wrote on comparative religion and, in particular, composed a tract entitled (in English translation) *The Philosphical Origins of Gentile Theology*, in which he pointed out numerous similarities between different religions and

---

[18]Newton, Clark Library ms, of 'Paradoxical Questions', quoted by Westfall (1980, p. 345).

[19]More to Sharp, 16 August 1680, quoted by Westfall (1980, p. 349).

de-emphasised the role of Jesus. This work was never published in his lifetime
and indeed remained unnoticed until comparatively recently.

Newton did not rise early, but slept through the time of morning church
services and tended to work through the evening ones, according to Humphrey.

3.10 Principia

Around 1679 Newton turned once again to the field of mathematics, re-reading
Descarte's *Geometry*. He evidently had developed an antipathy to the algebraic or
analytical approach and felt that all theorems should be proved using Euclidean
geometrical techniques, even if the inspiration for them might have come by
using analytical methods.

> I have often heard him censure the handling geometric subjects by al-
> gebraic calculations; and his book of Algebra he called by the name of
> Universal Arithmetic, in opposition to the injudicious title of Geometry,
> which Des Cartes had given . . . He frequently praised Slusius,[20] Barrow
> and Huygens for not being influenced by the false taste . . . Sir Isaac New-
> ton has several times particularly recommended to me Huygens's stile and
> manner. He thought him the most elegant of any mathematical writer of
> modern times, and the most just imitator of the antients . . . I have heard
> him . . . speak with regret of his mistake at the beginning of his mathe-
> matical studies, in applying himself to the works of Des Cartes and other
> algebraic writers, before he had considered the elements of Euclid with
> that attention, which so excellent a writer deserves.[21]

Cartesian geometry he referred to as 'the Analysis of the Bunglers in Mathematicks'.[22] Using geometrical methods, he extended Euclidean geometry to include concepts similar to algebraic limits, while in considering the interactions
of moving bodies, he rejected Descartes wholly 'mechanical' view, that bodies
could only affect each other by direct contact.

Hooke, by then Curator of Experiments of the Royal Society, tried to persuade Newton to take more interest in its work and, in a letter, reminded him of
some ideas that he [Hooke] had published several years before about planetary
orbits, that bodies gravitate towards each other, that they obey a law of inertia
which causes them to move in straight lines unless deflected by a force and that
the gravitational force is stronger the closer bodies are to one another. In these
remarks he was clearly coming quite close to the idea of universal gravitation.
Newton's luke-warm reply indicated that he had found these ideas interesting,
but concluded

> But yet my affection to Philosophy being worn out, so that I am almost
> as little concerned about it as one tradesman uses to be about another
> man's trade or a country man about learning, I must acknowledge my

---

[20]R.F.W. de Sluze (1622–1685), mathematician of Liège.

[21]Pemberton (1728, Preface, p. 2).

[22]Quoted by Westfall (1980, p. 380).

self avers from spending that time in writing about it wch I think I can
spend otherwise more to my own content & ye good of others . . . '[23]

Nevertheless, Hooke continued to correspond with Newton and even claimed,
correctly, but based on a false argument, that gravitational attraction obeyed an
inverse square law. This finally stimulated Newton at the end of 1679 to derive
the important theorem that Kepler's discovery that the planets move in ellipses
actually led directly to such a law! However, he characteristically put the proof
aside and did not tell other people of it.

In the early 1680s other able mathematicians were slowly becoming aware
of Newton's discoveries. They included his later rival in matters of reputation,
Leibniz. Apart from Hooke, Christopher Wren, and, above all, Halley were also
concerned with the problem of orbits. Finally, in August 1684, the genial Halley
visited Newton in Cambridge to ask him this question: What form would a
planet's orbit take if the attractive force towards the Sun followed an inverse
square law? As related by one of Newton's later disciples, the mathematician
Abraham DeMoivre,

> Sr Isaac replied immediately that it would be an ellipsis, the Doctor
> [Halley] struck with joy & amazement asked him how he knew it, why
> saith he I have calculated it, whereupon Dr Halley asked him for his
> calculation without any further delay, Sr Isaac looked among his papers
> but could not find it, but he promised him to renew it, & then to send it
> him . . . [24]

The result of Halley's question was a small treatise of nine pages called *De
Motu Corporum in Gyrum*, (The motion of bodies in orbit), written within the
short space of a few months, a masterpiece which showed that an elliptical orbit
implies an inverse square law and also that an inverse square law implies an orbit
in the shape of a conic section—in other words, it even included the parabolic
orbit followed by a comet.

The composition of *De Motu* was the stimulus that Newton needed to get
started on his masterpiece, the *Principia*. By the end of 1684 this work was in
full swing. For two years he thought of almost nothing else. All the partially
complete works, hints, and memoranda of the previous twenty years formed only
the basis for his masterful synthesis of mathematics and mechanics applied to
the description of nature. He took a Euclidean approach and built up a complete
system from the most elementary fundamental facts. Every step in the arguments
was examined carefully; matters which had previously been glossed over were now
investigated fully and every conclusion now justified with care and precision.
Concepts such as *mass*, *momentum*, and *force* now had to be defined properly.
Clarity on these concepts only emerged during the process of putting his thoughts
into logical order for the *Principia*. The main subjects it covered were the laws

---

[23]Newton to Hooke, 28 November 1679, quoted by Westfall (1980, p. 384).

[24]Newton's recollection, told to DeMoivre, Schaffner Collection, University of Chicago, ms
1075-7, quoted by Westfall (1980, p. 403).

of motion and of gravitation, giving the correct explanation of the tides. His intensity at this period was recorded by Humphrey Newton

> So intent, so serious upon his studies, y^t he eat very sparingly, nay, often-times he forgot to eat at all, so y^t going into his Chamber, I have found his Mess untouch'd of wch when I have reminded him, [he] would reply, Have I; & then making to ye Table, would eat a bit or two standing ... At some seldom Times when he design'd to dine in ye Hall, would turn to ye left hand, & go out into ye street, where making a stop, when he found his Mistake, would hastily turn back, & then sometimes instead of going into ye Hall would return to his Chamber again ... When he has sometimes taken a Turn or two [in the garden], has made a sudden stand, turn'd himself about, run up ye Stairs, like another Alchimedes, with an eureka, fall to write on his Desk standing, without giving himself the Leasure to draw a Chair to sit down in.[25]

The *Philosophiae Naturalis Principia Mathematica* (Mathematical Principles of Natural Philosophy) consists of three 'books'. The first (Fig 3.5) of these was presented to the Royal Society in April 1686 while the printing of the whole work was finished in June 1687. The financial risk and the effort of seeing the work through the press were undertaken by Edmond Halley, who was impelled by, and clearly understood, its revolutionary importance. This was a courageous gesture as his income from the Royal Society was not large and he was mostly dependent on rentals from houses that he owned (Cook 1998). In addition, Newton's irascibility and frequent threats to give up the whole project had to be contended with. Newton was incensed by Hooke's claims to priority concerning the inverse square law. Hooke had in fact done little more than guess it whereas Newton had proved outright that it fitted the facts. At this time he made his famous statement:

> Philosophy is such an impertinently litigious Lady that a man had as good be engaged in Law suits as have to do with her. I found it so formerly & now I no sooner come near her again but she gives me warning.[26]

Hooke's attitude disturbed Newton so much that he resented it all his life. In his original manuscript he changed a reference from 'distinguished Hooke' to plain 'Hooke' and his name was eliminated completely from a later edition.

Typically, Newton made no attempt to make his work easy to read, unlike Galileo. It was, of course, written in Latin, the universal scholarly language then understood by well-educated people in all European countries. He simply did not want to become involved in uninformed arguments. As a friend remarked: 'And for this reason, namely to avoid being baited by little Smatterers in Mathematicks, he told me, he designedly made his Principia abstruse ... '[27] Not sur-

[25] Humphrey Newton, Keynes ms 135, quoted by Westfall (1980, p. 406).

[26] Newton to Halley, June 1686, quoted by Westfall (1980, p. 448).

[27] Derham to Conduitt, 8 July 1733, Keynes ms 133, quoted by Westfall (1980, p. 459).

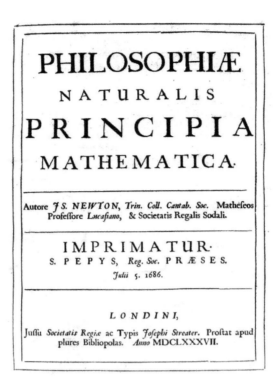

FIG. 3.5. Title page of the first edition of the *Principia* (1687). Samuel Pepys, the Secretary of the Navy, as President of the Royal Society, issued the official license to print.

prisingly, it was understood by very few laymen, let alone scientists, though the importance of its central ideas were clear to many who heard them.

3.11 Fame

The *Principia* was an immediate success: British mathematicians quickly recognised the advance that it represented. Partly through the efforts of Halley, copies were sent to the Continent where they were reviewed in the academic journals of the time. The advances it described were a revelation to almost all able mathematicians who saw it. Nevertheless, the Dutch 'natural philosopher' Huygens and the German mathematician Leibniz remained partially attached to Aristotelian ideas and could not accept Newton's notion of a force without a tangible agency, such as his law of gravity represented. However, his ideas were so powerful in their application that in the course of time they swept away such reservations. Newton's laws were triumphant for practical purposes and remain so to this day. They only break down under the most extreme conditions, where relativity and quantum mechanical principles must be applied.

Leibniz, later to become involved in disputes with Newton about claims of mathematical priority, reacted to the appearance of the *Principia* immediately, by publishing several papers that described his own views on refraction, motion in a resisting medium, and the theory of orbits in the journal *Acta Eruditorum*, published in Leipzig.

3.12   University politician

Newton's leap to fame occurred just at the time when the Roman Catholic King James II, to the great dismay of many sectors of society, inherited the throne of England. To consolidate his power, James set about placing Catholics in positions of influence wherever he could, and the University of Cambridge was a prime target. They soon rebelled, the immediate cause being an order by the King to confer the degree of MA on a Benedictine monk. Newton, the puritanical heretic, announced his support for his Anglican colleagues and became a leader of the opposition. He was not afraid to state his views clearly and became one of eight commissioners appointed by the University to answer a summons to London issued by the furious King. He took an active part in the drafting of position papers. In hearings before the Ecclesiastical Commission, a kangaroo court chaired by the powerful and dangerous Judge Jeffreys, they were told in no uncertain terms to conform. Their leader, the Master of Magdalene College, was deprived of his office.

The University was unrepentant and continued to disobey James' commands. Fortunately for them, the 'Glorious Revolution', which placed the protestant King William of Orange on the English throne, broke out a year later, in November 1688, and came to a rapid and successful conclusion. The supporters of the revolution were mainly from the 'Whig' party and their opponents were 'Tories'. The Glorious Revolution was a watershed in British politics, because thereafter the government was controlled by politicians rather than the monarchy.

Newton's outspoken courage in promoting the protestant cause made him known both to the government and to the University as a 'politically reliable' person. He was elected to represent the University in 1689 at the convention called to settle the Revolution, which afterwards became a parliament. The story goes that the only time he spoke in Parliament was to request that a window be closed! Afterwards he was appointed by Parliament as a commissioner to oversee tax collection from the University. The new King even intended to reward him with the headship of King's College, which was then vacant, but tactfully did not press the matter when the members of the college objected.

3.13   Change of character

Following his political adventures and the growth of his reputation as the greatest living British scientist, Newton began to cultivate a social life among influential people. He subscribed conspicuously to various 'good causes' and had his portrait (Fig 3.6) painted in London by Sir Godfrey Kneller, the leading portraitist of the time. He was evidently beginning to enjoy his fame.

FIG. 3.6. Newton's first portrait was painted by Sir Godfrey Kneller in 1689, when he was 46. It shows him two years after the publication of the *Principia* and about three years before his mental breakdown (Schuster and Shipley 1917).

During his stay in London on Parliamentary business he finally met the famous Christiaan Huygens face-to-face and also the philosopher John Locke. On closer acquaintance, the latter turned out to be another closet Arian. At one point (ca. 1690), Newton composed a pro-Arian pamphlet *An historical account of two notable corruptions of Scripture, in a letter to a friend* which he sent to Amsterdam for anonymous publication. Perhaps fortunately it was not actually published at the time, since its authorship would surely have leaked out and would have meant an abrupt end to his career.

3.14 Fatio

Another acquaintance that Newton made during his stay in London around 1689 was that of a promising young Swiss mathematician, Fatio de Duillier,

then twenty-five years old. They probably met for the first time at a meeting of
the Royal Society in June, when Huygens presented a talk on light and gravity.
Fatio was to become one of the first of Newton's 'disciples'.

They met again in July on a trip to the court at Hampton Court Palace.
Newton evidently found his company very agreeable: for example he asked Fatio
if there was space available at his lodgings which he could occupy during the
coming Parliamentary session. Over the following few years they saw a lot of each
other and Fatio came to share Newton's non-mathematical interests—Arianism,
alchemy, and biblical prophesies. Fatio also acted as an intellectual go-between
with Huygens.

Newton came to regard Fatio as his protegé and perhaps as an object of
platonic love. When Fatio wrote to say that he was seriously ill in London in
November 1692, following a visit to Cambridge, Newton became very concerned
and wrote telling him to get medical advice, with a promise of money if required.
As Fatio recovered, he was also offered an allowance if he would reside in Cambridge. Newton rarely showed much personal interest in other people and his
attention to Fatio seems to have been quite unusual.

Their close relationship ended in 1693. In May, Fatio met an alchemist who
convinced him to go into business to manufacture a mercury-based medicine
which would supposedly make his fortune. He asked Newton to help him financially towards this scheme and to come to London to give him advice. Newton
probably did go to London a few times during the following months, but his
actions there remain a mystery. One thing is clear: something caused him to
fall out with Fatio, and shortly thereafter his generally disturbed mental state
became so bad that he may be said to have suffered a nervous breakdown. He
remained rather cool towards Fatio in later years although the latter tried more
than once to repair the friendship.

3.15 Nervous breakdown

Newton's affliction, whatever its cause, reached its worst point in September
1693. He wrote rather paranoid letters to at least two of his friends for which he
afterwards had to apologise. One was to Samuel Pepys, Secretary of the Navy
and a member of the Royal Society, who he had often seen in London. Pepys had
invited him for a visit that he at first accepted and now wished to refuse:

> I am extremely troubled at the embroilment I am in, and have neither
> ate nor slept well this twelve month, nor have my former consistency of
> mind. I never designed to get anything by your interest,[28] nor from your
> acquaintance, nor by King James's favour, but am now sensible[29] that
> I must withdraw from your acquaintance, and see neither you nor my
> friends any more ... [30]

[28] ... I never schemed to take advantage of your influence ...
[29] ... I now feel ...
[30] Newton to Pepys, 13 September 1693, quoted by Westfall (1980, p. 534).

A few days later he wrote to John Locke:

> Being of opinion that you endeavoured to embroil me wth weomen & by
> other means I was so much affected with it as that when one told me you
> were sickly & would not live I answered twere better if you were dead.
> I desire you to forgive me this uncharitableness. For I am now satisfied
> what you have done is just & I beg your pardon for having hard thoughts
> of you ...[31]

This breakdown, which occurred when Newton was 51, could conceivably have
been associated with a kind of 'male menopause', and perhaps resulted from a
realization that his mathematical creativity was waning. He made few important
discoveries after this time even though he remained a competent mathematician.
Much of his later mathematical and scientific work consisted of polishing his
earlier accomplishments and defending his priority. Some have speculated that
his illness could have had a physical cause, namely the result of poisoning from
the heavy metals such as antimony, arsenic, lead, and mercury, that he used in
his chemical experiments (e.g. Spargo and Pounds 1979).

Fortunately, his friends realised that his behaviour was atypical and that
he was in a disturbed mental state. His touchiness had always been evident to
Whiston and Flamsteed, and even the amiable Locke had to say '...he is a nice
[overly precise] man to deal with, and a little too apt to raise in himself suspicions
where there is no ground' (quoted by Andrade 1947).

3.16   Opticks

Newton's greatest work, after the *Principia*, was his *Opticks*, printed in 1704. By
1695 he had prepared it in draft form but was reluctant, as usual, to publish.
He told the mathematician, John Wallis, that he was afraid publication would
cause him trouble. Wallis replied:

> You say, you dare not *yet* publish it. And why *not yet*? Or, if not now,
> when then? You adde, lest it create you *some trouble*. What trouble *now*,
> more than at another time? ... And perhaps some other may get scraps
> of ye notion, & publish it as his own; & then 'twil be His, not yours;
> though he may perhaps never attain to ye tenth part of what you are
> allready master of. Consider, that 'tis now about Thirty years since you
> were master of those notions about *Fluxions* and *Infinite Series* ... But if I
> had published the same or like notions, without naming you; & the world
> possessed of anothers *Calculus differentialis*,[32] instead of your fluxions:
> How should this, or the next Age, know of your share therein? ... I own
> that Modesty is a Vertue; but too much Diffidence (especially as the world
> now goes) is a Fault.[33]

---

[31] Newton to Locke, 16 September 1693, quoted by Westfall (1980, p. 534).

[32] This refers to the fact that the first actual publication of Newton's early ideas on the
calculus was, with proper attribution, in Wallis's *Opera*, dated 1693–1695.

[33] Wallis to Newton, 30 April 1695, quoted by Westfall (1980, pp. 523–524).

The consequences of his early reluctance to publish were to embitter his old age, when he became preoccupied with proving his priority. It seems that Hooke had remained his main bugbear, since he began to publish his optical and mathematical work only in 1704, after the latter's death. He took the opportunity to include his *Tractatus de quadratura*, concerning integration, and *Enumeratio ...*, concerning third-order curves, as appendices to the first edition of his *Opticks*, even though they represented work he had done thirty or more years before.

The importance of *Opticks* lay in its originality: unlike some of the results reported in the *Principia*, nobody could claim to have anticipated him. It was also written in English, in less formal language, and could be understood easily. Even though the most important results had been reported to the Royal Society in 1672, the new work had broad influence. Another of Newton's young disciples, Abraham DeMoivre, a Huguenot immigrant, saw the Latin version through the press.

3.17   Warden of the Mint

London life had appealed to Newton since his many visits there as a political representative of the University of Cambridge. At the time the *Principia* was being printed he had become intimate with Charles Montague (1661–1715), formerly a fellow-commoner undergraduate of Trinity College, who was a poet, an amateur of science, and President of the Royal Society. More importantly, he was a highly successful young Whig politician and something of a genius at economics. He was effectively the founder of the Bank of England. Appointed Chancellor of the Exchequer, or Minister of Finance, he faced the problem that the coins in circulation (there were no banknotes) were no longer worth their face value because criminals were cutting or filing small quantities of gold or silver from their edges. A committee of leading financiers and scientists including Newton was appointed to report on the problem. New machinery that produced an evenly milled edge was brought into use so that any deviation from standard size could instantly be recognised. In December 1695 it was decided that all existing coins would have to be melted down and replaced by the improved ones. The position of Warden of the Mint was offered to Newton in March 1696 and he was in office just over a month later. The genial Montague, once told by the arch-Tory Jonathan Swift that he was 'the only Whig in England I loved, or had a good opinion of',[34] informed Newton that the post was worth five to six hundred pounds 'and has not too much bus'nesse to require more attendance than you may spare'! In accordance with the morality of the times, Newton did not resign his Cambridge positions, which he treated as a sinecure, until five years later. In fact, the Wardenship of the Mint was itself usually so regarded. The motivation for Montague's action in appointing Newton is not very clear. While there may have been some altruistic aspects, in that he admired Newton's achievements very much, it would have been very unusual to give away a lucrative government

[34]Swift to Stella, 30 September 1710; see Swift (1908).

office in this manner. It is quite probable that Newton's clearly demonstrated loyalty to the Whig cause and to Montague himself was a factor. The French philosopher Voltaire suggested many years later that the attraction of Newton's vivacious niece, Catherine Barton, the daughter of a stepsister from his mother's second marriage, was the real reason but she and Montague probably did not meet until after the appointment.

Contrary to Montague's apparent expectations, as might have been predicted, Newton entered with zeal into his new task. He rapidly acquired a detailed mastery of the Mint's metallurgical processes, complex accounting system, and administration. He even took a leading part in the prosecution of counterfeiters. Suffice it to say that he accomplished the task of the recoinage with great success and entrenched his position very securely. His obvious competence enabled him to retain his job, essentially a political appointment, in spite of changes later on in government. At the end of 1699, the Master of the Mint died. This position had also been treated as a sinecure, but was a much more valuable one than the Wardenship as it entitled the holder to a commission on the total number of coins minted. Even though Montague was then out of office, Newton's technical expertise was sufficiently valuable to the government that he was appointed to the position. He now had an income of several thousand pounds in most years, about ten to twenty times his academic salary.

On 29 January 1696, Newton received a problem issued as an international challenge in the journal *Acta Eruditorum* by the Swiss mathematician Johann Bernoulli. This was to find the shape of the slope down which a body would slide the fastest from one given place to another. According to Conduitt, Newton started to work on it, tired from his day at the mint, at four o'clock in the afternoon 'but did not sleep till he had solved it wch was by 4 in the morning'.[35] He sent his solution anonymously to the Royal Society in London for publication. Bernoulli was not deceived and recognised his style 'tanquam ex ungue leonem' (as the lion is recognised by his print). Other solutions were provided by Leibniz, Johann Bernoulli, his brother Jakob, and the French mathematician de l'Hôpital. Newton may have solved the problem in the fastest time!

3.18   President of the Royal Society

After his move to London Newton at first took little interest in the Royal Society, in spite of having been elected a member of Council on two occasions. However, at the end of 1703, following the death of Hooke, he was elected its President, an office he continued to fill until his own death. Through his efforts, the meetings were enlivened with experimental demonstrations. He came to occupy a rather dictatorial position. Stukeley reported:

> Every thing was transacted with great attention and solemnity and de-cency; nor were any papers which seemed to border on religion treated

[35]Conduitt, Keynes ms 130.5, quoted by Westfall (1980, pp. 582–583).

FIG. 3.7. Isaac Newton. Mezzotint by John Smith dated 1712, from a portrait
   by Kneller of 1702 (SAAO).

without proper respect. Indeed his presence created a natural awe in the
assembly ... (Stukeley 1936, p. 80)

His friends in the early eighteenth century included William Whiston, his
successor as Lucasian Professor, and another young disciple, Samuel Clarke, the
translator of his *Opticks* into Latin, both of whom were Arians. Whiston was
foolish enough to publish his Arian views openly and was eventually removed
from his University position. Clarke trod a fine line but succeeded in keeping his
post, also at Cambridge. Others were infected with Arian ideas, even the ambitious Richard Bentley whose conscience bothered him so little that he occupied

the Regius Professorship of Divinity without any apparent qualms!

3.19   Relations with Flamsteed

Newton's relations with Flamsteed constitute what Westfall (1980) describes as
'the most unpleasant episode of Newton's life', an example of his sometimes
tyrannical behaviour in later years. Flamsteed, as Astronomer Royal and the
pre-eminent British observer of the time, was in possession of the best data on
the position of the Moon, to which Newton wanted to have access in order to test
his theory of its motion. In 1694–1695 Newton attacked this problem again—
his last major piece of scientific work. He approached Flamsteed for data and
was in fact given some on the condition that he did not show them to other
people and would allow Flamsteed to see the refined theory first. All he really
wanted in return was respect, something that Newton never had for him. Both
were very touchy men and each irritated the other. Further requests for data
eventually met with stalling. In the second edition of the *Principia* Newton in a
fit of petulance removed all references to Flamsteed. His attempt at improving
his theory proved a failure—one which he tried to blame on Flamsteed's tardiness
in supplying data.

Flamsteed's main aim in life was to complete a new catalogue of the stars. Indeed, since the practical way to measure the Moon's orbit was to plot it against
the 'fixed' or background stars, their positions had to be determined with as
much precision as possible. The publication of the Catalogue was likely to be expensive and in 1704 Flamsteed approached Queen Anne's consort, Prince George,
for funding. Unfortunately, Newton, as President of the Royal Society, managed
to obtain control of the funds promised and he used his power to browbeat
Flamsteed. Flamsteed's comprehensive plan for the publication of his *Historia
Coelestis* and an accompanying atlas was thrust aside in favour of an abbre-
viated edition. He resented that profit was to be made by a publisher, while
he had worked for thirty years on a miserly salary and had had to put £2000
of his own money into the work to purchase the instruments he had used and
to employ assistants. Fortunately, the printing work proceeded very slowly and
Flamsteed used the delays to work on his own version of the catalogue. In 1710
the Queen suddenly appointed Newton, obviously at his own instigation, and
other Fellows of the Royal Society, to be 'Visitors' to the Royal Observatory—
effectively a board of inspectors. 'Her Majesty' demanded that the catalogue be
finished forthwith, but Flamsteed claimed more time was needed to improve the
planetary tables. Newton nevertheless pushed ahead with his truncated version
of the catalogue, using Halley to complete the necessary calculations. He called
a meeting of the Visitors and Flamsteed in 1711 to show the latter who was
in control—very foolishly on his part, since Flamsteed knew just how to make
him lose his temper in public. Ostensibly, the Visitors were to check that the
instruments were in proper order and fit for continued use. Flamsteed told them
that they were in fact his own and did not fall under their authority. He went on

to say how Newton's insistence on publishing the catalogue his way had robbed
him of the fruits of his labours:

> at this he fired & cald me all the ill names Puppy &c. that he could
> think of. All I returned was I put him in mind of his passion desired him
> to govern it & keep his temper. This made him rage worse, & he told
> me how much I had receaved from ye Goverm^t in 36 years I had served
> ...I sayd nothing to him in return but with a little more spirit than I
> had hitherto shewd told them, *that God* (who was seldom spoke of with
> due Reverence in that Meeting) *had hitherto prospered all my labours &*
> *I doubted not would do so to an happy conclusion,* took my leave & left
> them.[36]

Flamsteed also had his politically powerful friends. When Queen Anne died
in 1714 and Newton's main protector Halifax (formerly Montague) the following
year, he managed to get hold of the unsold copies of the unauthorised *Historia
Coelestis*, which he burned (except for the parts he could use in his own edition)
as a 'Sacrifice ...to Heavenly Truth'.[37] He almost managed to finish his own
Historia Coelestis Britannica before his death in 1719; fortunately his former
assistants saw that it was completed.

3.20   Knighthood

In 1701 Newton stood in the Parliamentary election for a seat in Cambridge and
was elected. However, that Parliament was short lived, being dissolved following
King William's death the next year, and he was unwilling to stand again in the
election that immediately ensued. Halifax put pressure on him as a good Whig
when it came to the election after that, in 1705. He even persuaded the new
Queen, Anne, to visit the University to lobby on Newton's behalf. After dining
at Trinity College, she showed her approval by conferring a knighthood on him.
When we think of *Sir* Isaac, it is sobering to be reminded that his knighthood was
conferred as much for party political reasons as for his contribution to science!

In his later years, spent in London, he lived comfortably and fashionably in
a well-furnished house, even possessing two silver chamberpots. He kept a coach
and had as many as six servants—housekeeper, cook, maid, footman, etc. It is
clear that he could afford the life of an established member of the upper middle
class. The house that he occupied in St Martin's Street from 1710 is shown in
Fig 3.8. He paid £100 per annum in rent.

Stukeley (1752) reported that Newton enjoyed music, but opera was too much
for him: 'Sir Isaac said they were very fine entertainments; but that there was
too much of a good thing; it was like a surfiet at dinner. "I went to the last
opera," says he "The first act gave me the greatest pleasure. The second quite
tired me; at the third I ran away"'.[38]

---

[36]Flamsteed to Sharp, 22 December 1711, (Baily) quoted by Westfall (1980, p. 693).

[37]Flamsteed, (Baily) quoted by Westfall (1980, p. 696).

[38]Stukeley (1936, p. 14).

FIG. 3.8. Newton's house in St Martin's Street, London, which he occupied for fifteen years from 1710. Coincidentally, in 1774 it was purchased by the musicologist Charles Burney, a friend of William Herschel. It was demolished in 1910. (From F.J. Robotham, before 1922.)

Newton's niece, Catherine Barton, had taken up residence to 'keep house' for him in the closing years of the seventeenth century. She was a witty and charming young lady who rapidly became known in high society. By 1703 she was the intimate friend and probably the mistress of Charles Montague, since 1700 Earl of Halifax. She was the toast of the Kit Kat Club, the social centre of the Whig party. Even the caustic Swift remarked 'I love her better than any body here [London]', and often visited her.[39] Halifax died in 1715 and Catherine, to whom he left a substantial sum, married John Conduitt in 1717. Conduitt was of military background and was evidently wealthy, possibly as a result of having been commissary to the British army at Gibraltar for a number of years. Such posts were usually regarded as opportunities for self-enrichment.

3.21 Dispute with Leibniz

During his later years Newton no longer avoided controversy and became for a while an active participant in one. He became concerned to prove his priority in the development of calculus. Leibniz was equally determined. The issue was confused by the fact that both he and Leibniz hid behind others who ostensibly

[39]Swift to Stella, 24 March 1711; see Swift (1908).

were merely supporters. On Newton's side the mathematician Wallis tried to prove priority by publishing both of Newton's early *Epistolae* as part of his own *Opera* (Works) in 1699.

In the course of this propaganda war, Newton enlisted the support of the Royal Society, which in 1713 published an anonymous pamphlet entitled *Commercium epistolicum ...* or *The Correspondence of the Learned John Collins and Others Relating to the Progress of Analysis*, which contained copies of the correspondence with a commentary highly biased towards Newton's point of view. Modern research has shown that he himself was its author, a fact which he never disclosed during his lifetime.

This undignified argument was slowed down by Leibniz's death in 1716 but was nevertheless continued for some years afterwards by various supporters.

Leibniz's viewpoint was championed by the Swiss mathematician Johann Bernoulli, who tried to keep the dispute going long after other intermediaries had tired of it. After one of Bernoulli's attempts to provoke him in 1722, Newton told DeMoivre 'Since my youth I have hated disputes. Now that I am eighty years old I detest them'.[40]

3.22   Old age

In 1713, a second edition of the *Principia* was completed at the instigation of Richard Bentley, now Master of Trinity College, and under the care of the young mathematician Roger Cotes, who experienced Newton's querulous behaviour as his work progressed. It contained many corrections and new proofs, ending with a *General Scholium*, or explanatory note, against mechanical attempts, such as Descartes' vortex theory, to explain the motions in the heavens. For once he expatiated on his notion of an omnipresent God:

> This most beautiful system of the sun, planets, and comets, could only proceed from the counsel and dominion of an intelligent and powerful Being ... This Being governs all things, not as the soul of the world, but as Lord over all; and on account of his dominion he is wont to be called *Lord God*, παντοκρατωρ (pantocrator) or *Universal Ruler* ... He is omnipresent not *virtually* only, but also *substantially*; for virtue cannot subsist without substance. In him are all things contained and moved; yet neither affects the other: God suffers nothing from the motion of bodies; bodies find no resistance from the omnipresence of God. (Newton 1729)

As he had done many times before, Newton emphasised that he had not found the cause of gravity and would make no hypotheses: *Hypotheses non fingo*, as he famously put it in the Latin text. However, the universe had been put together by God and had continued without change thanks to the laws He had put in place.

Pirate copies of the *Principia* appeared in Amsterdam the following year. Characteristically, Cotes received no payment and scant thanks from Newton. A

---

[40]DeMoivre to Varignon, January 1722, quoted by Westfall (1980, p. 791).

third edition came out in 1726 under the editorship of Henry Pemberton, who was paid for his efforts.

> Though his memory was much decayed, I found he perfectly understood his own writings, contrary to what I had frequently heard in discourse from many persons. This opinion of theirs might arise perhaps from his not being always ready at speaking on these subjects, when it might be expected he should. But as to this, it may be observed, that great genius's are frequently liable to be absent [minded], not only in relation to common life, but with regard to some of the parts of science they are best informed of ... (Pemberton 1728, Preface, p. 3)

Newton probably continued to be an Arian but at the same time was active on commissions to promote the construction of new churches and the rebuilding of St Paul's cathedral in the City of London, following its destruction in the Great Fire of 1666. His private religious views were known only to a small circle of his friends. From time to time he visited the future King George II and Queen Caroline, who enjoyed his opinions on 'philosophy' and biblical chronology. While alone, it was said that he was hardly ever without a pen in his hand and a book before him.

Some time before his death, Newton reminisced to an unknown person on his lifetime of research:

> I don't know what I may seem to the world, but, as to myself, I seem to have been only like a boy playing on the sea shore, and diverting myself in now and then finding a smoother pebble or a prettier shell than ordinary, whilst the great ocean of truth lay all undiscovered before me.[41]

In his old age he became quite generous with cash gifts both to his extended family and to young mathematicians. Those whom he supported included James Stirling and Colin Maclaurin, both contributors to the development of mathematical analysis. He was by far the most financially successful member of his own family and was looked upon by his relations as a source of help when required.

Conduitt recorded that Newton was of 'middle stature' and quite plump in old age (Fig 3.9). His hair had been white from the age of 30, but he was not bald. He kept all his teeth but one until he died. He could still read without spectacles in 1726. His eating habits were moderate: Stukeley mentions several other details, such as that Newton's breakfast consisted of bread and butter with a tea made by boiling orange peel in water. He drank wine with dinner but otherwise only water. He would not eat rabbits because of the cruel way in which they were killed. He suffered from kidney stones in his last few years and also from some degree of incontinence.

His last illness involved a violent cough and agonising bladder pains. Offered the last sacrament of the Church by the Conduitts, he refused—probably a final defiant gesture of Arianism. He died on 20 March 1727.

---

[41] J. Spence, quoted by Westfall (1980, p. 863).

FIG. 3.9. Newton aged 82, in 1725, engraving from a portrait by Vanderbank (Lebon 1899).

After such a long and distinguished life, society at large showed their recognition of his contribution to the advancement of science and his body was laid in state in Westminster Abbey, where he was later buried. His pall-bearers were among the most important persons in the land, and included the Lord Chancellor. This show of appreciation for the life of an intellectual impressed Voltaire, then a visitor to England, who believed that such treatment would never have been granted in his native France, still at that time an absolute monarchy. In 1731 an ornate monument was erected above his tomb in the nave of Westminster Abbey. The concluding lines of the inscription read

Sibi gratulentur mortales

Tale tantumq extitisse
Humani Generis Decus.

'Let mortals rejoice that there has existed such and so great an ornament to the human race'.

References

Andrade, E.N. da C., 1947. Newton, in *Newton Tercentenary Celebrations*, Cambridge University Press, Cambridge, pp. 3–23.

Cook, A., 1998. *Edmond Halley: Charting the Heavens and the Seas*, Clarendon Press, Oxford.

Lebon, E., 1899. *Histoire Abrégé de l'Astronomie*, Gauthier-Villars, Paris.

Mitchell, O.M., 1856. *The Orbs of Heaven*, 6th edn, Nathaniel Cooke, London.

Newton, Isaac, 1672a. A Letter of Mr. ISAAC NEWTON, Professor of Mathematics in the University of Cambridge; containing his New Theory of Light and Colours . . . , *Phil. Trans. Roy. Soc.*, **6**, 80, 3075.

Newton, Isaac, 1672b. An Account of a New Catadioptrical Telescope, invented by Mr. Newton, F.R.S. and Professor of Mathematics in the University of Cambridge, *Phil. Trans. Roy. Soc.*, **7**, 81, 4004.

Newton, Isaac, 1728. *Optical Lectures Read in the Public Schools of the University of Cambridge*, Printed for Francis Fayram, London, p. 1.

Newton, Isaac, 1729. *Sir Isaac Newton's Mathematical Principles of Natural Philosophy and His System of the World* tr. Motte, Andrew, ed. Cajori, Florian, University of California Press, Berkeley, CA (1934).

Pemberton, Henry, 1728. *A View of Sir Isaac Newton's Philosophy*, S. Palmer, London.

Robotham, F.J., ca. 1922. *Story–Lives of Great Scientists*, Wells, Gardner, Darton, and Co. Ltd, London, undated.

Schuster, A. & Shipley, A.E., 1917. *Britain's Heritage of Science*, Constable & Co., London.

Spargo, P.E., and Pounds, C.A., 1979. Newton's 'Derangement of the Intellect'. New Light on an Old Problem, *Notes Records Roy Soc.*, **34**, 11–32.

Stukeley, William, 1936. *Memoirs of Sir Isaac Newton's Life, being some account of his family and chiefly of the junior part of his life*. From a manuscript dated 1752 in the possession of the Royal Society. Ed. White, A.H., Taylor and Francis, London.

Swift, Jonathan, 1908. *The Journal to Stella*, ed. Frederick Ryland, George Bell and Sons, London.

Turnbull, H.W. (ed.), 1959. *The Correspondence of Isaac Newton*, **1**, 74–75.

Westfall, R.S., 1963. Short-writing and the State of Newton's Conscience, 1662 (I), *Notes Records Roy Soc.*, **18**, 10–16.

Westfall, R.S., 1980. *Never at Rest: A Biography of Isaac Newton*, Cambridge University Press, Cambridge.

4

WILLIAM HERSCHEL: SURVEYING THE HEAVENS

*Then, speaking of himself, he said, with a modesty of manner which quite
overcame me, when taken with the greatness of the assertion:*
  *'I have looked further into space than ever human being did before me.
I have observed stars of which the light, it can be proved, must take two
million years to reach the earth.'*
  *I really and unfeignedly felt at the moment as if I had been conversing
with a supernatural intelligence ...* [1]

4.1 The Herschel family

From a note written by William Herschel himself,[2] it appears that his family
can be traced back to a brewer called Hans Herschel in the town of Pirna near
Dresden in Saxony, one of the eastern states of Germany. This Hans's second son,
Abraham, the grandfather of William, was born in 1651 and learned gardening
in the Elector of Saxony's gardens. He later became a landscape gardener near
Magdeburg, about 60 km west of Berlin. In spite of his manual occupation he
was interested in intellectual matters and knew something of arithmetic, drawing,
and music. His youngest son Isaac, born in 1707, the father of William Herschel,
showed an early talent for music and learnt to play the violin by ear. At first, he
had to earn his living by gardening, but he soon saved enough money to pursue
his real interest by taking formal lessons. His first musical job was in Potsdam,
the seat of the Prussian court, near Berlin, where he became an oboist. However,
the Prussian way of life was not to his liking and he moved on, obtaining a post
as a musician with the Foot Guards in Hanover.

Germany in the eighteenth century was a collection of small states, usually
ruled by princes. Political unity was still more than a century away. Each ruler
maintained a court, which gave encouragement to artists and musicians, according to his taste. The Herschel family were part of the infrastructure of this court
life, to which they contributed their skills in gardening and music. Their home
territory of Hanover was at that time connected with Great Britain through the
fact that they shared the same royal family. In fact, the Elector of Hanover,
George I (of England) had inherited the British throne in 1714.

Isaac Herschel married Anna Ilse Moritzen of the nearby town of Neustadt.
They had ten children, of whom six survived. The eldest was Sophie Elisabeth,
born 1733, who later married George Griesbach, a musician. Jacob, born 1734,

---

[1] Thomas Campbell, quoted by Lubbock (1933, p. 336).
[2] Dreyer (1912).

followed. Friedrich Wilhelm himself, later called Frederick William in England, was born 15 November 1738. Next came Alexander, born 1745, and Caroline Lucretia (Lina), born 16 March 1750. Dietrich, the youngest, was born in 1755. All members of the family are mentioned because they interacted frequently throughout their lives. Caroline was destined to serve as William's able astronomical assistant; the other male siblings became musicians.

4.2 Hanover years

In spite of their relative poverty, the Herschels maintained an interest in the world about them and were very conscious of the value of a good education. Isaac Herschel enjoyed astronomy and delighted in showing his children the night sky. Anna was, however, illiterate. Fortunately, the children were able to attend a school operated by the Hanover garrison. William turned out to be highly intelligent and quickly absorbed all that his schoolmaster could teach him. By the age of 15 he had to earn his keep, which he did by becoming an oboe player in the Hanoverian Guards. Part of the money he and his older brother Jacob received was used to pay a private tutor to teach them French. This man, a Mr Hofschlager, was kind and well-educated and encouraged his intelligent pupil to continue studying on his own. His little sister Caroline related how she was kept awake at night by the debates which took place between her father and brothers on all kinds of metaphysical and mathematical subjects. The two older brothers shared a bed. William would often find, after he had been discoursing at length on some erudite matter, that Jacob had fallen asleep.

When Frederick II of Prussia, known afterwards as Frederick the Great, became king in 1740 he started a series of wars which affected much of Europe. During this period, English and Hanoverian forces were involved in a lengthy campaign against the French, culminating in a victory at the Battle of Dettingen. Isaac had to accompany his band into battle but was stricken by asthma and rheumatism as a result of sleeping in wet conditions while on campaign. However, for a decade from 1746 there was peace. In 1756, while William was still a bandsman in the army, the Foot Guards were sent over to England to aid in defending that country against the French, who were thought to be on the verge of invading. They were quartered in Maidstone, Kent, where William had enough free time to learn English and get to know several musical families. The invasion never materialised and the troops after some time returned to Hanover.

William's chief souvenir of his visit was a copy of John Locke's *An Essay Concerning Human Understanding*. Having lived through the seventeenth-century revolutions in England, Locke (1632–1704) was against dogmatic religion, believing that a rational person will always temper his opinions with a little doubt. He advocated that people should act in their own long-term interests and should keep their passions under control. He believed in a rational world where everything moved like a giant clockwork along the lines of Newton's laws. He divided knowledge into a kind that is intuitive or can be demonstrated readily and another kind which was merely 'probable' or a matter of revelation. The latter had to be rejected if it clashed with the clear evidence of reason. Such were some of the ideas that influenced the young William Herschel and indeed many other admirably rational eighteenth-century figures.

In 1757 the Hanoverians had once more to go into action against the French but were routed at the Battle of Hastenbeck. Conditions in the army became very bad and even the musicians came under fire. At one point, Isaac Herschel advised his son to flee for his own safety, which he sensibly did. Isaac believed that this could not be regarded as a desertion because William, as a musician and a minor, had never been formally sworn in as a soldier. However, he made sure to ask their general for an official discharge when things settled down. After leaving the army, William and his older brother Jacob headed for Hamburg and took ship for England.

4.3   Wandering musician

The Herschel brothers arrived penniless in London but William quickly got piece-work as a music copyist. Jacob gave lessons and they sometimes found oppor-tunities to perform through the people they had previously met in Kent. It is evident that, while William was prepared to do any kind of work to bring in money, Jacob was very fussy. Caroline, stated after his death that Jacob was 'on the whole an excellent and sensible creature', but he was 'incorrigible where Luxury, Ease and Ostentation were in case'.[3]

After two years of this unsettled life, Jacob received an offer of a place in the Hanoverian court orchestra and decided to return home. William felt that he would get nowhere against the musical competition in London and accepted an offer to direct the small band of the Earl of Darlington's regiment, then stationed in Richmond, Yorkshire. For this band—two oboes and two French horns—he composed military music, but he once again found opportunities at private homes to better exercise his talent. He had to ride about Yorkshire under all weather conditions to get to these engagements. He was a welcome guest in the home of Sir Ralph and Lady Milbanke, where he wrote several symphonies. These are still occasionally played, but more as curiosities than for their musical originality. His style was intermediate between baroque and classical, written in the contemporary 'empfindsamer' or sympathetic manner, which was supposed to express feeling directly. All his spare time he used for studying languages and music theory. The latter led him to mathematics. On one memorable occasion he performed in Edinburgh, where he met the philosopher Hume and was invited by him to dinner.

His life was a lonely one and he found relief by writing frequently to his older brother in Hanover. Mostly they corresponded in German but when the opposite sex formed the subject, he wrote in French. In March 1761 he was quite smitten by two young ladies:

[3]C.L. Herschel, quoted by Lubbock (1933, p. 12).

one of them is the daughter of Mrs G. and the other is the daughter of Lady G--y. The first is the most beautiful person in the world, beauty itself personified; she is quite young and likeable, but not yet an adult and absolutely innocent. As to Lady G--y's daughter, she is already more advanced, as she is around 20 or 23; but of mediocre appearance and far from being beautiful, but *in compensation* she plays the guitar admirably und singt sehr lieblich darein.[4] She is extremely fond of the oboe and has several times invited me to spend a few weeks at the country house of My Lady her mother during the summer. She has something very attractive in her disposition and blushes at appropriate times; likewise she smiles and makes eyes ... But all that innocently ... To hell with your stories, you say. Well my dear [brother], to please you I will change the subject; let's come to Madamoiselle. Stop laughing, because it would be wrong not to be serious in speaking of such an admirable person. As one judges the feelings of one's friends better by the letters they write than by their conversation, I have got to know her quite perfectly because I often receive letters from her. Each letter surprises me by its good sense and justice; we have had several discussions on the subjects of friendship, feelings, knowledge of the human heart, virtue and several other matters, in which she displays more than common judgement. But what I admire most is that she knows how to write to me *as a friend*, frankly, without being any the less thoughtful or reserved.[5]

These little flirtation eventually had to come to an end. Presumably the various parties realised that their positions in life were too different for serious relationships to become possible.

Not surprisingly, when William came across the work of the stoic philosopher Epictetus he took his advice to heart! He lectured his brother (by post) on stoic philosophy:

> Let a man once know what sort of a being he is; how great the Being which brought him into existence, how utterly transitory is everything in the material world ... if one cannot control one's feelings, we have at least some power over them. We should therefore take pains to banish all those that disturb us and to retain the pleasant ones ... [6]

His religion at this stage of his life (21 years of age) seems to have been a kind of Deism, with a somewhat condescending attitude to Christianity:

> the Christian religion is very good and one must not oppose but rather do everything to encourage it.
>
> Anyone therefore who wishes to be a reformer would be acting very foolishly if he should begin meddling with religion. Does one not do a thousand things in order not to be out of the common and not to offend the custom of the country? One wears ruffles, one frizzes one's hair,

---

[4]and sings very sweetly to it.

[5]Herschel to Jacob Herschel, 29 March 1761, quoted by Lubbock (1933, p. 19); tr. ISG.

[6]Herschel to Jacob Herschel, 31 March 1761, quoted by Lubbock (1933, p. 25).

shaves one's beard, has silver candlesticks, coaches and horses, footmen and cooks, drinks tea twice a day, carries a sword in peace-time, sleeps till nine o'clock, says 'Your good health' when one drinks ...

[in French] And how many accidents there are which could finish us off in a moment! Four days ago, coming from Richmond, just as I was thinking of death and the order which one observes in all creation, I said to myself *All is in order*—at which very instant, by some accident, I fell violently from my horse. In getting up again, not having been hurt in the least, I said [in English] 'It would have been far better that thou hadst broke thy neck than that the laws of motion should have been altered for thy sake'.[7]

Like their clients among the upper classes, eighteenth-century musicians had to wear formal dress and practise the exquisite and artificial manners of good society. William easily fell into this way of life and was always known for his natural politeness and his willingness to entertain visitors however inconvenient their arrival might have been.

Early in 1762 Herschel moved to Leeds where he became director of the public concerts and rapidly built up a following for his musical talents, now mainly exercised on the violin or the organ. He began to feel more settled and his finances improved. Two years later he took a brief holiday during which he went to see his family in Hanover after an absence of nearly seven years. Of course, they were glad to see him again but for Caroline it was a particular joy. William had always been especially kind to her and was her favourite brother.

Soon after William's return to Yorkshire his father died (April 1767), leaving the remnants of the German family quite hard up. His own life continued as before, with a constant round of travels to Yorkshire towns and country houses. The winter was the season for living in towns and the summer was spent in the country. However, music was gradually ceasing to be his only interest. He was beginning to spend more and more of his spare time in mathematical studies.

4.4   Life in Bath

In 1766 he had moved to Halifax where he had applied for the post of organist at the parish church, which was just about to receive a new organ. As luck would have it, at the same time he received an invitation to move to Bath as organist of the 'Octagon' Chapel. However, he felt obliged to humour his Yorkshire friends by staying there for a few months longer. The Halifax organ post was to be decided by a practical musical examination. In those days, English organs did not have pedals, but Herschel managed to create a particularly sonorous effect by using small lead weights to hold down the bass notes while he played the other keys. With this subterfuge, he won the competition against a more dextrous rival! There is a well-known story that the organ builder, Johan Snetzler, said to have been Swiss, did not approve of the rival candidate, saying 'Te Tevel, te

---

[7]Herschel to Jacob Herschel, 12 May 1761, quoted by Lubbock (1933, pp. 26–27); part tr. ISG.

Tevel! He run over the keys like one cat; he will not give my piphes room for to Shpeak!' Herschel's slower but harmonious style was more to his liking: 'Aye, aye! tish is very goot, very goot indeed; I vill luf tish man for he gives my piphes room for to shpeak'.

However, in spite of an offer of a salary increase, Herschel departed as planned at the end of the year.

Bath, as its name implies, was a spa town and very popular among genteel invalids as a place to 'take the waters' in the eighteenth century. It had developed under the city-appointed Master of Ceremonies and arbiter of manners, Beau Nash, as a social centre where the fashionable could spend the winter 'season', the period of the year when the affluent socialised in the towns and cities and enjoyed entertainments such as the theatre, balls, and concerts. The atmosphere did not change very much as the eighteenth century wore on and the novels of Jane Austen, who lived there herself after 1801, give a faithful picture of Bath life.

The Octagon Chapel, where Herschel became the organist, was a somewhat snooty private church for those who could afford to pay to hear good music and avoid the disagreeable presence of common people. There were fires in the recesses for the sake of invalids and preassigned seats could be taken for a fixed number of months. Herschel rented a house to accommodate his numerous pupils and let part of it to his former landlord and friend from Leeds, a Mr Bulman, whose business had in the meantime collapsed. He found for the latter a job as clerk of the chapel. The inauguration of the organ was a special occasion celebrated by the performance of an oratorio, for which Herschel made the arrangements, importing well-known singers from London. His older brother Jacob was at the time in England and played the organ at the ceremony, while he himself led the band. For the occasion, Herschel trained a choir 'of carpenters and joiners' from scratch. He seems to have given up serious composition around 1769, although he mentions having composed glees, madrigals, and catches as late as 1779.

Life was relatively easy for Herschel during the next few years and his income increased steadily. He was able to help his younger siblings to get an education and find musical employment. Meanwhile, in Hanover, little 'Lina' was fretting. She had had smallpox as a child and was not considered pretty enough to be marriageable. Mrs Herschel was not as progressive as her husband had been and did not believe in educating her unfortunate daughter, whom she treated as a servant. Caroline for her part was unwilling to become a ladies' maid and was not well enough educated to be a governess. Eventually she was allowed to improve herself by attending a dressmaking school with several young Ladies of genteel families. One of the children she met there was later to become a member of Queen Charlotte's court (as Mrs Beckedorff) and a good friend to her.

The youngest brother, Dietrich, who had remained a musician in Hanover, was giving the family some anxiety because of the frivolous company he was keeping. He was accordingly invited to join the household in Bath, where he stayed about a year and William saw to it that he received a musical education.

Similarly, brother Alexander was falling into bad habits and was persuaded to
come over in 1770. William longed to help Caroline in some way and invited her
to Bath with the idea that she could be trained as a singer. He had to promise to
pay his mother the salary of a servant in exchange for her release. In the summer
of 1772 he made a trip to Hanover, via Paris, to fetch her. The eleven-day return
journey to England was not untypical of what an ordinary traveller might then
experience: they spent six days and nights on the open seats of a mail coach and
their ship lost its main mast in a storm, arriving in England almost wrecked.
They were landed from an open boat and were carried on the backs of sailors who
threw them 'like balls' onto the shore. They hired a cart to get to the London
stagecoach, but this overturned and left Caroline in a ditch! Eventually they got
to London and a comfortable inn. 'In the evening, when the shops were lighted
up, we went to see all what was to be seen in that part of London; of which I
only remember the optical shops, for I do not think we stopt at any other'.[8]

Once in Bath, Caroline soon settled into the role of housekeeper to William,
in addition to taking lessons from him and singing solo parts in the concerts that
he organised from time to time. She also trained the choirs that were needed for
the oratorios that he presented. She had 'no time to take care of [herself] or to
stand upon nicetys'.[9] She found the vapid world of Bath society not at all to her
taste and preferred the lessons that William gave her in basic mathematics to
the conversation of fashionable ladies. William's son J.F.W. Herschel later wrote
of her:

> She was attached during 50 years as a second self to her Brother, merging
> her whole existence and individuality in her desire to aid him to the entire
> extent and absolute devotion of her whole time and powers. There never
> lived a human being in whom the idea of Self ('der grosse Ich') was so
> utterly obliterated by a devotion to a venerated object unconnected by
> those strong ties of love and marriage which inspire such devotion in the
> female mind.[10]

4.5   Crazy about telescopes

By 1771 Herschel was earning about £400 per year, a very respectable income for
a professional person. He had to pay about £31.50 of this in rent on his house,
to which Mr Bulman contributed one-third. A servant could be had at the time
for about £10 per year, with board. In the spring of 1773 he laid out '£15.10.0[11]
for a handsome suit of clothes, it being then the fashion for gentlemen to be
very genteely dressed'.[12] More interestingly, his accounts also show that he was
beginning to buy astronomy books and tables. He also invested in a quadrant,
an instrument for observing the positions of stars.

[8]C. Herschel, quoted by Mrs J. Herschel (1879, p. 27).

[9]C. Herschel, quoted by Lubbock (1933, p. 57).

[10]J.F.W. Herschel letter to R. Wolf, 1866, quoted by Lubbock (1933, p. 58).

[11]Fifteen pounds ten shillings or £15.50 in modern terminology.

[12]Herschel, Memorandum for 1773, quoted by Lubbock (1933, p. 60).

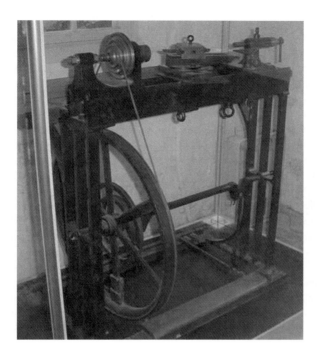

FIG. 4.1. Treadle lathe used by Herschel and his brother Alexander for making small telescope parts. The lathe is today in the William Herschel Museum in Bath (author's photograph).

Within a few months it was obvious that he could think of nothing other than building telescopes. His sister recorded:

> then it was to my sorrow I saw almost every room turned into a work-shop. A cabinet maker making a tube and stands of all description in a handsome furnished drawing-room; Alex putting up a huge turning ma-chine [lathe, see Fig 4.1] (which in the Autumn he brought with him from Bristol) in a bedroom for turning patterns, grinding glasses and turning eye-pieces &c ... I was to amuse myself with making the tube of pasteboard against the glasses [lenses] arrived from London, ... [13]

He started off by buying lenses from an optician in London and mounting them himself in tubes. In his quest for higher and higher magnifications, he bought lenses of longer and longer focal lengths, ending up with a tube that was 30 ft (9.1 m) long and almost impossible to keep pointed at a star. In September he hired a small Gregorian telescope (a design proposed by the mathematician James Gregory in 1663, see Fig 3.3). A Gregorian is a simple reflecting telescope using two mirrors and an eyepiece and, like all reflectors, is much shorter than a

[13]C. Herschel, quoted by Lubbock (1933, p. 65).

refractor (lens telescope) of similar power. The simplicity and good performance of this instrument encouraged Herschel to try making mirrors himself, according to the methods described in a treatise published by Robert Smith (1738), *A Compleat System for Opticks*. In September, he bought a job lot of tools and materials from an amateur mirror maker in Bath who had got bored with his hobby. As he proceeded in this practical manner he gradually came to understand the theory of the instruments he was making.

By the end of 1773 he had had some small mirror-blanks cast out of speculum metal and was learning how to polish them. Of course, he was fully occupied during the days with his concerts and lessons to private pupils. In the early months of 1776 he had finally achieved success in astronomical mirror-making. By July, he had constructed a 20 ft (6.1 m) reflector, namely one with that focal length, which was also about the length of its tube.[14] Its diameter was probably around 9 or 10 in. (23–25 cm). The instrument was extremely awkward to use because of its great length, and Herschel found himself experimenting on ways to improve telescope stands.

That winter he also became the conductor of the public concerts in Bath. He often had to make do with whatever singing talent he could find and composed songs of various kinds to suit the performers. He himself gave concerto performances on the oboe and the harpsichord. His brother Alexander performed solo cello items. But every leisure moment was spent on telescope work:

> and many a lace ruffle (which were then worn) was torn or bespattered by molten pitch, &c.; besides the danger to which he continually exposed himself by this uncommon precipitancy which accompanied all his actions . . .
>
> For my time was so much taken up with copying Music and practising, besides attendance on my Brother when polishing, that by way of keeping him alife [sic] I was even obliged to feed him by putting the Vitals by bits into his mouth;—this was once the case when at the finishing of a 7 feet [2.1m] mirror he had not left his hands from it for 16 hours together. . . . And generally I was obliged to read to him when at some work which required no thinking; and sometimes lending a hand, I became in time as useful a member of the workshop as a boy might be to his master in the first year of his apprenticeship.[15]

William undertook many family duties and seems to have borne most of the responsibilities that their father would have taken care of had he been alive. Caroline dreaded the 'empty-headed company' that their tenants, the Bulmans, invited to the house and her worst fears were realised when Alexander became engaged to one of them. After realising his imprudence, he turned to William to get him out of the situation. Perhaps as a result of embarrassment, the Bulmans decided to return soon afterwards to Leeds.

---

[14]Herschel customarily referred to his telescopes by their focal lengths, though later the diameter of the primary mirror or objective lens came to be preferred.

[15]C. Herschel, quoted by Lubbock (1933, pp. 67–68).

In July 1777 they received word that their little brother Dietrich had run away from home in Hanover with the idea of 'going to the East Indies in company with a young idler not older than himself'.[16] William immediately stopped his lathe, on which he was turning an eyepiece when he heard the news, packed his bag and hastened over to Holland where the miscreant was thought to be. He was too late. In the meantime, Dietrich had reached London but had fallen ill there. Caroline received a letter from an inn-keeper informing her of the situation and she passed the news on to her other brother, Alexander, then working in Bristol. He spent all his money, 'for the poor fellow was never famous of being an economist; especially where he thought he could assist a fellow creature in want',[17] in going to London, engaging a doctor and bringing Dietrich back with him. After his health was restored, William found work for him as a musician. He returned to Hanover two years later and mercifully soon married and settled down.

4.6 Recognition as an astronomer

In 1779, William and Caroline moved (something that they did several times while in Bath) to Rivers Street, where a newly formed 'Literary and Philosophical Society' was located. Their house had no garden and William had to carry his telescope outside and use it on the street in front of the house. One evening, while he was trying to find the heights of the mountains on the moon by measuring their shadows, a stranger stopped and asked him for a look through. The next day the same person called 'to thank me for [my] civility in showing him the moon'[18] and introduced himself as Dr William Watson Jr., a member of the Royal Society in London. They soon became friends and Watson introduced him to the rather short-lived Bath society to which he contributed some papers about electricity and optics, in 1780–1781.

A music pupil of Herschel's around this time later recounted how, on a cloudy January night, Herschel saw that the sky had suddenly cleared and rushed outside to 'apostrophize' a certain star.

> His lodgings resembled an astronomer's much more than a musician's, being heaped up with globes, maps, telescopes, reflectors, &c., under which his piano was hid, and the violincello, like a discarded favourite, skulked away in a corner.[19]

On another evening a stranger was brought to meet Herschel. They argued vehemently for several hours and, after he had been seen to the door, Herschel described him as 'a devil of a fellow'. It turned out that he was Rev Dr Nevil Maskelyne, the Astronomer Royal. They also struck up a friendship. Yet another

---

[16]C. Herschel, Memoir, quoted by Lubbock (1933, p. 69).

[17]C. Herschel, quoted by Lubbock (1933, p. 71).

[18]Herschel, Biographical Memorandum, quoted by Lubbock (1933, p. 73).

[19]Bernard, quoted by Lubbock (1933, p. 73).

professional visitor who realised that Herschel's work was worthy of serious attention was Dr Thomas Hornsby of Oxford. The latter introduced him to Alexander
Aubert, an amateur astronomer of Deptford who also became part of his circle.

In 1780, Watson thought it worthwhile to communicate two papers of Herschel's to the Royal Society, one about the variable star Mira Ceti and the other
on the heights of the mountains on the moon. Herschel had speculated in the
latter paper that the moon must be inhabited but he was politely asked to leave
such speculations out of the version to be printed. He complied rather reluctantly, arguing with Maskelyne that the moon was so similar to the earth that
'there is an almost absolute certainty of [its] being inhabited'.[20]

In March 1781, the Herschels moved again, this time to 19, New King Street
(Fig 4.2), where William was destined to make his most famous discovery. This
house is on sloping ground, with a garden behind it in which the telescopes
could be erected. It still stands today and is occupied by the 'William Herschel
Museum'.

4.7   Sweeping the sky

Herschel's approach to astronomy was something quite new. With fresh eyes,
he set about making the first comprehensive survey of the sky, looking at each
object under high magnification, a process he called 'sweeping'. Nowadays, whenever a new technology becomes available to astronomy, whether radio, X-rays or
gamma-rays, or a more sensitive detector is produced, one of the first things that
is done is to make a survey using the new technique, in the hope of discovering
new kinds of objects. Before Herschel, astronomers had concentrated on the positions of the so-called 'fixed stars' as well as the planets or wanderers. Progress
had consisted of doing this with more and more accuracy, and few had concerned
themselves with the possibility that there might be other interesting objects in
the sky.

Herschel's interest in surveying the sky arose from a decision he made to try
to measure the distances of the nearby stars directly by trigonometry, using the
earth's orbit as a baseline. There should be small effects due to parallax when a
nearby star is viewed against the background of distant stars, looking from two
places located six months apart on the earth's orbit around the sun. Attempts
had been made before to do this, but the errors of measurement had swamped
what was evidently a very tiny angle. The detractors of Copernicus and Galileo
had used the seeming absence of parallax as a cornerstone of their argument that
the earth is stationary, and thus the search to demonstrate its existence was an
important philosophical problem. In fact, success was achieved only in 1838, after
the end of Herschel's life, by Friedrich Wilhelm Bessel at Königsberg in Prussia
and Thomas Henderson in Cape Town, partly thanks to the development of
a new generation of specialised instruments. Nevertheless, Herschel's attempts
were to lead to unexpected discoveries of the greatest importance.

[20]Herschel to Maskelyne, ca. 1780, quoted by Lubbock (1933, p. 77).

FIG. 4.2. 19 New King Street, Bath, the house from which Herschel discovered the planet Uranus. Today, it contains the William Herschel Museum. (From Ball 1911.)

Newton's contemporary, Edmond Halley, had discovered that, over an interval of many years, there were some stars that had moved perceptibly against the background of the remainder. This 'proper motion' could only be detected after such a long interval of time because of the crude position-measuring technology then available. However, it did suggest that some stars might be nearer than others, and was an encouragement to those who were trying to measure parallax.

Herschel, following a suggestion made by Galileo and others after him, thought that he could determine the expected small apparent back-and-forth movements of parallax by finding cases where two stars looked very close together in the sky, so that only the differences in their positions had to be monitored. One member of an apparent pair was likely to be much closer to the observer than the other. For measuring their separations he devised a special eyepiece incorporating a micrometer and two threads. First, he had to find suitable close double stars to be monitored, and it is for this reason that he started his surveys. Before his work, the Cambridge physicist John Michell had speculated in 1767 and Christian Mayer, a Jesuit priest of Mannheim, in 1776, that some double stars might not merely be directional coincidences but could be pairs held together by gravity. Indeed, Herschel's precise observations were to lead in 1802 to a direct proof that many *were* indeed in orbit around each other (see Section 4.17).

The mounting of Herschel's telescopes was so awkward that they could not be pointed easily at different parts of the sky. Just like the sun, the stars and planets rise in the east and set in the west due to the rotation of the earth. He set up his telescopes to be able to look at a point on the meridian (a line running from north to south through the point in the sky directly overhead). In this way, a band of stars would pass through his field of view as the earth turned. The particular band depended on the angle upwards from the ground (the 'altitude') at which he set the telescope. Each clear night he would set the telescope to a slightly different altitude, so that, after a year (of clear nights!) he covered everything that could be seen from his latitude. To speed up the search, instead of working the whole night at a single altitude, he gave a small, rapid, up–down motion to the telescope to widen the band that he could monitor. Every time he saw a star or nebula he signalled his sister, whose job it was to note down the time from a precision clock and read off the telescope's altitude from a string-operated indicator. Any interesting object could easily be found again from this information and its position catalogued.

A few years later, in 1784, the French geologist Barthélemy Faujas-de-Saint-Fond was visiting Great Britain, making a point of looking up its most famous scientists and engineers. He went late one night to visit the Herschels and found them at work:

> I arrived at Mr. Herschel's about ten o'clock. I entered, by a staircase, into a room that was decorated with maps, instruments of astronomy and natural philosophy, spheres, celestial globes, and a large harpsichord.
>
> Instead of the master of the house, I observed, in a window at the farther end of the room, a young lady seated at a table, which was surrounded with several lights; she had a large book open before her, a pen in her hand, and directed her attention alternately to the hands of a pendulum clock, and the index of another instrument placed beside her, the use of which I did not know. She afterwards noted down her observations.
>
> I approached softly on tiptoe, that I might not disturb a labour, which seemed to engage all the attention of her who was engaged in it; and, having got close behind her without being observed, I found that the book she consulted was the Astronomical Atlas of Flamsteed, and that,

after looking at the indexes of both the instruments, she marked, upon a large manuscript chart, points which appeared to me to indicate stars ...

'My brother', said Miss Caroline Herschel, 'has been studying these two hours; I do all I can to assist him here. That pendulum marks the time, and this instrument, the index of which communicates by strings with his telescopes, informs me, by signs which we have agreed upon, of whatever he observes ...'[21]

4.8 Discovery of Uranus

On 13 March 1781, even before Caroline had made the move to the new address in King Street to join her brother, Herschel noticed a 'curious either nebulous star or perhaps a comet' in his field of view. He found that it was moving very slowly against the background stars. This was to turn out to be the planet Uranus.

It has been generally supposed that it was a lucky accident that brought this star to my view; this is an evident mistake. In the regular manner I examined every star of the heavens, not only of that magnitude but many far inferior, it was that night *its turn* to be discovered. I had gradually perused the great Volume of the Author of Nature and was now come to the page which contained a seventh Planet. Had business prevented me that evening, I must have found it the next, and the goodness of my telescope was such that I perceived its visible planetary disc as soon as I looked at it; and by the application of my micrometer, determined its motion in a few hours.[22]

News of the strange new object was passed to Hornsby at Oxford and Maskelyne at Greenwich for verification. The first thought it was a comet but the second immediately accepted that it was a new planet. Herschel quickly prepared a paper, called *Account of a comet* for the Royal Society, where it was read on 26 April 1781. He was received at the Society for the first time on 3 May.

Dr Watson later reported that Sir Joseph Banks had had a letter from Herschel 'containing further particulars of the *great* discovery and told me they were extreamely satisfactory'.[23] Banks was a naturalist who had accompanied Captain James Cook around the world and was to hold the Presidency of the Royal Society from 1778 to 1819. Charles Messier, the great French comet-hunter, on hearing the news, managed to view the 'comet' himself, with great difficulty, and sent his congratulations.

During the following summer, when the new planet could no longer be observed and the musical public had left Bath for the country, Herschel was free to resume his attempts to build bigger and better telescopes. Caroline heard of nothing but plans for telescope mountings and casting larger mirror-blanks.

---

[21]Faujas-de-St-Fond, B., Travels in England ... (1799), quoted by Ashworth (2003).

[22]W. Herschel, from an autobiographical essay quoted by Lubbock (1933 pp. 78–79).

[23]Watson to Herschel, 25 May 1781, quoted by Lubbock (1933, p. 85).

She was put to work as a human copying machine, laboriously transcribing the precious star catalogues that her brother was able to borrow from time to time.

> [this] kept me employed when my Brother was at the telescope at night;
> for when I found that a hand sometimes was wanted when any particular
> measures were to be made with the Lamp micrometer &c. and a fire
> kept in, and a dish of Coffe necessary during a long night's watching; I
> undertook with pleasure what others might have thought a hardship.[24]

Experiments on the right mixtures of copper, tin, and other elements to use for making mirror blanks were continually under way. The speculum metal had to be able to take a polished surface of high reflectivity but also had to be resistant to shocks such as might arise from a sudden change of temperature or while manhandling a mirror into place in a telescope. The main activity that summer of 1781 was an attempt at constructing his largest mirror yet. A foundry was set up in the garden-level basement of No 19, New King Street. The mould for the mirror was made from horse-dung, which had to be pounded and sieved in huge quantities. William, Alexander, and Caroline all took part, with further assistance from Dr Watson. Finally, in August, the great day arrived, but the mould leaked and the mirror cracked at the first attempt. William's rather deadpan account of what happened is quoted first:

> When everything was in readiness we put our 537.9 pounds of metal into
> the melting oven and gradually heated it; before it was sufficiently fluid
> for casting we perceived that some small quantity began to drop through
> the bottom of the furnace into the fire. The crack soon increased and the
> metal came out so fast that it ran out of the ash hole which was not lower
> than the stone floor of the room. When it came upon the pavement the
> flags [stones] began to crack and some of them to blow up, so that we
> found it necessary to keep a proper distance and suffer the metal to take
> its own course.[25]

A somewhat more graphic account was provided by Caroline:

> both my Brothers, and the caster and his men were obliged to run out
> at opposite doors, for the stone flooring (which ought to have been taken
> up) flew about in all directions as high as the ceiling. My poor Brother
> fell exhausted by heat and exertion on a heap of brickbatts'.[26]

Telescope activities had to be put aside as another hectic musical season started. By the end of August, the planet could once again be seen. Writing about 'this comet, or planet as some are disposed to call it', Hornsby of Oxford told Herschel that 'Mr Lexell, a very celebrated mathematician' had determined its orbit and found that it was sixteen times the distance of the earth from the sun.[27] This calculation was the confirmation that was needed to prove its

[24]C. Herschel, quoted by Lubbock (1933, p. 87).

[25]Herschel, Records, quoted by Lubbock (1933, pp. 89–90).

[26]C. Herschel, quoted by Lubbock (1933, p. 90).

[27]Hornsby to Herschel, 14 October 1781, quoted by Lubbock (1933, p. 93)

planetary nature, which was then generally accepted. Anders Johan Lexell (1740–84), of Swedish origin, was Professor of Astronomy in St Petersburg. The famous French mathematician, Pierre-Simon de Laplace (1749–1827), also calculated an orbit independently.

In November 1781, the Royal Society decided to present its Copley Medal to Herschel. He was advised of the award by Sir Joseph Banks (the President) and warned to do something about naming the new planet:

> Some of our astronomers here incline to the opinion that it is a planet and not a comet; if you are of that opinion it should forthwith be provided with a name and our nimble neighbours, the French, will certainly save us the trouble of Baptizing it.[28]

At the time of the presentation he said:

> Your attention to the improvement of telescopes has already amply repaid the labour which you have bestowed upon them; but the treasures of the heavens are well known to be inexhaustible. Who can say but your new star, which exceeds Saturn in its distance from the sun, may exceed him as much in magnificence of attendance? Who can say what new rings, new satellites, or what other nameless and numberless phenomena remain behind, waiting to reward future industry.[29]

Herschel's new planet was the first one of the three that have been discovered in recorded history, all the others having been part of folklore dating from time immemorial. It is interesting to note that Herschel was not the first person to spot it—Uranus had been recorded as a star on no fewer than twenty occasions between 1690 and his discovery of its planetary nature (Clerke 1893, p. 96).

In December 1781, a paper of Herschel's called *On the Parallax of the Fixed Stars* was read to the Royal Society. Dr Watson reported to him that some of his notions had not gone down well with the astronomers present. He had claimed to be able to estimate the angle between the two members of a double star in terms of the diameter of either star. He also used exceedingly high magnifications, for example of 6450 times, in his telescopes, much higher than are ever used today. Viewed in a large telescope with good optics, the image of a star under such extraordinarily high magnification is seen as large numbers of small pieces, called speckles, continually jumping around, thanks to the movements of small pockets of air in the earth's atmosphere. Herschel claimed to see perfectly round images: these can only have been the result of some optical defect of his mirrors. It is probably true, however, that a practised observer of double stars can make out, among the moving speckles, whether a star is a single one or a close double. Herschel, in his own defence, stated to Dr Watson:

---

[28]Banks to Herschel, 15 November 1781, quoted by Lubbock (1933, p. 95).

[29]Sir Joseph Banks on presenting the Copley Medal of the Royal Society to William Herschel for his discovery of a new planet, 1781.

Seeing[30] is in some respects an art, which must be learnt. To make a person see with such a power is nearly the same as if I were asked to make him play one of Handel's fugues upon the organ. Many a night have I been practising to see, and it would be strange if one did not acquire a certain dexterity by such constant practice.[31]

A few years later, in 1786, Herschel found himself sitting next to the eccentric bachelor Henry Cavendish at dinner at the house of their mutual friend, Mr Aubert. Cavendish was an early scientific chemist famous for having shown that water is composed of hydrogen and oxygen and later, in 1798, for providing the first experimental determination of Newton's gravitational constant. Although a member of an aristocratic family that included the Dukes of Bedford and a very rich person indeed, he dressed plainly in old-fashioned clothes and had a reputation for extreme taciturnity. Sir John Herschel, William's son, was fond of telling an anecdote that arose on this occasion: After having said nothing for some time, Cavendish suddenly addressed his neighbour, Herschel:

'I am told that you see the stars round, Dr. Herschel'.
'Round as he button', he replied.

Cavendish then kept silent until, near the end of the dinner, he again opened his lips to say in a doubtful voice:

'Round as a button?'
'Exactly, round as a button', repeated Herschel, and so the conversation ended.[32]

This story was given as an example of Cavendish's eccentricity, but it could well be that he found Herschel's 'perfectly round' stars inexplicable, as indeed they were.

4.9 King George III

Early in 1782, Herschel's friends Watson and Banks started agitating for a state income from King George III so that he would no longer have to spend most of his time earning money as a musician. King George was interested in astronomy and had a private observatory in Kew, where he employed an astronomer. One of his equerries arranged that Herschel should go to London in May 1782 to have an audience with His Majesty, who received him very graciously. During the interview, the King asked after his musician brother Alexander and gave him permission to attend a court concert where he could hear his in-laws, the Griesbachs, play. He had to stay in London until such time as the King dismissed him. His telescope was set up at Greenwich so that Maskelyne, as Astronomer Royal, could compare it with his own ones.

[30]By seeing, Herschel meant observing; today 'seeing' refers to the steadiness of the atmosphere at a particular time.

[31]Herschel to W. Watson Jr, 7 Jan 1782, quoted by Lubbock (1933, p. 101).

[32]Quoted by Lubbock (1933, p. 102).

Herschel's visit to London was something of a triumph. He spent much of his time with his friends Aubert and Maskelyne and was complimented by the latter for having introduced him to high magnifications. He wrote of his experiences to his sister, ending with 'Lina, I tell you all these things. You know vanity is not my foible, therefore I need not fear your censure'.[33] His next letter reported how, at the royal concert, the King had talked to him for half an hour and had asked George Griesbach to play a solo concerto for his benefit.

> I am introduced to the best company. To-morrow I dine at Lord Palmerston's, and the next day with Sir Joseph Banks, &c. &c. Among opticians and astronomers nothing now is talked but of *what they call* my Great discoveries. Alas! this shows how far they are behind, when such trifles as I have seen and done are called *great*. Let me but get at it again! I will make such telescopes, and see such things ... that is, I will endeavour to do so'.[34]

At last, on 3 July, the King and the Royal Family had him demonstrate his telescopes in Windsor. The King enjoyed himself. The following night, the sky was cloudy but Herschel had an artificial Saturn ready for demonstration purposes! Herschel's friends pushed him to make application to the King for a pension, which he eventually did. The King responded positively and granted him an amount of £200 per year, on condition that he should reside near Windsor. Alexander Aubert sent his congratulations:

> How great! How generous! How Kingly! this encouragement of science is, in his Majesty; may God bless him for it and make every matter contribute to his happiness; there never was a more deserving, a more virtuous, a more benevolent sovereign.[35]

Herschel wrote an official letter to Banks on the subject of a name for the new planet:

> Sir, By the observations of the most eminent Astronomers in Europe it appears, that the new star, which I had the honour of pointing out to them in March, 1781, is a Primary Planet of our Solar System ...
>
> In the fabulous ages of ancient times the appellations of Mercury, Venus, Mars, Jupiter, and Saturn, were given to the Planets, as being the names of their principal heroes and divinities. In the present more philosophical aera, it would hardly be allowable to have recourse to the same method, and call on Juno, Pallas, Apollo, or Minerva, for a name to our new heavenly body. The first consideration in any particular event, or remarkable incident, seems to be its chronology: if in any future age it should be asked, *when* this last-found Planet was discovered? It would be a very satisfactory answer to say, 'In the Reign of King George the Third'. As a philosopher then, the name of GEORGIUM SIDUS presents

[33] Herschel to C. Herschel, undated, quoted by Lubbock (1933, p. 115).

[34] Herschel to C. Herschel, 3 June 1782, quoted by Lubbock (1933, p. 116).

[35] Aubert to Herschel, 31 August 1782, quoted in Lubbock (1933, p. 120).

itself to me ... But as a subject of the best of Kings, who is a liberal protector of every art and science;—as a native of the country from whence
this Illustrious Family was called to the British throne;—as a member of
that Society, which flourishes by the distinguished liberality of its Royal
Patron; and, last of all, as a person now more immediately under the
protection of this excellent Monarch, and owing everything to his unlimited bounty;—I cannot but wish to take this opportunity of expressing
my sense of gratitude, by giving the name Georgium Sidus ... to a star,
which (with respect to us) first began to shine under His auspicious reign
(Herschel 1783).

Meanwhile, Johann Elert Bode (1747–1826) of Berlin had suggested the name
'Uranus'. In writing to Herschel he apologised for pre-empting him by saying 'had
I been in your situation I should have felt it proper to do as you have done'.[36]
Bode's name was the one that stuck. Even in England, there were some that
preferred it, but others, including Herschel himself, continued to refer to the
'Georgian Planet'.

Several of Herschel's friends felt that the King's 'unlimited bounty' could
have stretched to more that £200 per annum; however, he did receive further
special grants of several thousand pounds and a big telescope order from His
Majesty. After all, the Astronomer Royal's salary was only £300 per annum.
Herschel, always polite, did not himself reveal the amount of his pension outside
his family except to Sir William Watson Senior, the father of his friend. 'Never
bought Monarch honour so cheap' was Watson's tart comment.[37] Others were
simply told that 'the King had provided for him'.

4.10   Motion of the Sun

As mentioned, Edmond Halley had been able to detect the 'proper motion' of
some stars against the background of the more distant ones and Herschel, having almost finished his third 'sweep' or survey, published a paper about changes,
whether in position or brightness, that had occurred since Flamsteed's *British
Catalogue* or *Historia Coelestis Brittanica*, based on observations made about
a hundred years earlier. Some of the apparent changes in position could be explained as being the 'reflection' of the movement of our own sun (and solar system
with it). He deduced that the 'Apex' or direction of this motion is towards the
star λ (lambda) of the constellation Herculis, surprisingly close to the currently
accepted direction,[38] and published his result in 1783.

---

[36]Bode to Herschel, July 1783, quoted by Lubbock (1933, p. 123).

[37]C. Herschel, Reminiscences, quoted by Lubbock (1933, p. 133).

[38]This motion is now known to be a by-product of the rotation of the sun and nearby stars
around the centre of the Milky Way galaxy. Viewed from outside, young stars move together
in relatively orderly circular orbits in the plane of the Milky Way, while older ones have more
random orbits out of the plane that do not show much rotation. These differences give rise to
the apparent 'solar motion', which is a relative one.

By this time, people were asking how the stars could be stable in the sky, considering the power of gravity to pull them together. Newton had hardly discussed this question at all and merely supposed that God had taken care of things by balancing their attractions in a systematic way. The problem had since been taken up by others, such as Tobias Mayer and Joseph de Lalande. Just as he went to press, Herschel received a tract from Alexander Wilson, Professor of Astronomy in Glasgow, *Thoughts on General Gravitation, and Views Thence Arising as to the State of the Universe*, which discussed the stability of the stars as a result of motion about a central point. In some ways this was the beginning of the concept of the Milky Way as a galaxy.

4.11 Construction of the Heavens

Before beginning his fourth survey of the sky at the end of 1783, with a new telescope of 18.7 in. (47.5 cm) aperture and 20 ft (6.1 m) focal length, Herschel had become interested in the 'construction of the heavens'. He believed the Heavens, or Universe, to consist of various '*nebulous and sidereal strata* (to borrow a term from the natural historian [i.e. geologist])'. His first paper on the subject (Herschel 1784) could 'as yet give a few outlines, or rather hints. As an apology, however, for this prematurity, it may be said, that the end of all discoveries being communication, we can never be too ready in giving facts and observations, whatever we may be in reasoning upon them'.

Charles Messier had published his famous lists of nebulae in the year 1774, with supplements in 1780 and 1781. His intention had been to provide a list of objects that might be mistaken for comets, his own main interest. Most nebulae are extremely faint and difficult to discern in small telescopes. When the moon is full they often cannot be seen at all against the the sky background. Encouraged by his friend Alexander Aubert, Herschel found with his much more powerful telescopes that many of them, which had previously been thought of as being nothing more than luminous patches of light, could be resolved into stars. Soon after he started looking for nebulae, he could report that he had found hundreds more. 'It is very probable, that the great stratum, called the milky way, is that in which the sun is placed'. Much of his fourth survey, which ended in 1802, was to be devoted to cataloguing nebulae and clusters.

In 1785, he published his second and most famous cosmological paper *On the Construction of the Heavens*.

> if we would hope to make any progress in an investigation of this delicate nature, we ought to avoid two opposite extremes, of which I can hardly say which is the most dangerous. If we indulge a fanciful imagination and build worlds of our own, we must not wonder at our going wide from the path of truth and nature; but these will vanish like the Cartesian vortices, that soon gave way when better theories were offered. On the other hand, if we add observation to observation, without attempting to draw out not only certain conclusions, but also conjectural views from them, we offend against the very end for which only observations ought to be made; I will endeavour to keep a proper medium. (Herschel 1785)

FIG. 4.3. William Herschel in 1785. Based on a portrait painted by Lemuel
   Abbott for William Watson Jr. (SAAO).

This paper contains an account of his 'gages', where he looked in many different directions in the sky and counted the numbers of stars he could see in his
eyepiece. By assuming that all stars are of about the same intrinsic brightness
(a rather bad assumption in actuality) and equally scattered along the line of
sight, he could estimate how far the Milky Way must extend in every direction
and use this information to make a map of it (see Fig 4.4).

In the same paper he reported the discovery of a new class of object, the
Planetary Nebulae. These were so-called because they looked like stars or small
planets with rings (now known to be gaseous) around them.

4.12   Minor discoveries

In 1787 Herschel reported that he had found two satellites of the 'Georgian
planet'. Part of his success in seeing these exceedingly faint objects he ascribed
to a new form of telescope that he had invented, called the 'front view' (see
Fig 3.3). Traditionally, reflecting telescopes had contained a secondary mirror
to bring the image out somewhere convenient for placing an eyepiece. However,
speculum metal is an inefficient reflector and perhaps 40% of the light was lost

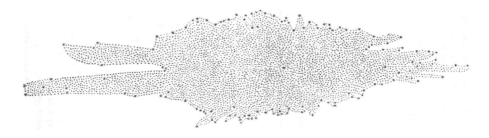

FIG. 4.4. Cross-section of the Milky Way according to Herschel (1785). The position of the solar system is given by the asterisk near the middle. The distances to the borders in each direction are estimated from star counts, assuming that all stars are equally luminous. The fork on the left side is the apparent rift in Sagittarius.

because of this extra mirror. Herschel's idea, practicable only with quite large telescopes, was to tilt the primary and put the eyepiece at the end of the tube, towards the side. With the long focal lengths that Herschel used, this off-axis arrangement had almost no effect on the quality of the image.[39]

4.13 Datchet, Windsor and the 40-ft telescope

Herschel was already forty-three years old when, thanks to his Royal pension, he was able to give up his musical career. He played the organ in Bath for the last time on Whit Sunday 1782 and by that summer he and Caroline were settled near Windsor. He rented a house that had a large number of outbuildings suitable for use as workshops in the village of Datchet. It was almost a ruin, having served as a hunting lodge 'for some Gentleman' and had not been occupied for years. Looking for a spot in the garden on which to erect a 20-ft telescope, Alexander narrowly escaped falling down a well that was hidden by the long grass! Caroline as usual had to take care of the housekeeping:

> I ... saw at once that my Brother's scheme of living cheap in the country (as jokingly he said) on Eggs and Bacon would come to nothing; for at Bath I had the week before bought from 16 to 20 eggs for 6d, here I could get no more than five for 4d, Butcher's meat was $2\frac{1}{2}$d and 3d per pound dearer; and the only butcher at Datchet would besides not give honest weight and we were obliged to deal, all the time we lived there, at Windsor. O dear! thought I, what are we to do with 200 a year! after Rent and Taxes are deducted ... and there being upwards of 30 Windows[40] ...[41]

---

[39]Some later giant telescopes, such as the 200-in. at Palomar, are big enough that the observer can be placed in a small cage on the main axis of the primary (see Fig 9.8).

[40]A tax was payable on windows!

[41]Caroline Herschel, quoted in Lubbock (1933, pp. 134–135).

William frequently went to Windsor in the evenings to show the King and his circle objects through his 7 ft telescope. During the days he was continuously busy with improvements and pressed poor Caroline into helping him. In her usual not-complaining-but-complaining way she wrote:

> I had the comfort to see that my brother was satisfied with my endeavours in assisting him when he wanted another person, either to run to the Clocks, writing down a memorandum, fetching and carrying instruments, or measuring the ground with poles ... [42]

Good workmen were impossible to find locally and all fine work on the lathe had to be done by Herschel himself or by his brother, when he came from Bath during the summer low season for music. The telescopes had been put together hastily after the move so that no clear nights would be lost. On one evening when the wind was strong William had just got off a ladder when the whole structure came crashing down. 'Fortunately it is a cloudy evening so that I shall not lose time while I repair the havock that has been made' (quoted by Bennett 1976). On another occasion, Caroline had a nasty accident in the dark when everything was covered with snow. She fell on a hook attached to the base of a telescope and it went into her right leg:

> my Brothers call 'make haste' I could only answer by a pitiful cry 'I am hooked'. He and the workmen were instantly with me, but they could not lift me without leaving near 2oz. of my flesh behind. The workman's wife was called but was afraid to do anything, and I was obliged to be my own surgeon by applying aquabaseda and tying a kerchief about it for some days; till Dr Lind hearing of my accident brought me ointment and lint and told me how to use it. But at the end of six weeks I began to have some fears about my poor limb and had Dr Lind's opinion, who on seeing the wound found it going on well; but said, if a soldier had met with such a hurt he would have been entitled to 6 weeks' nursing in a hospital.
>
> I had however the comfort to know that my Brother was no loser through this accident, for the remainder of the night was cloudy and several nights afterwards afforded only a few short intervals favourable to sweeping ... [43]

A Portuguese acquaintance, referred to by the Germanized name Von Magellan, visited Herschel at Datchet one fine winter's evening and found him observing at a temperature of 13 °F or –10 °C. He normally observed the whole night through, only stopping to warm up once every 3 or 4 hours. He worked in the open air rather than a dome because he found that the telescope had to be at the same temperature as its surroundings to obtain the clearest images.

His favourite telescope was the 'large' 20-ft mentioned in Sect 4.11, having a mirror 18.7 in. in diameter (as opposed to the small one whose mirror was only 12 in. or 30 cm in diameter). This he finished in 1783 (Bennett 1976). It saw

[42]C. Herschel, quoted in Lubbock (1933, p. 136).
[43]C. Herschel, quoted by Lubbock (1933, p. 137).

service for many years; half a century later his son used it to survey the Southern Hemisphere from Cape Town.

4.14   Telescope business

From about 1785 onwards, Herschel operated a telescope-making business which enjoyed huge success, as everybody wanted to have one of the powerful new instruments. In 1781 he had made mirrors for two members of the Bath Philosophical Society and in the following years he supplied optics to J.H. Schröter of Lilienthal in Germany as well as his friend Alexander Aubert. However, it was the King who encouraged him to go into business on a more serious basis:

> The goodness of my telescopes being generally known, I was desired by the King to get some made for those who wished to have them. Getting the woodwork done by His Majesty's cabinet maker, I fitted up five 10 feet (3m) telescopes for the King, and very soon found a great demand for 7 feet (2.1m) reflectors. This business in the end not only proved very lucrative, but also enabled me to make expensive experiments for polishing mirrors by machinery.[44] (See Fig 4.5)

As usual, Alexander did all the lathe-work and William figured the mirrors. His usual production rate was about five instruments per year. Many were ordered by dilettantes, but some were used by established astronomers (Spaight 2004). Unfortunately, very few advances in astronomical science resulted. Speculum metal mirrors were hard to maintain because they tarnished rapidly and had to be re-polished every few months; their figures could easily be destroyed if this work was done by unskilled hands.

In 1785, Herschel's health broke down as a result of the dampness and fogs that prevailed at Datchet, which was marshy from its closeness to the Thames. According to his son, he used to rub his face and hands with a raw onion whenever the nearby land was flooded, in order to keep off the infection of the ague. Eventually, he did however succumb and he consulted Dr William Watson, Senior, Physician to the King, for advice. He was prescribed a medicine called Red Bark and told to take mutton broth and bread for nourishment, with wine negus and toast for 'between whiles'.

This precipitated a move to Windsor proper. Soon afterwards, Herschel applied to the King through the President of the Royal Society for a grant towards the construction of a large telescope of 30 or 40 in. (76 cm or 1 m) diameter. 'His majesty fixed upon the largest', i.e. the 40-in. and provided an amount of £2000. He later made a second grant of the same amount and an annual one of £200 for maintenance. Caroline's essential contributions were recognised by a separate salary of £50 per year 'the first money in all my lifetime I ever thought myself at liberty to spend to my own liking ...'[45]

---

[44]From Herschel's notebook 'Memorandums', quoted by Dreyer (1912, p. xlv).

[45]C. Herschel, quoted by Mrs John Herschel (1879, pp. 75–76).

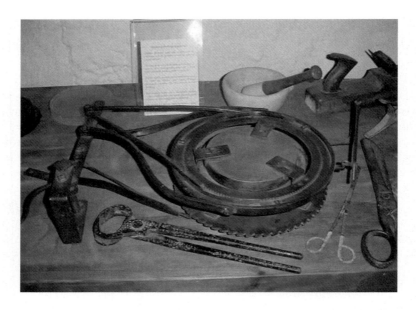

FIG. 4.5. A grinding-cum-polishing machine and other tools used by Herschel. The mirror being ground is the circular disc in the centre, which is face down above the tool. Between them would be the grinding paste: the mirror would become concave and the tool convex as the handle to the left was used to move the mirror back and forth over the tool. The ratchets would have given a small rotary action after each stroke. (Author's photograph, taken in the William Herschel Museum, Bath.)

Work started on the 40-ft (12.2 m) tube for the 40-in. telescope in April 1786. Several artisans 'including a very ingenious Smith', a turner, a 'brassman', and a whole troup of labourers were kept busy, altogether thirty or forty people. Herschel himself was so dextrous at turning and forging that one of his workmen asked him what he had been brought up to do. 'To fiddling' was his reply.

In July 1786 he had to make a trip to Hanover to deliver a telescope to the University of Göttingen as a present from the King. He made sure that Caroline had plenty of work to keep her busy during his absence, but she was able to spend her evenings searching the skies for comets with a small reflector. On 1 August she discovered a comet which she could verify the following night. She communicated her find to the Royal Society and to Alexander Aubert. Ultimately she discovered seven more.

Unfortunately, the landlady at Windsor proved too grasping and the whole operation had to be moved again in the following year, this time to Slough. According to Caroline, Herschel observed until dawn in Windsor on his last night there and could start again the following evening in Slough. This was to become their permanent home. As usual, it took time to find a workman sober

FIG. 4.6. Model of Herschel's 40-ft telescope in the Deutsches Museum, Munich (Author's photograph).

and reliable enough to operate the sweeping mechanism. Many entries such as 'Lost a Neb by the blunder of the person at the handle' were to be found in the observing record.

Christoph Papendiek, a flautist at court, and his wife Charlotte, Assistant Keeper of the Wardrobe to the Queen, had lived in the house previously. Mrs Papendiek noted in her diary about Herschel: 'His first step, to the grief of everybody who knew the sweet spot, was to cut down every tree, so that there would be no impediment to his observations of the heavenly bodies'.[46]

During the construction of the enormous telescope tube, 40 ft long and 4 ft 10 in. (1.5 m) diameter (see Fig 4.6), out of rolled iron sheets, the King and Queen as well as other members of the Royal Family visited the Herschels to inspect progress:

> Before the optical part was finished many visitors had the curiosity to walk through it, among the rest King George III. And the Archbishop of Canterbury following the King and finding it difficult to proceed, the King turned to give him the hand, saying: 'Come my Lord Bishop, I will show you the way to heaven'. This was in the year 1787, August 17th ...[47]

[46]Charlotte Papendiek, Journal, quoted by Lubbock (1933, p. 146).
[47]from a letter of Caroline to Mrs John Herschel, quoted by Lubbock (1933, p. 157).

The mirror was polished by hand by a team of twelve men holding it in a frame with twelve handles and imparting polishing strokes according to Herschel's instructions.

The first serious observations were made in August 1789 and immediately resulted in the discovery of a sixth satellite of Saturn, although Herschel soon realised that he had seen this body two years previously in one of his other telescopes. However, a dramatic announcement was expected in order to justify the effort and expense that had been entailed. In fact, the 40-ft telescope was to prove too awkward to use on a regular basis, partially because it took considerable time to prepare for work and required two workmen to help handle it. Re-polishing of its spectrum was time-consuming and labour-intensive. In practise, it was used only rarely, mainly for searching for new planetary moons. Nevertheless, it stood for over fifty years at Slough and became something of a landmark. A paper describing it in great detail was presented to the Royal Society in 1795. The exact technique that Herschel used in the production of mirrors he regarded as a trade secret and he never published it.

4.15   Marriage

Herschel had met a couple, John and Mary Pitt, who lived at Upton, the parish near Slough to which he himself belonged. Mary Pitt, born Baldwin, was the daughter of a wealthy London merchant. He got on well with Pitt, who was a cultivated and literate person 'with a well-chosen library' where they spent considerable time together in conversation. Unfortunately, John Pitt died in 1786. Mrs Papendiek was a good friend of Mrs Pitt and noted in her diary that

> [she] poor woman, complained much of the dullness of her life, and we did our best to cheer her, as did also Dr Herschel, who often walked over to her house with his sister of an evening, and as often induced her to join his snug dinner at Slough.
>
> Among friends it was soon discovered that an earthly star attracted the attention of Dr Herschel. An offer was made to Widow Pitt, and accepted. They were to live at Upton, and Miss Herschel at Slough, which would remain the house of business.[48]

Mary soon realised that if they got married William would continue to spend most of his time with his sister observing at Slough and decided that the idea was, perhaps, not such a good one after all. Herschel 'expressed his disappointment', but explained that he could not give up astronomy, for which he had trained Caroline to be a most efficient assistant. Furthermore, she was indefatigable and would make any sacrifice in his interest. After some time, Mary relented and the future partners came to an agreement that they would maintain two establishments, with a footman to go between them and that Caroline was to have an apartment of her own over the workshops. According to Hoskin (2003b),

[48]Charlotte Papendiek, Journal, quoted by Lubbock (1933, p. 174).

at the time of their marriage Mrs Herschel was worth at least £10,000, with the expectation of receiving an equivalent amount when her mother died.

They were married on 8 May 1788 in Upton church, with Alexander and Caroline Herschel as witnesses. Mary's brother gave her away and Sir Joseph Banks 'acted as friend of the Doctor'. The honeymoon was spent in the house at Slough, but six weeks afterwards, cards were sent around from Upton and that was where visitors could offer their congratulations.

Mrs Papendiek visited them, with baby and nurse, one fine afternoon, after, at the last minute, putting on her white gloves under a yew-tree in Upton churchyard, and 'was ushered into the well-known tent, where cake and wine were presented'. Rather crossly she mentions that she 'hoped for some assistance home, but none was offered, so we walked back again, and I was not so fatigued as might have been expected'.[49]

William's marriage was a heavy blow to Caroline, who was dependent on him for almost everything in her life, and she resented giving up her monopoly on caring for him. Her diaries for this period were torn up, so we do not have a first-hand account of her personal tragedy. The new Mrs Herschel was evidently a calm and kind person who enjoyed life without being over-intellectual. She travelled with her husband and had a circle of good friends. Many years later, her daughter-in-law, Mrs John Herschel, wrote:

> I could wish to bear personal witness to the graceful and dignified gentleness of Lady Herschel's manner, even in old age, which were the faithful indications of the genuine kindness and amiability of her nature. Miss Herschel's good sense soon got over the startling innovation of an English lady-wife taking possession of her own peculiar fortress, and she who gladdened her husband's home soon won over the entire affection of the tough little German sister.[50]

At the time of their marriage, William was 49 and Mary 38. The novelist and diarist Fanny Burney (1752–1840), who knew many of the eighteenth century's famous figures, such as the lexicographer Samuel Johnson and the statesman Edmund Burke, was, with her musicologist father, a friend of Herschel's. On meeting the newly-weds at a tea party she wrote:

> Dr Herschel was there, and accompanied them (the Miss Stowes) very sweetly on the violin; his new-married wife was with him and his sister. His wife seems good-natured; she was rich too! and astronomers are as liable as other men to discern that gold can glitter as well as stars.[51]

In fact, Herschel's telescope business brought in a good income, even if it involved a lot of hard work. Dreyer (1912), in his *Biographical Introduction to Herschel's Works*, added up such telescope sales as he knew of and came to a total of £14,743, but it is clear that he did not know of them all (Spaight 2004).

[49]Charlotte Papendiek, Journal, quoted by Lubbock (1933, p. 176).

[50]Mrs John (Lady) Herschel, quoted by Lubbock (1933, p. 177).

[51]Diary of Fanny Burney, quoted by Lubbock (1933, p. 177).

His personal effects were eventually to be valued in his will at £6000 and he left £25,000 in '3% Reduced Annuities' to his son, besides £200 to his brother Dietrich and several other large legacies.[52].

4.16   Social life

Marriage led to a more relaxed lifestyle. After his fourth sweep of the heavens he gave up laborious survey work and concentrated on particular classes of objects. He also seems to have taken more interest in laboratory experiments, which could be pursued during the daytime. The extreme care that had gone into his sweeps is illustrated in a paper he read to the Royal Society in 1799 *On the Power of penetrating into Space by Telescopes* ... First, it took about twenty minutes for his eyes to reach maximum sensitivity in the dark. If he accidentally watched a bright star passing through the eyepiece this dark adaption process might have had to be repeated. Then, the weather conditions had to be perfect to get optimum results:

> In order therefore to calculate how long a time it must take to sweep the heavens, as far as they are within the reach of my 40-feet telescope, charged with a magnifying power of 1000, I have had recourse to my journals, to find how many favourable hours we may annually hope for in this climate. It is to be noticed, that the night must be very clear; the moon absent; no twilight; no haziness; no violent wind; and no sudden changes of temperature; then also, short intervals for filling up broken sweeps will occasion delays; and, under all these circumstances, it appears that a year which will afford 90, or at most 100 hours, is to be called very productive. (Herschel 1800)

Herschel, always polite to an extreme, was imposed upon by many visitors, partly because of his proximity to the Court at Windsor, to whom he became almost a tourist attraction. A register of visitors still exists and is displayed at the William Herschel Museum in Bath. For example, twenty or more British and foreign nobles, bishops, or scientists turned up each month during the summer of 1788. A few years later the musician Joseph Haydn was one such casual visitor. His 'First London Notebook', a compendium of travels and unreliable gossip, contains under 15 June 1792 the entry:

> I went from Windsor to [Slough] to Doctor Hershel [sic], where I saw the great telescope ... In his younger days Dr Hershel was in the Prussian service as an oboe player. During the seven-years' war he deserted with his brother and went to England, where he supported himself as a musician for many years: he became an organist at Bath, but gradually turned more to astronomy. After having provided himself with the necessary instruments, he left Bath, rented a room near Windsor, and studied day and night. His landlady was a widow, fell in love with him, married him, and gave him a dowry of £100,000. Besides this he has a yearly pension

---

[52]Quoted by Sime (1900, p. 212).

of £500 from the King, and his wife, at the age of 45, presented him with a son this year, 1792. Ten years ago he had his sister come, and she is of the greatest assistance to him in his observations. Sometimes he sits for 5 or 6 hours under the open sky in the bitterest cold weather.[53]

The story of the 'dowry' from his rich 'landlady' may have been current at the time, but seems to have no relation to the facts of the matter. Nor was his wife 45 when John Herschel was born, but she was about 42, a dangerous enough age to have had a child.

The Director of Paris Observatory, Jérôme le François, called Lalande[54], was a strong admirer and wrote a letter to him in 1787 addressed simply to 'Monsieur Herschel, le plus célèbre astronome de l'univers, Windsor'. This contact was followed by several visits in 1788 and a thank you note:

> Dear Sir and Colleague
> I have arrived in London completely taken, completely penetrated by your kindness, by your lessons, by your affability, by your genius, by your marvels. My first task is to thank you, as well as Mrs H. and the kind Miss, who I have had so much pleasure in seeing busy with you in a way which will serve me as a model for our frivolous French girls ... [55]

In a further note sent before he returned to France, Lalande stated 'I will never forget especially the night of 5 August: I have told everybody that I have never passed one as agreeable, not even excepting those spent in making love'.[56] Lalande named his niece, who was born on the day that Caroline discovered one of her comets, after her.

Dr Charles Burney, the musicologist, was an occasional visitor. In September 1797 he mentioned in a letter how he had dropped in on Herschel:

> The good soul was at dinner but came to the carriage himself, to press me to alight immediately and partake of his family repast; and this he did so heartily that I could not resist ...
> We soon grew acquainted, I mean the ladies [Mrs Herschel, her mother, Miss Wilson, daughter of the Glasgow astronomer, and Caroline Herschel] and I, for Herschel I have known very many years; and before dinner was over all seemed old friends just met after a long absence. Mrs Herschel is sensible, good-humoured, unpretending, and obliging; the Scots lady sensible and harmless; the little boy [J.F.W. Herschel] entertaining, comical, and promising. Herschel, you know, and everybody knows, is one of the most pleasing and well-bred natural characters of the present age, as well as the greatest astronomer.[57]

[53]Haydn, J., First London Notebook, tr. Landon (1959, pp. 254–255).

[54]Lalande had adopted the aristocratic title 'de Lalande' under the *Ancien Régime* but later prudently contracted it to the more plebeian-sounding Lalande following the Revolution.

[55]Lalande to Herschel, 8 August 1788, quoted by Lubbock (1933, p. 217), tr. ISG.

[56]Lalande to Herschel, quoted by Hoskin (2003c, p. 171), tr. ISG.

[57]Burney's Memoirs, quoted by Lubbock (1933, p. 291).

FIG. 4.7. William Herschel in 1794, from a pastel by John Russell (1745–1806). Herschel is shown holding a diagram of 'The Georgian planet with its Satellites' (from *The Scientific Papers of Sir William Herschel*, 1912), Vol. 1.

If Herschel was 'one of the most pleasing characters of the present age', Burney himself must sometimes have been hard to tolerate: he had written a 'Poetical History of Astronomy', which he read to Herschel in many lengthy sessions. Herschel evidently listened politely. But, Burney wrote:

> He made a discovery to me, which had I known it sooner, would have overset me, and prevented my reading any part of my work. He said that he had almost always had an aversion to poetry, which he regarded as the arrangement of fine words without any useful meaning or adherence to truth; but that when when truth and science were united to these fine words he liked poetry very well.[58]

Time was available for private musical events. Herschel played the piano as well as other instruments and Mrs Papendiek records having taken part in singing one of his catches at a party in her own house where his nephews, the Griesbachs, also performed.

[58]Burney, quoted by Sime (1900, p. 202).

Herschel and his wife took holidays for a month or so during the summers, when the nights were too short for serious work. Sir William Watson, who continued to live at Bath, and Alexander Aubert remained special friends whom he visited many times. Patrick Wilson, the son of Alexander Wilson, encouraged him to visit Glasgow, which he duly did. They toured the Lake District, climbing many of the mountains. The journey of 1792 was mainly to see the factories of the industrial revolution; one visit was to the Soho (Birmingham) works of Boulton and Watt. James Watt, who is best-known for his improvements to the steam engine, was one of the inventors and entrepreneurs who constituted the Lunar Society of Birmingham, so called because they could meet conveniently (in the absence of street lights) only at the time of full moon. These men were the leaders of the movement towards industrialisation. Watt himself could reflect in old age 'My inventions are giving employment to the best part of a million of people, and [have] added many millions to the national riches' (Smiles 1878, p. 405). As a result of their friendship, Herschel was asked some years later to appear as an expert witness in a patent infringement case, Boulton and Watt v. Bull (1793). The furthest point of this particular trip was Glasgow, where he was entertained by the city and the university, receiving an honorary doctorate from the latter—his second, having received one from Edinburgh six years before.

In 1802 Herschel undertook a long-planned visit to Paris to see Lalande, Méchain, Delambre, Laplace (who had been one of the first to calculate the orbit of Uranus), Messier, and other astronomical correspondents. For once, France and England were not at war. It was the period when Napoleon was consolidating his power and carrying out internal reforms, such as the introduction of his new legal code. The journey took over twelve days, including two spent at Dover waiting for good weather before crossing the Channel. Nearly all the eminent scientists in Paris were keen to meet him, and he them. A high point of this trip was a visit to Napoleon, whose official title was at that time 'First Consul', but he was already, in fact, a dictator. Among the company were Count Rumford (Benjamin Thomson),[59] a displaced American scientist who had taken the wrong side during the Revolution, and Laplace. Napoleon sat down and invited Herschel to do so also, but he preferred to remain standing with the others, who had not been so honoured. The First Consul asked 'a few questions relating to Astronomy and the construction of the heavens to which I made such answers as seemed to give him satisfaction'. Herschel was unfavourably impressed by an argument that Napoleon had with Laplace:

[59] Benjamin Thompson (1753–1814) was born in Massachusetts before the American Revolution but acted as a spy for the loyalists and was eventually forced to flee. He was later a distinguished scientist who worked on the theory of heat. He was made Count Rumford by the Elector of Bavaria, for whom he was a consultant on military matters. He was famous to the general public for having invented a new kind of efficient stove and was the first to show the equivalence of heat and energy. In 1805 he married the widow of the French chemist Lavoisier. A newspaper of the time remarked 'by which nuptial experiment he obtains a fortune of £8000 per annum—the most effective of all the Rumfordizing projects for keeping a house warm'. He was one of the founders if the Royal Institution in 1799.

The difference was occasioned by an exclamation of the first Consul, who asked in a tone of exclamation or admiration (when we were speaking of the extent of the sidereal heavens): 'And who is the author of all this!' Mons. de Laplace [evidently a deist like Herschel] wished to shew that a chain of natural causes would account for the construction and preservation of the wonderful system. This the first Consul rather opposed.[60]

This exchange he recalled eleven years later in a conversation with the poet Thomas Campbell: 'The First Consul did surprise me by his quickness and versatility on all subjects, but in science he seemed to know little more than any well-educated gentleman, and of astronomy much less than, for instance, our own king ... I remarked his hypocrisy in concluding the conversation on astronomy by observing how all these glorious views gave proofs of an Almighty wisdom'.

4.17   Later discoveries and interests

Infrared radiation

One of Herschel's most important discoveries was published when he was 62 years old, namely that of infrared radiation. In making observations of the sun, he had had to use filters made of coloured glass to reduce its heat and brightness, which he found were not related to each other. Intrigued by this discovery, he tried the experiment of passing a beam of sunlight through a prism and onto a table where there were three thermometers arranged side-by-side so that only the middle one received the sunlight. The tiny heating effect of the light could be separated from changes in the temperature of the surrounding air by comparing them. He found that the heating was greatest at the red end of the spectrum and that it was even greater past the end of the visible light. This was the discovery of infrared radiation, reported to the Royal Society in 1800 in a paper called *On the rays which occasion heat*. Banks, the Society's President, wrote 'I hope you will not be affronted when I tell you that highly as I prized the discovery of a new planet, I consider the separation of heat from light as a discovery pregnant with more important additions to science'.[61] Henry Cavendish and Count Rumford concurred.

He was able to show that infrared rays could be refracted and reflected like visible light but could not find any heat in rays beyond the violet end of the spectrum. Stimulated by his infrared discovery, Johann Wilhelm Ritter of Jena found in the following year that ultraviolet rays were very effective in blackening silver chloride—a property that turned out later to be the key to photography.

Nebular theory

Herschel, like Johann Heinrich Lambert, the philosopher Immanual Kant, Pierre Simon Laplace, and other eighteenth-century scientists, felt that there ought to be an evolutionary sequence present in the skies. He was a supporter of the

[60] Herschel, Diary, quoted by Lubbock (1933, p. 310).
[61] Banks, quoted by Hoskin (2003a).

'nebular hypothesis', which suggested that stars formed out of nebulous material and then drew together into clusters. Eventually the clusters would amalgamate to form even larger bodies which contained so many stars as to appear seamless. These would then be 'Island Universes'. He thought, for example, that the Milky Way's nebulosities would eventually collapse into clusters of stars. These ideas had a long lifetime, surviving to the end of the nineteenth century, mainly because they were somewhat 'nebulous' themselves and impossible to prove or disprove. Neither were they altogether wrong. Herschel, in his later life, tried to classify all his nebulae into such an evolutionary scheme.

Physical binaries

In 1802 Herschel announced another great discovery—that many double stars are, in fact, binaries in elliptical orbits revolving around each another, bound together by gravity just like the planets of our own solar system and obeying Kepler's laws. Thus the gravity that operates in the solar system was also at work between the distant stars. During the following two years he summarised his twenty-five year observational programme of some fifty double stars, placing the truth of his important discovery beyond doubt.

Stellar brightness and variable stars

Two other areas that Herschel contributed to were the study of variable stars and the measurement of brightnesses. The idea that stars might differ radically from each other in intrinsic brightness had not yet taken hold at the start of his career. However, his discovery that the binary stars could be physically close to each other but markedly different in their properties showed that they could not all be identical.

4.18   Doubtful speculations

Some less successful speculations were concerned with the phenomenon called 'Newton's rings', the colouring effects that are visible in thin films, such as when oil forms a layer on water. Herschel attempted to explain them on Newton's corpuscular theory of light rays (1807–1810). In fact, they had already been analysed perfectly successfully using the wave theory of light by Thomas Young, although this work was somewhat ignored at the time. In cases like this, Herschel's speculations at times retained an amateur flavour. He did not distinguish as clearly as he might have between hard fact and guesswork.

Herschel's friend Alexander Wilson of Glasgow believed that the sun was covered by luminous clouds with occasional gaps—the sunspots—through which the solid surface beneath could be seen. The clouds could be replenished by comets, which were thought to be made of a special luminous matter, falling in from time to time. Herschel thought that the sun, like the moon, should be inhabited. He speculated that the radiation of the sun fluctuated enough to affect the weather on earth and even suggested that the price of wheat could be correlated with the number of sunspots. This work was received with considerable criticism, even ridicule.

4.19 Asteroids and the Celestial Police

In 1772, Bode of Berlin enunciated a law (partly anticipated by Johann Titius) that the spacings of the orbits of the planets had a certain regularity, except that there seemed to be a bigger than expected gap between Mars and Jupiter. Naturally, there was speculation that it should be occupied by a missing planet. Herschel's discovery of Uranus extended the sequence and made the gap seem even more mysterious. Franz Xaver von Zach, an early proponent of international collaboration, set up a special group, who came to be known as the 'Celestial Police', to look for the missing body. Although Herschel was one of the 'policemen', it was Giuseppe Piazzi of Palermo who found (in 1801) the first body in the right place, a rather tiny planet which received the name of Ceres. Herschel used his powerful telescopes to examine it. Three similar objects were found within the next few years and it was Herschel who proposed that they should be called 'asteroids', because they seemed small enough to be star-like.

4.20 John Frederick William Herschel

On 9 March 1792, Mary Herschel gave birth to their only child, a boy, who was christened John Frederick William (see Fig 4.8). His godfathers were Sir William Watson and Count Komarzewski, a Polish noble who had become another close friend.

John Herschel, as the only child of William, and indeed the only male Herschel of his generation, was very much cossetted. He was sent to Eton, a famous school attended by many upper-class boys, which was nearby in Windsor, but was soon withdrawn as it was felt to be too rough. He eventually became a student in mathematics at Cambridge and graduated 'First Wrangler', or top of his class. With his friends, who included Charles Babbage, the computer pioneer, he helped to modernise mathematical teaching at the University. The cumbersome notation that Newton had used for the calculus had continued in use and had cut Cambridge off from the mainstream of mathematics. They set up the 'Analytical Society' to rescue Cambridge from its 'dot-age', a reference to Newton's notation for differentials in calculus.

John Herschel obtained a Fellowship at St John's College, Cambridge, in 1813. The question soon arose as to what his future occupation would be: his father suggested the Church but he inclined towards the law. This led to a sharp but revealing letter from the seventy-five-year old to his son, one of the few occasions when he set aside his extreme politeness:

Nov. 8, 1813

My dear Son

You request my opinion upon a subject which is of so much con-sequence, and upon which so much may be said that it will hardly be possible to enter into it properly by letter writing; but what is more, you have already given your opinion of it, and to ask mine seems to be a mere matter of form. You say—I cannot help regarding the source of church emolument with an evil eye.—The miserable tendency of such a

FIG. 4.8. John Herschel, son of William and Mary. (From Ball 1911.)

sentiment, the injustice and the arrogance it expresses, are beyond my conception.

You mention the necessity of a mode of acquiring a living, and that a man should have some ostensible means of getting his bread by the labour of his head or hand, as if you already found yourself in confined circumstances, or had no expectation of assistance from your parents, which you know is not a just idea of your situation.

You say that Cambridge affords you the Society of persons of your own age and of your way of thinking, but know, my dear Son, that the company and conversation of old, experienced men of sound judgement, whose way of thinking will often be different from your own, would be much more instructive, and ought to be carefully frequented.

But now let us compare the end proposed and the means of obtaining it in the two liberal professions of which it seems the choice is to be made. Your own words with regard to the church were, 'The path was wide and beaten, and I only had to pursue it'. Can anything be a higher commendation of it? Such a path must surely lead to happiness, or else it would never be so wide and so beaten. You admit that you had only to pursue it. Very true, for you are in it and have all the time been in it; nay, you are already far advanced in it, having not only carried off the highest honours of the University but being already in possession of a fellowship which ensures your success in pursuit of the path.

How is it about the path of law? It is crooked, tortuous and precarious. It is also beaten, but how many have miserably failed to acquire an *Honest* livelihood? You are not in that path, and it is almost too late to enter it;

your studies have been of a superior kind.

You say the church requires the necessity of keeping up a perpetual system of self-deception, or something worse, for the purpose of supporting the theological tenets of any particular set of men ... A philosopher could not scruple to recommend the forbearing disposition of Epictetus, and would as little hesitate to inculcate the Doctrine of Christ to love your neighbour as yourself and to act towards him as you would wish he should act towards you.

Now cast a look upon the Law and Lawyers. Is it not evident that at least one half of them do act against their better knowledge and conscience? ...

On the other hand, a Clergyman may get his bread, and always act conscientiously and do good to every person with whom he has any connection ...

A Clergyman ... without the least derangement of his ostensible means of livelihood, has time for the attainment of the more elegant branches of literature, for poetry, for music, for drawing, for natural history, for short pleasant excursions of travelling ... He may also be a happy family man, a husband, a father, and with a paternal fortune added to his other means he will have all the real enjoyments of life within his reach.

With my best wishes for your happiness in which all the family joins, I remain, my dear son,

Your affectionate father,

WM. Herschel.[62]

William's attitude was typical of a 'latitudinarian' Anglican of the eighteenth century, before the 'evangelicals' arrived to make the established church take religion more seriously. John nevertheless did attempt to become a lawyer but in 1816 he gave up and went home to work with his father on astronomical matters. He was to become highly successful in continuing his father's work and made important contributions to many other areas, such as photography, but his career is beyond the scope of this book.

A touching description of Herschel senior in a more typical benign mood was written by the poet Thomas Campbell in a letter to an unknown correspondent, just a few weeks before the incident referred to:

Brighton, Sept 15, 1813

I wish you had been with me the day before yesterday, when you would have joined me, I am sure, deeply in admiring a great, simple, good old man in Dr Herschel ...

Now for the old Astronomer himself; his simplicity, his kindness, his anecdotes, his readiness to explain and make perfectly perspicuous too his own sublime conceptions of the universe are indescribably charming. He is 76, but fresh and stout, and there he sat nearest the door at his friend's house, alternately smiling at a joke, or contentedly sitting without share or notice in the conversation. Any train of conversation he follows

---

[62]Herschel to J.F.W. Herschel, 8 November 1813, quoted by Lubbock (1933, pp. 349–350).

implicitly; anything you say he labours with a sort of boyish earnestness to explain ...

Then, speaking of himself, he said, with a modesty of manner which quite overcame me, when taken with the greatness of the assertion:

'I have *looked further into space than ever human being did before me.* I have observed stars of which the light, it can be proved, must take two million years to reach the earth.'

I really and unfeignedly felt at the moment as if I had been conversing with a supernatural intelligence ...

After leaving I felt elevated and overcome, and have, in writing to you, made only this memorandum of some of the most interesting moments of my life.[63]

In May 1816, Herschel was made a knight of the Hanoverian Guelphic Order by the Prince Regent, later George IV (1762–1830), then reigning in the place of his father George III, who had become insane in old age.

Near the end of Sir William's life he was asked to become the first President of the Astronomical Society, later the Royal Astronomical Society. His son John was one of its founder members. They had felt the need for a specialised body and had hoped that the Duke of Somerset would take on the position, but Sir Joseph Banks was afraid that it would detract from the older Royal Society and went so far as to persuade the Duke, a personal friend, not to accept.

In the years before he died, Herschel became progressively more frail and suffered frequent illnesses. One of his last trips to London was to have his portrait painted by William Artaud (see Fig 4.9). On 15 August 1822, he had to take to his bed and on the 25th he expired. He was buried in the church at Upton. In his epitaph, composed by the Provost of Eton, are the words 'Coelorum perrupit claustra', meaning 'He broke through the barrier of the Heavens'.

4.21 Last years of Caroline

Caroline felt that her brother's death was in many ways the end of her own existence. To increase her loneliness, the court of Queen Charlotte, which was the centre of her social life, dissolved in 1818 on the death of the Queen, two years before her husband George III. His successor, George IV, continued Caroline's £50 state pension, which was augmented by £100 per annum from William's estate. She retired to Hanover, where she expected to be happier, soon after his death. In 1835 she and Mrs Somerville[64] were made honorary members of the Royal Astronomical Society. Only three women received such a distinction before Margaret Huggins and Agnes Clerke were awarded it in 1903.[65] She lived to the age of 97, dying on 9 January 1848. She was a lively if somewhat caustic old soul who was well-remembered in astronomical circles. She had the satisfaction of seeing the completion in May 1838 of her nephew's famous expedition to the

---

[63]Thomas Campbell to a friend, 15 September 1813, quoted by Lubbock (1933, pp. 335–336).

[64]A noted female mathematician

[65]The third was Miss Anne Sheepshanks, one of the Society's benefactors.

FIG. 4.9. Portrait of William Herschel in June 1819, painted by William Artaud (1763–1823). (Frontispiece of *The Scientific Papers of Sir William Herschel* (1912), Vol. 2).

Cape of Good Hope. In recognition of this work, he was made a baronet (a hereditary knight) at the time of Queen Victoria's coronation.

References

Scientific Papers of Sir William Herschel Knt. Guelph., LL.D., F.R.S., 1912. 2 vols., ed. J.L.E. Dreyer, Royal Society and Royal Astronomical Society, London.

Ashworth, William B., 2003. Faujas-de-Saint-Fond visits the Herschels at Datchet, *J. Hist. Astr.*, **34**, 321–324.

Ball, Sir Robert S., 1911. *Great Astronomers*, cheap edition, Pitman, London.

Bennett, J.A., 1976. 'On the Power of Penetrating into Space': The Telescopes of William Herschel, *J. Hist. Astr*, **7**, 75–108.

Clerke, Agnes M., 1893. *A Popular History of Astronomy During the Nineteenth Century*, 3rd edn., Adam & Charles Black, London and Edinburgh.

Dreyer, J.L.E., 1912. (see *Scientific Papers . . .*), above.

FIG. 4.10. Portrait of Caroline Herschel, aged 79, painted by Tielemann in 1829. (From *The Scientific Papers of Sir William Herschel* (1912), Vol. 2, facing p. 651.)

Herschel, Mrs. John, 1879. *Memoir and Correspondence of Caroline Herschel*, 2nd edn., John Murray, London.

Herschel, W., 1783. A Letter from William Herschel, Esq. F.R.S. to Sir Joseph Banks, Bart, P.R.S., *Phil. Trans. Roy. Soc.*, **73**, 1–3.

Herschel, W. 1784. Account of some Observations tending to investigate the Construction of the Heavens, *Phil. Trans. Roy. Soc.*, **74**, 437–451.

Herschel, W. 1785., On the Construction of the Heavens, *Phil. Trans. Roy. Soc.*, **75**, 213–266.

Herschel, W., 1800. On the Power of penetrating into Space by Telescopes; with a comparative Determination of the Extent of that Power in natural Vision, and in Telescopes of various Sizes and Constructions; illustrated by select Observations, *Phil. Trans. Roy. Soc.*, **90**, 49–85.

Hoskin, M., 2003a (ed.) *Caroline Herschel's Autobiographies*, Science History Publications Ltd, Cambridge.

Hoskin, M., 2003b. Herschel's 40ft Reflector: Funding and Functions, *J. Hist. Astr.*, **34**, 1–32.

Hoskin, M., 2003c. *The Herschel Partnership*, Science History Publications Ltd, Cambridge.

Landon, H.C. Robbins, 1959. *The Collected Correspondence and London Notebooks of Joseph Haydn*, Barrie and Rockliff, London.

Lubbock, C.A., 1933. *The Herschel Chronicle: The Life Story of William Herschel and his Sister Caroline Herschel*, Cambridge University Press, Cambridge.

Sime, James, 1900. *William Herschel and His Work*, T. & T. Clark, Edinburgh.

Smiles, Samuel, 1878. *Lives of the Engineers. The Steam Engine. Boulton and Watt*, new revised edn. John Murray, London.

Smith, Robert, 1738. *A Compleat System of Optics*, 4 vols., Cornelius Crownfield, Cambridge.

Spaight, J.T., 2004. "For the Good of Astronomy": The Manufacture, Sale, and Distant Use of William Herschel's Telescopes, *J. Hist. Astr.*, **35**, 45–69.

5

WILLIAM HUGGINS: CELESTIAL CHEMICAL ANALYSIS

It was just at this time [1859], when a vague longing after newer methods of observation for attacking many of the problems of the heavenly bodies filled my mind, that the news reached me of Kirchhoff's great discovery of the true nature and the chemical constitution of the sun from his interpretation of the Fraunhofer lines.

This news was to me like the coming upon a spring of water in a dry and thirsty land. Here at last presented itself the very order of work for which in an indefinite way I was looking—namely, to extend his novel methods of research upon the sun to the other heavenly bodies. A feeling of inspiration seized me: I felt as if I had it now in my power to lift a veil which had never before been lifted; as if a key had been put into my hands which would unlock a door which had been regarded as forever closed to man—the veil and door behind which lay the unknown mystery of the true nature of the heavenly bodies ... (Huggins 1897)

The names of William Huggins and his wife Margaret are not as well-known today as they once were. Nevertheless, this couple played an important role in the development of astrophysics, especially in the analysis of starlight. William Huggins's life largely coincided with the reign of Queen Victoria, who was on the throne from 1837 to 1901 and gave her name to an age. It was a time characterised by immense self-confidence and national pride. As a result of the industrial revolution and the economic expansion that it brought, the possession of private wealth and cultural interests became widespread: a spirit of optimism permeated society.

In the seventeenth and eighteenth centuries there had been very few posts for astronomers, the main ones being at the Royal Observatories of England and France and at a few universities. The state observatories had been established to improve navigation at sea, essential to the working of powerful navies and expanding merchant marines. Their employees spent most of their time 'grinding the meridian'—making observations of objects as they passed the north–south line through the zenith to determine their positions in the sky. Very little time was available for more speculative work.

In the nineteenth century there arose a number of gifted amateurs who were amongst the first to apply the newly made discoveries in physics and chemistry to celestial matters. Most were rich individuals who had turned to astronomy after having made their fortunes or who had benefitted from substantial inheritances. They included the Irish landowner William Parsons, third Earl of Rosse, who was famous for his giant telescope 'The Leviathan of Parsonstown' and having first made out the spiral form of certain galaxies (though he did not, of course, know that is what they were). The brewer William Lassell built large telescopes and discovered Triton, the satellite of Neptune, and Hyperion, a satellite of Saturn. The printer Warren de la Rue was a pioneer of solar photography; the machine-tool builder James Nasmyth, inventor of the steam-driven pile-driver, built the eponymous Nasmyth telescope (the precursor of the present generation of large telescopes), and Richard Carrington, from another brewing family, discovered terrestrial effects related to sunspot activity. In fact, it became quite fashionable for serious-minded rich people to have private observatories. Even a moderately well-off person might own a small telescope for looking at the Moon and planets. The founding of the Royal Astronomical Society in 1820 brought together those with an interest in astronomy and helped to increase enthusiasm for the subject.

5.1   Early years

William Huggins was born in Gracechurch Street, close to the Bank of England in the centre of the City (business district) of London, on 7 February 1824. He was the only child of a mercer[1] and linen draper, William Thomas Huggins, and his wife, Lucy, née Miller. Like many other shopkeepers of their time, the Huggins family lived in the building that housed their business, as did their assistants. They were well enough off to have one or two domestic servants.

William was christened on 24 August 1824 in a 'dissenting chapel', called Pinners Hall. Dissenters were protestants, in the Huggins's case Congregationalists, who did not belong to the official Church of England. Members of such minority religious groups were still discriminated against in some areas, such as in getting government employment. As a result, they were more than usually conscious of the need for their children to be well educated if they were to have successful careers. Before the age of 10, William was sent to a small school in Great Winchester Street for a short time. He took extra lessons in classics and mathematics from a clergyman and also attended lectures in the 'Adelaide Gallery', a sort of precursor technical school.

When the City of London School opened in January 1837, Huggins was one of its first pupils. It was an enlightened institution, offering a curriculum similar to a present-day academically orientated high school. However, his time there was brought to an end by an attack of smallpox, after which his parents took him away, fearing further infectious diseases. When he was about 15, his mother brought him to Paris where he obtained a kit for making Daguerreotypes, the earliest commercial process for photography, which had only just been invented. According to his biographers, Mills and Brooke (1936), during his late 'teens he took lessons at home with private tutors, in the hope of studying later at Cambridge University. This would not have been possible for a dissenter. Either this information is incorrect or he intended to swallow his pride and join the official church. He was good at languages, supposedly being fluent in French,

---

[1] Mercers were dealers in expensive cloths such as silk.

German, and Italian as well as Spanish, Swedish, and Hebrew. Music was not neglected: he had violin lessons and possibly piano as well.

Mills and Brooke (1936) state that he was already at this time an active amateur scientist, building his own apparatus for chemistry, optics, physics, electricity, and photography. At first he seems to have been most interested in microscopy and plant physiology. It is recorded that he bought his first telescope in 1842 for £15, when he would have been eighteen years old. This was about the annual income of an unskilled labourer. The smoky skies of central London and its many tall buildings were by no means ideal for astronomy. Furthermore, he would have had to use his telescope from the roof of his home or through an open window.

As a good Victorian, William 'denied himself idle amusements' and seems to have had a circle of rather serious friends. However, he did not adopt the strict Calvinistic ideas of the Falcon Square Congregational Church, to which the family then belonged. In later life his wife was to describe him as 'Christian unattached'.

5.2 Shopkeeper

Just at the time when he would have gone to university his father became seriously ill. To complicate matters, the family business then needed close attention because they suspected that their foreman was stealing from them. William had to give up any idea of studying at Cambridge. Instead, he was forced to take over the shop. However, the pursuit of business was interspersed with holidays on the continent, which he enjoyed with friends. We know of two of these trips, undertaken in 1850 and 1854. The earlier one was to Belgium, Germany and Switzerland. He travelled with a rather superior attitude towards foreigners. He referred in his travel diary to the German habit of smoking 'which has called down upon the devoted nation the sarcasm and rebuke of every English traveller, as a selfish and disgusting habit, which sacrifices another's comfort to the gratification of one's own prurient appetite (an appetite too, not natural, but forced and learned)' (Mills and Brooke 1936, p. 16). His travels in 1854 were to Paris, the Loire, and the Pyrenees. He was still at the age of 30 very much a bachelor. 'I cannot describe the ladies, as I never notice them'. At Eaux Bonnes, near Pau, he remarks 'It is a very fashionable bathing place and much reknowned for diseases of various kinds, especially those of the skin. Even celibacy, says our witty guide book, is often cured here. *I* very much doubt whether the remedy is not found much worse to bear than the disease itself'. He was also averse to wine: 'The little relish I have for wine is certainly not fostered by being told how the labourers wash their hands and feet, and . . . but I forbear'.

While working in London he had sufficient time to pursue interests in microscopy and astronomy, joining the Royal Microscopical Society in 1850 and the Royal Astronomical Society in 1852. These societies were in fact dominated by amateurs, at least in membership numbers, during the nineteenth century. No qualification was required for entry to the Royal Astronomical Society except the recommendation of two existing members, called 'Fellows'.

In 1853 he bought a second-hand Dollond refractor (lens-based telescope) with an aperture of 5 in. (13 cm) for £110. Dollond was a long-established London firm which had held the original patent on the achromatic lens, a great improvement on the simple lenses used by pioneers such as Galileo, because it made use of two different kinds of glass to bring all the colours to the same focus. The new price of the object glass had been 200 guineas (£210), so he must have driven a hard bargain.

5.3   Independence

In the following year, 1854, at the age of 30, Huggins made the decision to sell his business and the lease on his Gracechurch Street property and move to a neighbourhood south of built-up London. His new house was at 90 Upper Tulse Hill Road, at 177 ft (54 m) above sea level. Being a little higher than the rest of the city, it escaped some of the notorious smog generated by coal-burning. It was small and two-storied, with a front garden and a larger back one. Soon after the move, his father died but his mother lived on with him for a number of years. He was evidently comfortably off, though not exactly wealthy, and had no anxiety about the necessities of life. They could afford a cook and a female servant.

Astronomy now became his main occupation and he set about constructing a serious observatory which connected directly to the upper floor of his house. He hired a carpenter to construct a building raised 16 ft (4.9 m) above the ground on iron pillars so that his view would not be obstructed by the houses of his neighbours. The Dollond telescope was in a chamber of 12×18 sq. ft (3.7×5.5 m^2) with a dome on top. A transit telescope, by Jones, for measuring star positions, occupied a separate room. An astronomical clock by Arnold completed his instrumental setup (André and Rayet 1874). The telescopes were mounted on concrete pillars for stability. He was sufficiently proud of his observatory to publish a description in *Monthly Notices of the Royal Astronomical Society* (Huggins 1856).

He started his new calling rather like an enthusiastic schoolboy, with an observing notebook whose title page he embellished in Hebrew ('Look to the Heavens and count the stars'; Genesis), Latin ('Phenomena Observed by Will. Huggins through his Telescope . . . ') and Greek ('One star differs from another in glory'; 1 Corinthians). Its early pages were filled with conventional observations of the surface features of the planets such as many other amateurs were then occupied with. His notebooks still exist and are to be found in Wellesley College, Massachusetts, USA, to which they were bequeathed by his widow.

In 1858 Huggins sold his 5-in. Dollond and bought an 8-in. (20 cm) lens from another Fellow of the Royal Astronomical Society, the 'eagle-eyed' Rev William Rutter Dawes, a well-known observer of double stars who was reputed to have the most acute vision of anybody for precise visual measurements. The 8-in. lens had been made by an American painter turned optician, Alvan Clark of

FIG. 5.1. William Huggins around 1870 (SAAO).

Cambridge, Massachusetts, whose workmanship had impressed the ultra-critical Dawes. In fact, it was Dawes's glowing endorsements that ultimately led to the success of Clark's business. His telescopes are even now prized for their excellent optics. Huggins's newly acquired lens was placed in a tube and mounted by the English telescope maker Thomas Cooke of York. It is evident that from this point on his observing became more seriously scientific, although for some time he continued to receive advice from Dawes on suitable planetary projects to undertake. Meanwhile, he was looking around for a really challenging project with which he might be able to make a scientific name for himself.

5.4 'A spring of water in a dry and thirsty land'

The year 1859 saw the publication of Darwin's *The Origin of Species*, which revolutionised biology and challenged conventional thought in many other areas. In October of the same year, the Heidelberg scientists Gustav Kirchhoff and Robert Bunsen (Fig 5.2) published a paper which caused another revolution,

FIG. 5.2. Robert Bunsen (left) Gustav Kirchhoff (right) (SAAO).

this time in physics. They explained for the first time the nature of the Sun's spectrum as first described in detail by Josef von Fraunhofer in 1814. He had found that when a narrow[2] beam of sunlight falls onto a prism and the emergent rays are looked at through a telescope, the primary colours are seen to be crossed by narrow *dark* lines (see Fig 5.6, second panel from top).

By the time of Bunsen and Kirchhoffs' work it was already known that each element produces a unique set of *bright* lines when burned (see Fig 5.6, second panel from bottom, for the characteristic lines of hydrogen). In fact, this had become a method for telling which elements are present in an unknown substance; i.e. it was becoming possible to use spectroscopy for chemical analysis. Kirchhoff and Bunsen had now shown that when pure white light, whose spectrum is simply that of the rainbow colours, is allowed to pass through the relatively cool vapour of an element, the same lines as before are seen in the spectrum, but now as *dark* ones superimposed on the rainbow colours. In other words, the explanation of the lines seen by Fraunhofer in the sun's spectrum had now been found. They originate when light from the hot surface is absorbed by the vapours of elements high up in the cooler parts of its atmosphere.

Surprisingly, Kirchhoff was a traditionally precise Germanic professor who believed that the main facts of nature had already been established and that all that remained was to elaborate the known correlations between facts and to make more accurate measurements. He is said never to have missed a lecture, except once—'I regret to announce', he said on a certain Thursday at the conclusion

---

[2]Newton's beams were not narrow enough for him to have seen the Fraunhofer lines.

FIG. 5.3. Henry Roscoe (left) William Allen Miller (right) (SAAO).

of one of his lectures, 'that circumstances prevent my meeting you to-morrow'. The 'circumstances' were that he was going to get married, and the honeymoon lasted from Friday till Monday, when he was at his desk again. 'New discoveries almost worried Kirchhoff', whereas Bunsen was somewhat absent-minded and had 'the appearance of a prosperous farmer, and a somewhat cynical but at the same time good-natured smile'. He remained interested in new ideas even in retirement' (Schuster 1932).

The apostle of Bunsen and Kirchhoff in England was Henry Roscoe (Fig 5.3), Professor of Chemistry in Owens College, Manchester, the forerunner of the University of Manchester. He had been a student of Bunsen. He spread the word to the scientific community through George Gabriel Stokes, the leading theoretical physicist of Cambridge University, and a powerful figure in the Royal Society. He also gave public lectures on the subject in 1861 and 1862 at the Royal Institution. This organisation had been founded in 1799 by Count Rumford, born Benjamin Thompson (see Section 4.16) for research and had become the host for popular lecturers on scientific subjects. The new method of chemical analysis that Roscoe described had almost immediately (in 1860) led to the discovery of the elements caesium and rubidium by Bunsen in Germany and of thallium by William Crookes in England.

In his later years, as will be described, Huggins was to write an article very full of self–pride on the rise of the 'New Astronomy', meaning spectroscopic astrophysics, which he largely associated with his own career, for the magazine

The Nineteenth Century (Huggins 1897).[3] He had this to say on his introduction
to spectroscopy:

> I soon became a little dissatisfied with the routine character of ordinary
> astronomical work, and in a vague way sought about in my mind for
> the possibility of research upon the heavens in a new direction or by
> new methods. It was just at this time, when a vague longing after newer
> methods of observation for attacking many of the problems of heavenly
> bodies filled my mind, that the news reached me of Kirchhoff's great
> discovery of the true nature and the chemical constitution of the sun
> from his interpretation of the Fraunhofer lines.
>
> This news was to me like the coming upon a spring of water in a dry
> and thirsty land. Here at last presented itself the very order of work for
> which in an indefinite way I was looking—namely, to extend his novel
> methods of research upon the sun to the other heavenly bodies. A feeling
> of inspiration seized me: I felt as if I had it now in my power to lift a
> veil which had never before been lifted; as if a key had been put into my
> hands which would unlock a door which had been regarded as forever
> closed to man—the veil and door behind which lay the unknown mystery
> of the true nature of the heavenly bodies. . . .
>
> It was just at this time that I happened to meet at a soirée of the
> Pharmaceutical Society, where spectroscopes were shown, my friend and
> neighbour, Dr. W. Allen Miller, Professor of Chemistry at King's College
> [London], who had already worked much on chemical spectroscopy. A
> sudden impulse seized me to suggest to him that we should return home
> together. On our way home I told him of what was in my mind, and
> asked him to join me in the attempt I was about to make, to apply
> Kirchhoff's method to the stars. At first, from considerations of the great
> relative faintness of the stars, and the delicacy of the work from the
> earth's motion, even with the aid of clockwork, he hesitated as to the
> probability of our success. Finally he agreed to come to my observatory
> on the first fine evening . . . (Huggins 1897)

In fact it was that very neighbour, William Allen Miller (Fig 5.3) of 103
Tulse Hill Road who in January 1862 had given the lecture that he described.
Miller was a well-established scientist, Professor of Chemistry in King's College,
London, and had been the Treasurer and a Vice-president of the Royal Society.
He was an expert on spectroscopy and had for many years been interested in the
use of spectral lines for chemical analysis. He must have realised straight away
the significance of Kirchhoff's discovery. In his lecture, he mentioned clearly that
the dark lines had been seen in the spectra of stars by Fraunhofer himself. Like
so many discoveries, Kirchhoff's work had built on what had gone before and
represented a kind of grand synthesis of existing trends: many scientists had already shown interest in the possibilities of spectroscopy for chemical analysis. In
short, it was an idea whose time had come. Kirchhoff had also mapped the Solar

---

[3]Huggins's rather selective recollections of his collaboration with Miller, as well as of the
contributions made by others, are dealt with in an interesting critical article by Becker (2001).

spectrum. Miller himself was working at that time on metallic spark spectra in the laboratory. Huggins was talking to the right person. One had the spectroscopic experience and the other knew how to work a telescope. A collaboration was proposed: on the next clear night, Miller visited Huggins's observatory and they got to work.

To place a spectrograph on a telescope and get it to operate properly was no easy matter. The conditions in an observatory dome, open to the sky, are much more hostile than those in a laboratory. Because the telescope has to track a star and be pointable in many different directions, any instrument that is attached to it must be lightly constructed but at the same time highly stable mechanically. It must be impervious to changes in temperature. It has to be ready to work the moment there is a gap in the clouds. To make comparisons of stellar material with terrestrial substances in the way suggested by Bunsen and Kirchhoffs' work it was necessary to generate laboratory spectra in the dome at the same time that a star was being observed and to display them simultaneously or nearly so. The star spectra were extremely faint and the observer had to work in near-darkness to maintain his eyes at maximum sensitivity. Huggins designed a new kind of spectrograph, one that could be exchanged for the eyepiece of his telescope, with two prisms to disperse the light into its primary colours. Between them, Huggins and Miller assembled the necessary apparatus and learned how to use it effectively (Fig 5.4).

> Then it was that an astronomical observatory began, for the first time, to take on the appearance of a laboratory. Primary batteries, giving forth noxious gases, were arranged outside one of the windows; a large induction coil stood mounted on a stand on wheels so as to follow the positions of the eye-end of the telescope, together with a battery of several Leyden jars; shelves with Bunsen burners, vacuum tubes, and bottles of chemicals, especially of specimens of pure metals, lined its walls. (Huggins 1897)

Huggins and Miller were by no means the only people to have realised the potential of spectroscopic observations. It rapidly became a 'hot' subject. Giovanni Batista Donati and Father Angelo Secchi in Italy and Lewis Rutherfurd in the United States were examining stellar spectra almost simultaneously. As a consequence, Huggins and Miller needed to publish their work as quickly as possible if they were to get credit for it. Their first paper appeared in *Philosophical Transactions of the Royal Society*, a logical choice because Miller was a Fellow. Their referee was the astronomer Romney Robinson.[4] He particularly commended their use of comparison spectra to calibrate the wavelengths of the spectral lines, an important improvement on Secchi's approach.

Huggins was an adept pupil and learned the techniques of spectroscopy rather quickly from Miller. He was very soon conducting original research and writing

---

[4]Robinson, R.R., 1862. Royal Society 'Referees' Reports 1863–1865' No 121, addressed to G.G. Stokes, 21 June 1862 (Royal Society).

FIG. 5.4. The interior of Huggins's dome as it appeared in the 1860s. The telescope is his 8-in. Clarke/Cooke and his first spectroscope (see Fig 5.5) is
shown mounted on the eye end. In the foreground is an induction coil, used
to generate the high voltages needed to make spark spectra for comparison
purposes. (Woodcut by J. Swain, courtesy Prof R. French, Whitin Observatory, Wellesley College, Massachusetts.)

scientific papers on his own. He rapidly became self-assured. At the Royal Astronomical Society meeting of 10 April 1863, the aged but still formidable Astronomer Royal, George Bidell Airy, presented the spectroscopic work of one of
his junior staff members. Huggins was scathing in his comments, pointing out
that the Greenwich spectra lacked sensitivity and were not well calibrated in
terms of wavelength. On the other hand, when he was questioned about his own
achievements by Charles Pritchard, later professor of astronomy at Oxford, he

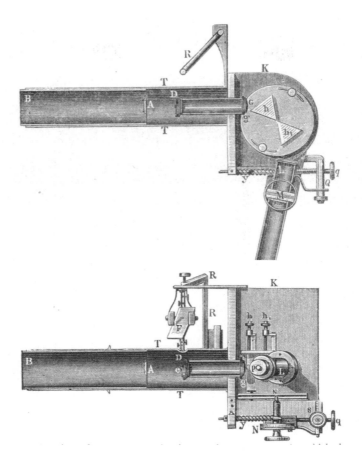

FIG. 5.5. Plan (top) and elevation (bottom) of Huggins' first spectroscope. The tube B to the left was inserted into the telescope and the star was focussed onto the slit at D. The light passed through the collimator lens g into the prisms h. The spectrum was viewed through the small telescope L. The arrangements around R were for introducing light from a comparison lamp. (From Lockyer 1874.)

was able to answer with confidence.

In fact, careful calibration was an important contributor to Huggins and Millers' success. They built up spectral atlases of the terrestrial elements in the laboratory using a special six-prism spectrograph to achieve high resolution. Whereas Kirchhoff had measured spectral lines relative to an arbitrary scale on his apparatus, Huggins used sparks in air between platinum electrodes to generate standard comparison spectra. This meant that others could easily reproduce the scale he used. Unfortunately, early attempts to record the spectra more ob-

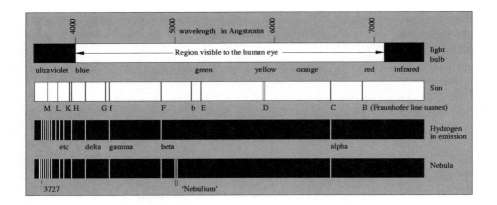

FIG. 5.6. Examples of spectrographic images. A pure hot source, such as a household light bulb, gives a rainbow-like continuous spectrum as in the top panel. The sun's spectrum, next from the top, is crossed by dark lines, caused by cool absorbing gases high in its atmosphere. Its most important lines were found and named by Fraunhofer. The third spectrum shows the lines of hot hydrogen gas. The last spectrum is that of a typical nebula, showing hydrogen as well as the mysterious 'nebulium' lines to the right of H-beta and the 3727 Å line. A wavelength scale in Ångstrom units is given at the top of the diagram. Blue is on the left and red on the right.

jectively by photography were frustrated by the insensitivity of the plates.

5.5   The 'Riddle of the Nebulae'

William Huggins's greatest discovery occurred late in 1864 during his work on the so-called 'nebulae'. After April 1864 Miller 'had not sufficient leisure to continue working with [him]', as he was careful to emphasise in his retrospective.

> working alone, I was fortunate in the early autumn of the same year, 1864, to begin some observations in a region hitherto unexplored; and which, to this day, remain associated in my memory with the profound awe which I felt on looking for the first time at that which no eye of man had seen, and which even the scientific imagination could not foreshadow.

Early interest in nebulous objects stemmed from their nuisance value to observers of comets, which they resembled and were often mistaken for. Even today, the term *nebula* covers more than one kind of object. The word arose during earlier observations by Herschel and others and was used to mean any object that appeared 'nebulous', meaning 'like a cloud'. As telescopes had grown in light-gathering power, astronomers such as the Earl of Rosse, whose 72-in. (1.83 m) telescope had the most powerful light-grasp of all, found that many of the nebulae could be resolved into individual stars and they argued that ultimately *all* nebulae would be found to be similar. Others felt that some of them consisted of an unknown form of luminous matter.

We do not know what encouraged Huggins to study the nebulae, which are extremely faint and were particularly difficult and unpromising objects to examine spectroscopically by eye. Nobody knew what they were, and certainly any understanding of galaxies lay far in the future. When Huggins started looking at them with his spectroscope he found that in some cases their light was not as difficult to see as expected because, to his and everybody else's great surprise, it turned out to be concentrated into just a few bright lines instead of showing the faint near-continuous, rainbow-like spectra crossed by narrow dark lines that the sun and stars did (see Fig 5.6, bottom panel). Of course, he checked and re-checked his apparatus to make sure this new phenomenon was not the result of an error or artefact arising from some mis-adjustment. Soon, he was able to satisfy himself that what he had seen was absolutely real.

The first nebula to show line emission was a 'planetary' in Draco—the name planetary had been given by William Herschel to a class of small, usually disc-like, nebulae surrounding hot stars, but having nothing in reality to do with planets. It was not long before Huggins examined many other nebulae and found that a good fraction of them showed the bright emission lines. Some, however, resembled the 'Andromeda Nebula', now known to be a neighbouring spiral galaxy somewhat similar to the Milky Way, and could just barely be seen to have weak star-like spectra. The obvious resemblance of the emission-line spectra to the earthly spectra produced when electricity was passed through a gas at low pressure proved that some nebulae had to consist of gases while the others really *were* agglomerations of faint and distant stars.

Huggins lost no time in sending an account of his discovery to the Royal Society, a much more selective organisation than the Royal Astronomical Society. Its fellowship—the 'FRS'— was by then only open to scientists of merit. To be able to write these letters after one's name conferred a certain distinction. His paper was refereed favourably by G.G. Stokes, then the secretary of the Society, who ordered that it was to be printed 'at once' because of its 'interest and importance'. Huggins was overjoyed, seeing this as his first step towards scientific fame, which he desired to an inordinate degree. He was at that time very friendly with Norman Lockyer, who was just then becoming interested in astronomy (they later became enemies). He persuaded Lockyer to write up his discovery in the popular journal, *The Reader*, to which he was a frequent contributor. Such a direct craving for publicity was then fairly unusual for a scientist, though common enough in today's more competitive atmosphere.

In his retrospective article of 1897 for the *Nineteenth Century*, full of poetic quotations and purple passages, he had this to say on the nebular discovery:

> On the evening of the 29th August, 1864, I directed the telescope for the
> first time to a planetary nebula in Draco. The reader may now be able to
> picture for himself to some extent the feeling of excited suspense, mingled
> with a degree of awe, with which, after a few moments of hesitation, I

put my eye to the spectroscope. Was I not about to look into a secret
place of creation?

I looked into the spectroscope. No spectrum such as I expected! A
single bright line only! At first I suspected some displacement of the
prism, and that I was looking at a reflection of the illuminated slit from
one of its faces. The thought was scarcely more than momentary; then
the true interpretation flashed upon me ...

The riddle of the nebulae was solved ... Not an aggregation of stars,
but a luminous gas. (Huggins 1897)

In spite of this exaggerated description, which can hardly be regarded as
a realistic record of his thoughts at the actual time, concerned as he would
have been with the complicated practicalities of the observation, it was a really
important discovery. Colonel Edward Sabine, the President of the Royal Society,
or 'PRS', proclaimed that in consequence of Huggins's work 'a totally different
view opened'. Using his 8-in. telescope he had answered a question which had
baffled the possessors of the largest instruments. His efforts were rewarded by
election to Fellowship of the Royal Society in June 1865. In 1866, he was awarded
the Society's Royal Medal for his discoveries and in 1867 he and Miller jointly
received the Royal Astronomical Society's Gold Medal for the same work. He
was also elected Secretary of the Royal Astronomical Society in the same year.
He was to remain a member of its Council until his death in 1910. Recognition of
the importance of the new discoveries by the broader scientific community was
virtually immediate. His gratification can only be imagined.

5.6   Nova T CrB

Huggins soon found himself in an enviable position. He had the best equipment
for making precise spectroscopic measurements and had a telescope immediately
available for use in his own home. It was by no means easy for a rival to duplicate
the apparatus, specialised knowledge, and observing skills that he had acquired.
The whole sky was waiting to be looked at. People automatically turned to
him with interesting new ideas. As many astronomers have done following the
development of a new instrument, he set about observing everything that seemed
to offer the promise of new discoveries.

An example of a special object that was brought to his attention was the
'Nova' or stellar outburst of May 1866. An amateur astronomer of Tuam in the
West of Ireland, John Birmingham, had spotted an apparently new bright star
in the constellation 'Corona Borealis', or Northern Crown. This star is known
today as the variable T CrB. He had written to *The Times* to announce his
discovery but took the precaution of notifying Huggins directly as a prominent
astronomer, just in case his letter was not published. There was no time to be
lost and indeed Huggins lost none in turning his telescope to the 'new' star.
Its spectrum was most unusual, combining bright emission lines with a normal
continuous background and dark absorption lines. Huggins interpreted his finding
as being a star with a nebula around it—a 'star on fire' as he put it. W.A.

Miller was invited around to confirm his result and, on the very next night, he reported it to the Royal Society. The discovery was published in a popular science journal and also reported to the British Association which met that August in Nottingham. The Association's yearly gathering attracted scientists of all disciplines and also drew a large following from the public. His address (Huggins 1866) was published as a pamphlet by William Ladd, an instrument maker who offered stellar spectroscopes for sale.

> The more precise statement of what occurred during our observations, as made afterwards from the pulpit of one of our cathedrals—'That from afar astronomers had seen a world on fire go out in smoke and ashes'— must be put down to an excess of the theological imagination. (Huggins 1897)

5.7 Comets

Comets, members of our solar system, are sometimes bright and obvious to the general public, in whom in ancient times they created great fear. They were natural targets for the early spectroscopists. The Italian astronomer Giovani Batista Donati had just asserted that they could possess light of their own, but Huggins was able to go further:

> I had myself, in the case of three faint comets, in 1866, in 1867, and January 1868, discovered that part of their light was peculiar to them [i.e. was not simply reflected sunlight], and that the light of the last one consisted mainly of three bright flutings [bands in their spectra]. Intense, therefore, was the great expectancy with which I directed the telescope with its attached spectroscope to the much brighter comet which appeared in June 1868. (Huggins 1897)

This last was known as Brorsen's comet and Huggins was able to make a direct comparison of its spectrum with a laboratory spectrum of 'olefiant gas'. He was able to show by this means that it contained hydrocarbons. However, it is interesting that he once again felt the need to invite Miller to confirm his observations.

5.8 Frustrating interlude

During a solar eclipse in August 1868, the extended corona, or outer atmosphere, of the sun, usually impossible even to glimpse because of being overwhelmed by the brightness of its photosphere or surface, was seen to have a spectrum consisting of bright lines. The French astronomer Jules Janssen and Huggins's English rival Lockyer observed the edge of the uneclipsed sun shortly afterwards through a spectroscope tuned to the bright red line of hydrogen and were able to detect the smaller prominences that appeared there, then described graphically as 'red flames'. For this they were awarded a special medal. Huggins was obviously unhappy at having missed this discovery and even tried to imply that spectroscopic observation of the prominences had been nothing very special:

> About the time that the news of the discovery of the bright lines at the
> eclipse reached this country, in September, I was altogether incapacitated
> for work through the death of my beloved mother. We had been all in all
> to each other for many years. The first day I was sufficiently recovered
> to resume work, December 19, on looking at the sun's limb with the
> same spectroscope I had often used before, now that I knew exactly at
> what part of the spectrum to search for the lines, I saw them at the first
> moment of putting my eye to the instrument. (Huggins 1897)

Typical of Huggins's enthusiasm in exploiting new instruments, he tried many
times to detect the sun's extended corona outside a total eclipse. This is a much
more difficult task than observing the prominences because the corona is fainter
and more spread out. He had hoped that by filtering out all but the red light
he would be able to reduce the blinding effect of the bright background. He put
several years of effort into this project, including sending a special telescope to his
friend Sir David Gill who observed under the clearer skies of the Cape of Good
Hope, but the apparently successful observations were found to be false. His
idea was a fundamentally sound one, but the amount of background reduction
required meant that the method had to await the development of narrow-band
filters in the second half of the twentieth century before it became usable.

With Miller he tried very hard in 1868 to detect heat radiation from the stars,
without success. For this work he used an early form of infrared detector called
a thermopile, which converts heat into electricity. The signals it produced were
far too feeble to measure satisfactorily. They could not get consistent results and
eventually had to give up. The instrumental difficulties had proved too great.
Sensitive infrared observations only became possible during the following century.

5.9   Radial motion of the stars

It had been known for a long time that stars show 'proper motion', i.e. foreground
objects appear to move very slowly across the background formed by the more
distant ones. This was clear evidence for movement in two dimensions, the third,
along the line of sight, being unobservable. The detection of such motion, called
'radial velocity', was first claimed by Huggins.

In 1842 the Austrian physicist Christian Doppler had shown that the pitch
of the sound emitted by a moving object, as heard by a stationary observer, is
affected by the speed of the movement. The classical example is the increase in
pitch of an express train whistle as it approaches a person standing on a railway platform and its decrease as it retreats into the distance. Doppler had suggested that a wavelength shift should be noticeable in starlight and even thought
(wrongly) that the different colours of stars could be due to this effect. In fact,
the wavelength change in starlight is very small, requiring a large spectrograph
to detect it. Father Angelo Secchi, working from the observatory of the Roman
College, the very institution controlled in Galileo's time by Cardinal Bellarmine,
had already tried to see the wavelength shift, without success. Huggins consulted
James Clerk Maxwell about the amount of change to be expected. Maxwell was

the theoretician who had put together the experimental knowledge of electricity that had accumulated before the mid-nineteenth century into a great synthesis called today 'Maxwell's Equations.'[5] It was clear that any effect would be so small as to be beyond the accuracy of existing equipment. In 1868 Huggins set about improving his spectroscope to be able to make the necessary very fine measurements. Using an instrument containing many prisms, he made comparisons between the positions of the lines in the spectrum of Sirius, the brightest star in the sky, and laboratory spectra. Still, he could not be certain. He altered his spectroscope so as to display the laboratory spectrum on both sides of the star spectrum. He could then measure any small displacements more exactly. By mid-1868 he was able to submit a paper to the Royal Society, in which he claimed that Sirius was receding at 47 km/s. It was refereed by the Astronomer Royal, still Airy, who was sceptical but nonetheless in favour of publication. Airy questioned whether the terrestrial line was really caused by the same substance as the celestial one and whether reliable measurements could be made at all of a star which never rises very high in the sky, as seen from England. In fact, Huggins's visual measurements were not confirmed by later observers—Sirius is actually approaching us, not going away as Huggins thought. However, the obvious need for greater sensitivity in these measurements led to support by other scientists for a new project that Huggins had in mind—the acquisition of a larger telescope.

5.10 New facilities

The sensational discoveries of the mid-1860s led to widespread recognition of the value of astronomical spectroscopy and the need for better facilities. Until then, astronomers had few expenses besides salaries after the initial provision of a telescope: now, all kinds of additional apparatus were going to be required. Rev Charles Pritchard, the remarkable President of the Royal Astronomical Society who became Professor of Astronomy at Oxford at an age when most people would have thought about retiring, made mention of the new trend when he handed over the Royal Medal to Huggins and Miller in 1867. Huggins thereafter began to propagandise his friends in high places with a view to getting a larger instrument. He even thought of spending part of his own meagre fortune to construct a 36-in. reflector. Telescope makers such as the Grubbs of Dublin and Cooke of York hoped, of course, for a lucrative contract. Howard Grubb paid Huggins a visit in April 1868 and they discussed the various possibilities. Together they sought the support of Thomas Romney Robinson, the Director of Armagh Observatory in Ireland, a great booster of his fellow-countrymen. Robinson put forward a formal proposal for a new telescope to the Royal Society. He asked for £2000, a huge sum by the standards of contemporary scientific grants. At that time, the annual budget dispensed by the Royal Society, the main source of grant money

---

[5]One of his predictions was the existence of radio waves, later produced experimentally by Heinrich Hertz.

for research, on behalf of the government, totalled only £1000. In an argument reminiscent of 'It's only money, darling', Robinson informed the Council of the Society that concern over cost was

> not the way in which persons who represent the Intellect and Knowledge of a Nation like ours, look at such matters. The power of the Royal Society over men's minds does not rest on the amount of its annual income or its balance in the bank. Its real wealth is whatever its animating influence or its helping hand has added to the treasures of science; its real power consists in the conviction of our countrymen that it is a mighty instrument of the highest and brightest progress, that its motives are as generous as its acts are beneficient and noble.[6]

G.G. Stokes was another supporter (also, incidentally, Irish-born and Robinson's son-in-law). He threw his weight behind the proposal. Fortunately, just before that time, the Royal Society had received a large bequest, which was eventually channelled into the telescope project. The decision was an easy one. The telescope would be the property of the Royal Society but would be lent semi-permanently to Huggins.

When planning how to use the new telescope, Huggins thought of moving out of London completely, but he was reluctant to give up the convenience of the big city as well as the home that he had turned into an observatory. He sold his Clark/Cooke telescope in August 1869 and set about changing his dome to a turret of 18 ft (5.5 m) diameter to accommodate the new instrument (Fig 5.7). The alterations took from the autumn of 1869 to February the following year. According to M.J. Perrotin (1881), Director of Nice Observatory, who made a tour around European observatories in 1880, the building then had three floors. A 6-horsepower steam engine, looked after by Huggins himself, drove Siemens and Holz generators. These were placed on the ground floor. The intermediate floor housed his laboratory equipment. Huggins became worried that to move the larger turret would tax his limited strength and feared that he might have to engage an assistant. Also, the new instrument actually consisted of two separate telescopes, interchangeable only with considerable effort—a refractor of 15 in. aperture (38 cm) and a reflector of 18 in. (46 cm). The changeover took several hours of work. The light-gathering power of the reflector was probably about the same as the refractor, because of the poor reflectivity of the speculum-metal mirrors that were still in use. However, a reflector is by its nature a truly achromatic telescope: it focuses rays of all wavelengths equally well, and speculum metal is an equally efficient (but not a *very* efficient) reflector in the ultraviolet and infrared. Although silver-coated glass mirrors, which were much more effective for visible light, were then coming into use, they did not reflect the ultraviolet, to which early photographic materials were most sensitive. For this reason, the choice of speculum metal was to prove a good one. Even though the contract was placed

---

[6]T.R. Robinson to Royal Society, quoted in 'Minutes of Council', 21 January 1869, Royal Society Library.

FIG. 5.7. The exterior of Huggins's observatory after its enlargement around
    1870 to accommodate his new telescope. (From Huggins and Huggins 1899.)

in April 1869 for a delivery before the end of the year, Grubb did not complete
it until February 1871. The refractor was at the time one of the largest in the
world. A new spectrograph with much increased resolving power was also supplied by Grubb. Huggins enjoyed showing off his new acquisitions. On 15 April
1871, he recorded that Janssen, Lewis Rutherfurd, Roscoe, and Pritchard had
dinner with him. In July, Dom Pedro, the indefatigable Emperor of Brazil, paid
the observatory a visit. He later conferred on Huggins the Brazilian Order of the
Rose.
    During the negotiations for the new telescope, on 30 September 1868, accord-
ing to Huggins's notebook, 'My intensely beloved mother died. From the above

date to Nov 5 I was quite unable to observe'. He took a brief trip to Brighton with a friend in October to help him get over his loss. Only two years later, on 30 September 1870, his (by then) only occasional collaborator, William Allen Miller, died rather suddenly, but the need for his expert help had long been overcome.

In 1870 the Royal and Royal Astronomical Societies promoted an expedition to Oran to observe a solar eclipse predicted for the Eastern Mediterranean. After much procrastination, the Royal Navy at the last minute provided the astronomers with a ship. Huggins was in charge of one of three parties that set sail on 5 December 1870. He brought with him an 'automatic' recording spectroscope of his own invention. This was a device for recording the positions of spectral lines as quickly as possible during the very short period—a few minutes—of a total eclipse. It was still the pre-photographic era and the new apparatus enabled a visual observer to set a crosswire quickly on a spectral line and record its position by pricking a card with a sharp pointer. His group included several well-known amateurs, as well as the established scientist William Crookes who we recollect had discovered the element thallium by spectroscopic means in 1870 (he later devised the 'Crookes Tube'—a precursor of today's television tube). As it turned out, they experienced a rough voyage and the weather in Oran itself was bad at the critical time. Crookes kept a diary of the voyage that indicates Huggins's swollen-headed leadership was none too popular;

> On returning to the ship we had a committee meeting. Little Huggins's bumptiousness is most amusing. He appears to be so puffed up with his own importance as to be blind to the very offensive manner in which he dictates to the gentlemen who are cooperating with him, whilst the fulsome manner in which he toadies to Tyndall must be as offensive to him (Tyndall)[7] as it is disgusting to all who witness it. I half fancy there will be a mutiny against his officiousness. (Quoted by Becker 1993, p. 319)

Again, on the rough voyage back, Crookes mentions with evident pleasure:

> Eating was almost impossible, for nearly all one's attention was needed to keep the meal, &c., on the plate, and ourselves on the benches. Huggins being small and not very careful, disappeared once, plate and all, under the table. (Quoted by Becker 1993, p. 319)

Apparently Huggins had had enough. He was afterwards reluctant to visit Dublin to inspect his finished telescope. Grubb cynically wondered whether it was rough seas or the fear of Fenian revolutionaries that was deterring him (Glass 1997, p. 67).

Also in 1870 Huggins attempted to see if a strong magnetic field affected the lines emitted by a substance. He was able to detect no change; however, with more sensitive apparatus such an effect was found later by the Dutch physicist Peter Zeeman in 1896. This is an interesting example of Huggins's rather eclectic

---

[7]John Tyndall (1820–1893), well-known Irish-born physicist.

approach which, given the fact that astronomical spectroscopy was a new and relatively unexplored technique, sometimes led to unexpected results.

5.11 Witness at a séance and other activities

Sometime in 1871, Huggins and an eminent lawyer were invited by Crookes to be objective witnesses of occult phenomena. Crookes, though a highly reputed scientist, was a firm believer in the existence of 'psychic' forces, of which he was an articulate defender. The 'medium' that Huggins was invited to see in action was Daniel Douglas Home, one of the most successful practitioners (Pearsall 1972). Even then, spiritualism was regarded sceptically by almost all scientists. Indeed, few were willing to waste their time on it. One can only think that Huggins was hoping to witness an interesting new phenomenon that he could investigate scientifically.

The demonstrations took place in a large gaslit room in Crookes' own house. Home caused an accordion to play and fly about, apparently spontaneously, within a cage placed under the table around which those present were sitting. In another part of the room was a horizontal wooden beam with one end resting on a table and the other suspended from a spring balance. The reading of the spring balance was seen to increase from 3 pounds to $6\frac{1}{2}$ pounds, supposedly by the application of psychic force. The feet and hands of the medium were watched closely throughout and no movement on his part could be detected.

Huggins was able to detect no fraud and afterwards, at Crookes' request, made a cautious statement in writing:

> Dear Mr Crookes
> Your proof [of a printed account of the seance] appears to me to contain a correct statement of what took place in my presence at your house. My position at the table did not permit me to be a witness to the withdrawal of Mr. Home's hand from the accordion, but such was stated to be the case at the time by yourself and by the person sitting on the other side of Mr Home.
> The experiments appear to me to show the importance of further investigation, but I wish it to be understood that I express no opinion as to the cause of the phenomena which took place. (Crookes 1870)

Although many mediums were shown to be tricksters who used the techniques of magicians, such as sleight-of-hand and distraction, Home seems to have escaped being exposed. He made it clear before a séance that his 'powers' were sometimes not at their best and presumably was unwilling to take risks when the witnesses could observe him too closely.

In August 1872, when the English nights were short, Huggins made an extensive trip to Sweden, Denmark, and Germany. He visited the famous spectroscopists Ångstrom, Thalén, D'Arrest, and Vogel amongst others.

He had also acquired a mastiff to keep him company. To this dog he gave the name Kepler. This animal had an incredible antipathy to butchers, in which

he followed his father and grandfather. Huggins was so interested in this apparent example of inherited behaviour that he wrote to Charles Darwin about it.
Darwin (1873) later published Huggins's description of the dog's behaviour in
Nature because 'The ... letter seems to me so valuable, and the accuracy of the
statements vouched for by so high an authority ... '

In February 1874, he severely sprained his ankle and was unable to observe
for several weeks. 'Afterwards I found my eyes painful (probably from too much
reading during the time I was kept in bed by my foot). I have frequently used
the telescope, but did not think it prudent to make any observations'.

5.12   An able and enthusiastic assistant

Fortunately, the need to engage an assistant to help with the operation of the
new telescope did not materialise. It appears that he first met Margaret Lindsay
Murray, born in 1848, at a musical evening held in the home of mutual friends,
the Montefiores. Later, he met her again in Dublin when visiting Howard Grubb,
the maker of his telescope. She was only twenty-seven, and thus twenty-four years
younger than he, when they married on 8 September 1875 at her home parish
church of Monkstown, near Dublin. He had acquired 'an able and enthusiastic
assistant', as he put it in *The New Astronomy* (Huggins 1897). As was usual
in those days, where property was involved, a trust was established before the
wedding to protect Margaret's assets. Otherwise, like any other woman, she
would have lost her rights to her husband on marriage.[8]

Margaret came from a fairly wealthy background. Her grandfather Robert
Murray, like her grandmother, Isabella Webster, had gone from Scotland to
Ireland in 1825, where he had risen to become the chief officer of a bank, with
a salary of £2000. His income must have been one of the highest in Dublin at
the time. Their oldest son, John Majoribanks Murray, who was to be Margaret's
father, was educated in Scotland at the Edinburgh Academy and later became
a solicitor (attorney) in Dublin. He married another Scot, Helen Lindsay, who
was to be Margaret's mother.

At the time of Margaret's birth, the Murray family lived in Monkstown, close
to the harbour of Kingstown (now Dun Laoghaire). Margaret Lindsay Murray
had one sibling, Robert Douglas Murray, who became a barrister (advocate) in
Dublin. Their mother died in 1857 when Margaret was only nine years old and
her father married again after a decent interval.

Margaret related in later life that she had been taught to recognise the constellations by her grandfather Robert. As a child of thirteen her aunt had presented her with the book of *Telescope Teachings* by Mary Ward, a cousin of Lord
Rosse. She was a diligent pupil: two other books, which she still had at the end
of her life, were given to her as prizes when she was fifteen and sixteen years old,

[8]Information from the will of Margaret Lindsay Huggins, kindly provided by Mrs M.T.
Brück.

for being top of her class (Whiting 1915b).[9] She had a small telescope that she used to project an image of the Sun to observe sunspots. It is said that she became interested in astronomical spectroscopy by reading an article in a popular magazine called *Good Words*, which praised the work of William Huggins. She constructed an astronomical spectroscope herself, with which she could study the Fraunhofer lines (Whiting 1915a).

Her wedding present from William was a sixteenth century Italian cameo in sardonyx (a type of quartz with layers of different colours). It showed the Biblical Flight into Egypt on one side and the Madonna on the other. Another gift was a miniature direct-vision spectroscope, only about 5 cm long, from the instrument maker Adam Hilger.

Having a bright and lively young wife somewhat softened William's serious outlook on life. They occasionally took time off to holiday on the Continent. In 1881,[10] they attended the famous Passion Play at Oberammergau in Bavaria. 'In later years they ceased to go abroad, rather to Mrs Huggins' regret, in order that William Huggins might indulge in his favourite sport of fishing at Benson on the Thames or in a wherry on the Norfolk Broads, for he was an expert pike fisherman, and loved the beauty of the river and the quiet that is essential to the angler' (Mills and Brooke 1936, p. 43). Perhaps it was while William was fishing that Margaret made the sketches of Norfolk scenes that can be seen in her surviving sketchbooks at Wellesley College.

5.13   A Victorian household

After his mother's death, Huggins had begun to collect artistic objects. Given that Margaret was an artist, their house soon began to resemble a museum. Her friend Alice Donkin described in an obituary (Donkin 1915) the appearance of the household and Margaret herself as they were when she first became acquainted with the Hugginses in 1882:

> The visitor, ushered into the drawing-room through doors panelled with old Flemish or German paintings of saints and kings, waited amongst Venetian furniture, curtains of rare old brocade, old pictures and Venetian glass . . . Then the hostess would come in, looking like the informing spirit of the place in a dress designed by herself with regard only to convenience and picturesque effect in colour and ornament; for in her belief, 'entire reasonableness and beauty are wedded.' Her striking face was framed by waving hair, black, but already broadly streaked with white, and worn short—this for convenience when lying back in the chair from which she watched the stars. Her manner to strangers was almost timid in its singular gentleness and courtesy; but her bright hazel eyes

[9]These were, according to some notes by S.F. Whiting of Wellesley College, Macauley's *Essays* and Prescott's *Conquest of Peru*. I was unable to find either book in the Wellesley Collections.

[10]The date probably should have read 1880, as the Passion Play normally took place only in the decadal years.

soon betrayed the unlimited possibilities of energy and keenness beneath
... (Donkin 1915)

Margaret loved medieval artefacts and spent her holidays sketching in places
with historical associations. After visiting Pevensey, the landing-place of William
the Conqueror in 1066, she wrote 'I did enjoy working out the scenery of the
Norman Conquest ... I have a stone from Pevensey beach; and who can say that
the great William did not kick that particular pebble? (Donkin 1915)
In her own words:

> This house is as distinct an artistic creation as any of Burne Jones' pic-
> tures for instance. So it has again and again been recognised by painters
> and the gifted generally. William's art is born in him—no one ever taught
> it to him: his taste may always be trusted. He only went wrong in not
> wearing his hair long, and in having a moustache, and in not wearing
> velvet coats in the afternoons. But he has improved in these respects.
> (Mills and Brooke 1936, p. 37)

We have from the bequest of Margaret Huggins to Wellesley College a list
of some of the interesting and beautiful things that they owned. Indian em-
broideries, antique jewellery, an eighteenth century lantern clock, a medieval
statuette of Christ the Good Shepherd ... She enjoyed hunting for under-valued
antique bric-a-brac which she spent many hours in restoring with her own hands.
She declaimed against collectors who went no further than Christie's auction
rooms, saying they could never have the pleasure of recognising by real expert
knowledge some old and valuable piece of work put aside on some dusty shelf.
The library at 90 Upper Tulse Hill Road contained many art books. The house
was full of stained glass, brasswork, enamelling, needlework, lace, and carvings.

Music also formed a part of their lives together. Margaret persuaded William
to take up the violin again. He bought himself a Stradivarius[11] dated 1708 and
was accompanied by his wife at the piano or organ. 'No more powerful antidote
against the wearying effects of the excessive and anxious activity of our age
is to be found than in what may be called the stimulating rest which music
affords'. She nevertheless described William as 'always rather an intellectual than
a perfervid player'. Sometimes they persuaded a friend to play his cello with them
in trios. Margaret wrote a monograph on the sixteenth century violin-maker Gio
Paola Maggini of Brescia (Huggins 1892). They also wrote together a paper on
technical aspects of the violin.

Another early visitor to the newly married Hugginses was Charles Piazzi
Smyth, the Astronomer Royal for Scotland. The house had already taken on the
appearance that visitors never failed to comment on. In June 1876 he noted:

> 19 June. Off to Huggins. Far out among Clapham's bowery roads and gar-
> dens, a long way, too, from railway station; a small house, small garden in
> front, large behind, small rooms and long and narrow staircases, but all

[11]Now known as the 'Huggins Stradivarius', and owned by the Nippon Music Foundation.

filled to overflowing with the most exaggerated ideas of medieval furniture: the painted glass at door however reproducing with the group from the Bayeux Tapestry *isti mirent* at the comet; a sun with red prominences and a nebula, and their respective spectra. Fernery and palmhouse, though small, grandly successful. The dogs, the big yellow mastiff Kepler and the little barky black terrier Tycho Brahe. Access through house to the observatory dome. carpeted floor, and large, with space for laboratory as well as driving clock, ... induction coil etc ... He has much liking for his old pieces of apparatus and making up things himself, and was once beyond everyone else in sp. But now, even with Mrs H.'s assistance, he must see that others do not pass him in refinements of mechanics. (Quoted by Brück 2003)

Early in their marriage, William introduced Margaret to a lady friend, Hannah Pipe, the founder of Laleham School for girls. Hannah was to become a new mother to Margaret, who always referred to her as 'My madre'. They frequently went on holidays and outings together. Hannah evidently appreciated the artistic side of her character (McKenna-Lawlor 1998).

Margaret enjoyed gardens and nature. It was she who tended the garden at Tulse Hill, for example. One of her sketches was of a dovecote, and she designed another as a Christmas present for Madre. A bookplate of her design featured an elaborate sundial which stood on a pillar in the garden.

5.14 Advent of photographic spectra

In 1873, Lockyer (later to discover the element helium in the solar spectrum) drew attention to the fact that two American amateurs, Lewis Rutherfurd and Henry Draper, had succeeded in photographing the spectrum of Sirius. Photography offered the convenience of a permanent record that did not require subjective interpretation at the eyepiece but allowed a spectrum to be examined at leisure. In fact, Huggins and Miller had made attempts of this kind in 1862, but the results were very unsatisfactory because of the insensitivity and the inconvenience of 'wet plate' technique when used on a telescope which might have to be orientated at any angle. During long exposures they tended to dry out and lose their sensitivity. A new process, involving gelatine-based plates, became available in 1871, but required some improvement before it was sufficiently sensitive for astronomical purposes.

By 1876, the dry plates were good enough to be used for recording astronomical spectra. They were particularly sensitive to ultraviolet light (the invisible rays beyond the violet end of the spectrum). As mentioned, Huggins's speculum metal reflecting telescope was ideal for this kind of investigation. However, the glass prism of his spectrograph did not pass ultraviolet light and had to be exchanged for a crystalline one of 'Iceland spar' (calcium carbonate). Similarly, the camera and collimator lenses had to be of rock crystal (quartz). With this arrangement he was able to observe the ultraviolet lines of Sirius (Huggins 1879) and other hot stars. He found several new ultraviolet lines of hydrogen in Sirius that were detected only later in the laboratory. For the next few years, the

Hugginses were to devote their efforts to photographing visible and ultraviolet spectra.

The driving clockwork of the telescope, one of the first that Howard Grubb had made, was not really precise enough to keep a star well-centred on the slit of the spectrograph. Such a drive is necessary to counteract the apparent movement of heavenly objects from east to west across the sky, caused by the rotation of the earth. An accurate drive was becoming an essential feature of bigger telescopes by that time. Huggins put pressure on Grubb to do away with his simple governor control system and to produce something better. When the latter developed an ultra-precise technique for gear-cutting some seventeen years after he had delivered the telescope, he re-worked and greatly improved Huggins's motor which was by then 'in a sad state' (Glass 1997, p. 138).

In 1881, the Hugginses were able to photograph the ultraviolet spectrum of a bright comet—'It was under a great tension of expectancy that the plate was developed, so that I might be able to look for the first time into a virgin region of nature, as yet unexplored by the eye of man'. The spectrum once again exhibited the 'fluted' (we would now say banded) spectra characteristic of hydrocarbons but also showed reflected sunlight with its characteristic absorption lines. One of the bands was demonstrated by George Liveing and James Dewar of Cambridge to be due to the cyanogen radical—a compound of carbon and nitrogen. This discovery ultimately led to some degree of hysteria, fomented by the press, during the 1910 appearance of Halley's comet. Sensationalists played on the fear that cyanide gas would poison the atmosphere when the comet made its closest approach to the earth!

A later important discovery made by the Hugginses (in 1890) was of a series of bands, now called the 'Huggins bands', in the far-ultraviolet spectra of the bright blue stars Sirius and Vega. These bands, although they did not realise it, originate in the earth's atmosphere rather than in the stars. Their cause, the absorption of starlight by terrestrial ozone, was only discovered much later on (Fowler and Strutt 1917).

Margaret wrote some notes about her role as assistant to William:

> The work in winter in favourable weather would begin about 6 p.m. and would continue until 9.30 or 10 p.m. Or if we began later than 6 we might work later. In this work I observe while William looks after the clock [she means here the massive clockwork motor which drove the telescope], dome, etc. When we first began, our exposures on each star had to be very long. I have, I think, worked on one for about three hours. But in our later work from three quarters to one and a half hours would be about the time. I had to teach myself what to do by degrees: at first I had my difficulties, but now my eyes are trained and are very sensitive. Also my hands respond very quickly and delicately to any sudden necessity. I can go and stand well at good heights on ladders and twist about well. (Astronomers need universal joints and vertebrae of india rubber). In our star work, having got my star into good position, I look at it, say, for three seconds, then close my eyes, say for six or seven seconds, then look

again for three seconds. By looking in flashes in this way I rapidly judge as to whether the star has been kept where it should be by certain means, and if not, rapidly adjust. With certain atmospheric or clock conditions I might look quite differently; say look five seconds and not three seconds, and so on. As I observe, I direct William as to what I need and he moves me bodily on my ladder, so that I am not disturbed more than is necessary . . .

Here [in the laboratory] the work may be of various kinds. It may be photographic, in which case I should help in arranging instruments, keeping the light right, and so on, if we are working on the sun. If working electrically, I should work batteries, fix electrodes, and be generally handy. I may take a turn at mixing up chemicals, pounding, weighing, dissolving, boiling—in short be a jack of all trades. If we work with our steam engine, William attends to it, *i.e.*, gets up steam. If he is at the engine while anything is working in the laboratory, I am always on duty to watch things, and ring a bell at once if steam must be turned off.

When needful I dust and wash up the laboratories, for no housemaid is allowed into those sacred precincts. I am a capital scientific housemaid. His pet bottles I let William deal with himself, as he will then know where to find them. I used to clean up the steam engine, but I have not for some time felt so able as I used for housemaiding, and William now gets a man to give it a rub now and then. But I understand about it and can do it at need. One is interesting with a lump of engineer's waste in one hand and some nasty oily stuff in a can in the other. (Quoted by Mills and Brooke 1936, pp. 38–42).

5.15 Further spectroscopic forays

In 1882 Huggins had Grubb re-arrange the telescope mounting so that both reflector and refractor could be kept permanently mounted. No longer would the changeover be a time-consuming operation involving the use of a crane. He also had a new spectrograph made that offered more resolution and had a mirror-like slit. The image of the star on the slit could be monitored by means of a periscope and the telescope could be guided more easily to keep the star centred correctly. This arrangement, now standard, seems to have been first devised by Huggins.

5.16 The mystery of 'Nebulium'

Two of the lines in nebular spectra, around 4959 and 5007 Å in modern parlance, defied all the attempts of the early astrophysicists at finding a terrestrial counterpart. At first Huggins thought they could be due to nitrogen, then, as more refined measurements showed a definite lack of coincidence, he thought their source might be the same element Doppler-shifted due to the motion of the nebula. Later, magnesium became a candidate, championed especially by Lockyer. It was eventually shown conclusively by James Keeler of Lick Observatory in California that it could be none of these. In fact, its explanation had to await the physicist Ira S Bowen who, only in 1927, was able to show that it was caused by so-called 'forbidden' emission from oxygen under the rarefied conditions existing in the vastness of space, circumstances that could not be duplicated in a terrestrial laboratory.

5.17   Eminent Victorian

The great advances then occurring in spectroscopic astronomy were chronicled by the Irish author Agnes Mary Clerke in her highly successful book *A Popular History of Astronomy During the Nineteenth Century*, first published in 1885 (Clerke 1893). Very soon after it appeared, a friendship bloomed between the Hugginses and Miss Clerke. In her books she 'never lost an opportunity for singing their praises' (Brück 2002). She attempted to have the hydrogen spectral line series that Huggins had observed in 1879 named after him but, in spite of her efforts, they are today associated with the amateur Swiss mathematician Johann Balmer, who in 1885 found the the formula that related their wavelengths. Miss Clerke's *History* eventually ran to four editions and must have helped to build up public knowledge of the Hugginses' work.

In 1891 Huggins was President of the British Association for the Advancement of Science, which each year held a great jamboree of scientists and those interested in science in some British or Empire city. The young American astronomer George Ellery Hale (see Section 6.6) was then his guest. That year the Association met in Cardiff. Huggins's name appeared often in the press and he became quite a national figure.

Observations continued to be made at Tulse Hill. In 1892 the spectrum of Nova Aurigae was measured and showed evidence for an expanding hydrogen shell.

In 1897 Hale invited Huggins to participate in the dedication of the Yerkes Observatory of the University of Chicago. Huggins composed an address to be used on the occasion but in the end did not feel able to make the long journey. The manuscript for the address was turned into the article *The New Astronomy ...* (Huggins 1897), already quoted from, his self-centred view of the history of spectroscopic astrophysics. It was published in *The Nineteenth Century*, a magazine that appealed to the serious reader. In this work he painted a romanticised portrait of himself as the distinguished originator of astrophysics, proceeding uninterruptedly from one astounding discovery to another. Later, he used extracts from this article in the introductory sections of his *Scientific Papers* (Huggins and Huggins 1909). The writers of his obituaries later on made rather uncritical use of it. This is, for example, how he described his early days of discovery:

> This time was, indeed, one of strained expectation and of scientific exhaltation for the astronomer, almost without parallel; for nearly every observation revealed a new fact, and almost every night's work was red-lettered by some discovery. (Huggins 1897)

On the occasion of the diamond Jubilee of Queen Victoria in 1897, Huggins was knighted, in the Order of the Bath, for 'the great contributions which, with the collaboration of his gifted wife, he had made to the new science of astrophysics'. Margaret was presented at court, wearing a gown given to her by Madre. For a Victorian wife, this was generous recognition indeed!

The apotheosis of the Hugginses stellar spectroscopic work occurred in 1899 with the publication of *An Atlas of Representative Stellar Spectra* (Huggins and Huggins 1899). This was an impressive, large-format, lavishly produced tome with generous margins, issued as Volume I of the *Publications of Sir William Huggins's Observatory*. Its authors were Sir William Huggins, KCB (followed by his many honorary degrees and honorary memberships of foreign scientific societies) and plain Lady Huggins. Apart from its scientific contents, it was embellished with drawings and medieval-style capital letters created by Margaret. In it was a history of the Tulse Hill observatory and its instruments. They also presented their ideas on the evolution of stars. These are now only of historical interest, since Huggins believed that the general character of a star's spectrum was dependent more on the atmospheric pressure at its surface than on its temperature, a view that is opposite to the present one.

From 1900 to 1905, Sir William was President of the Royal Society. He used his presidency to emphasise the importance of science to the national interest. In many ways, his outlook was shared on the other side of the Atlantic by George Ellery Hale. His portrait by the Hon. John Collier (Fig 5.9), painted in 1905, shows him as he wished to be remembered. A copy painted by the artist himself was presented to Margaret. As he got older, Huggins became more and more involved in science policy issues and no longer pursued observational astronomy so energetically. He was very conscious of the fact that Britain was dropping behind the Unites States and Germany economically, and was among the many scientists who advocated better technical and scientific education. His 'swan song' as President of the Royal Society, 'Science in Education', was delivered to a packed audience. Margaret Huggins wrote 'Well, Sir William has closed his five years of office brilliantly. This is the general verdict—a brilliant and memorable Presidency brilliantly brought to a close'.

> *Sciencia vinces*, [science conquers]—whether it be on the field of battle, on the waves of the ocean, amid the din and smoke of the workshop, or on the broad acres under the light of heaven; and assuredly, in the future, even more than in the past, not only the prosperity, but even the existence of the Empire will be found to depend upon the 'improvement of Natural Knowledge',[12]—that is, upon the more complete application of scientific knowledge and methods to every department of industrial and national activity. (Huggins 1906, pp. 116–117)

Among their last scientific investigations was an examination of the spectrum produced by a radioactive element—radium—in air. Pierre and Marie Curie had visited London in 1903 to lecture at the Royal Institution and were later awarded the Davy medal of the Royal Society. In their investigations, the Hugginses found

[12]Part of the full title of the Royal Society as given in a Royal warrant of 1663.

FIG. 5.8. A cartoon of Huggins by the famous 'Spy' [Sir Leslie Ward] appeared in *Vanity Fair* in April 1903.

that the glow produced was from the nitrogen in the surrounding air, excited by the particles emitted by the radium.

In 1906, Sir William published another book, *The Royal Society or, Science in the State and in the Schools*, containing his Presidential Addresses to the Society. This was a handsome red-bound volume that matched his later *Scientific Papers*. Honours continued to be bestowed on him: he was one of the first twelve people admitted to the newly formed Order of Merit in 1902, whose numbers are restricted to twenty-four. Other famous scientists admitted at the same time were Lord Kelvin, Lord Lister and Lord Rayleigh.

5.18   Margaret Huggins and education

Always possessing wide interests besides astronomy, Margaret Huggins (Fig 5.10) continued her interest in educational matters—for example, her acquaintance with the educator Hannah Pipe had existed from the early years of her marriage—and believed that women should compete on an equal level with men:

FIG. 5.9. William Huggins's Royal Society portrait by Hon. John Collier, finished in 1905 (From Huggins and Huggins 1909). The original hangs today in the Royal Society and Margaret's copy is at the National Portrait Gallery, London.

I find that men welcome women scientists provided they have the proper knowledge. It is absurd to suppose that anyone can have useful knowledge of any subject without a great deal of study. When women have really taken the pains thoroughly to fit themselves to assist or to do original work, scientific men are willing to treat them as equals. It is a matter of sufficient knowledge. That there is any wish to throw hindrances in the way of women who wish to pursue science I do not for a moment believe. (MLH in 1905, quoted by Newall 1916)

FIG. 5.10. Photographic portrait of Margaret Lindsay Huggins. (From Huggins and Huggins 1909)

She conceived a great admiration for Wellesley College, a famous womens' college situated in Wellesley, Massachusetts, to which she left many of her possessions. She probably met the founder of the College's Whitin Observatory, Sarah Frances Whiting, some years before its opening. On 22 September 1899 she wrote:

> As we couldn't be present—would that we could—I am sending you a portrait of Sir William which I thought perhaps might interest the students. I hesitated about adding one of myself—but Sir William and I do

all our work together and I decided that perhaps mine should go too. At least I hope it may show my interest in your work.[13]

5.19   A place of pilgrimage

Along with her work as assistant to William, Margaret kept up her artistic interests. Each year, their friends received Christmas cards based on her watercolour sketches. She wrote and illustrated several books. Apart from the life of Maggini already referred to, she wrote an introduction to a life of Stradivarius, she illustrated a life of Sir Philip Sidney and she wrote an article on the astrolabe (Huggins 1894). The last work combined her interests in medieval artefacts, art and science. It was illustrated with her own diagrams and with photographs of the fine Persian astrolabe that they owned. (Later, this instrument belonged to George Ellery Hale.) One of the authorities she quoted was Chaucer, who himself had written a treatise on the astrolabe in 1391.

As they grew older, the Hugginses welcomed astronomical visitors from many countries, one such being the spectroscopist Heinrich Kayser of Bonn, who wrote of them in an article for the 'Scientific Worthies' series published in *Nature*:

> Far from the noisy centre of London, in Upper Tulse Hill, there is a quiet house. The welcome accorded to one who has the good fortune to enter at its hospitable doors reminds one of Philemon and Baucis, and the visitor amidst the artistic decorations of the house feels transplanted into another world. In the garden extending far behind the house one sees the astronomical observatory with its dome, and recognises that this is the house of a man of science, not an artist. (Kayser 1901)[14]

The erratic American astronomer T.J.J. See, in his tribute of 1910, says

> Tulse Hill was long a place of pilgrimage for men of science visiting London, and many Americans will recall the hearty greeting they received from the venerable sage and his devoted wife, and the unostentatious simplicity with which they entertained their visitors. Sir William's house was full of books and beautiful pictures, and he seemed to live in his library, laboratory and observatory, when not mingling in social life or attending scientific meetings in the city. (See 1910)

Similarly, from George Ellery Hale, whose admiration for a kindred spirit shines through in his obituary of 1913:

> One ...thinks of Sir William in his home on Upper Tulse Hill, where, with the sole aid of his learned and devoted wife, he carried on his study and research. None who has ever entered that quiet library, its shelves overflowing with books, can ever forget the sight of the aged philosopher in his natural environment. (Hale 1913)

---

[13]Wellesley College Archives, Department of Astronomy, 'Souvenir Whitin Observatory'.

[14]Philemon and Baucis were an aged couple in Greek mythology who were the only people to entertain Zeus, the father of the gods, and his messenger Hermes when they travelled through Phrygia disguised as ordinary mortals.

Annie Jump Cannon, the famous Harvard classifier of spectra, wrote:

> To anyone who has crossed, for the first time, the great ocean and felt the
> strangeness of foreign lands, nothing was ever more delightful than a visit
> to this warm-hearted hostess. Who can ever forget an afternoon at Tulse
> Hill? Such was my pleasure in 1908. Immediately upon entering the door,
> one felt somewhat as if removed from the materialistic present into an age
> in which we are wont to think life was more romantic and sentimental. To
> pass through the portals ornamented by the Halley Comet transparencies,
> and then into the reception-room by the door-way of Flemish medieval
> paintings, and be greeted as an old-time friend by Sir William and Lady
> Huggins, surely warmed the heart-strings of the traveller from afar. The
> charm of such a visit can never be forgotten; the inspection of the ob-
> servatory and instruments used at the very beginning of the science of
> Astrophysics, the walk in the garden, the cheeriness of the tea hour, the
> spirit of the house with its many interesting antiques, and, above all, the
> remarkable personality of the host and hostess. (Cannon 1915)

5.20　Last years of William ...

In 1908, at the age of 84, Huggins decided to give up observing and return the
telescope that the Royal Society had built for him. He asked that it be sent to
Cambridge. Sir Howard Grubb recollected the funereal scene:

> I shall not easily forget the scene that presented itself, as we mounted
> the stairs to the well-remembered Tulse Hill Observatory.
>
> The equatorial had been partially dismantled; all the numerous parts
> and attachments had been removed and were scattered all over the floor,
> which was encumbered and littered with axes and various parts of the
> instrument, some of which had already been placed in packing cases;
> and in the midst of this litter, wrapped in a large cape and seated on a
> packing case, was Sir William himself, and his faithful collaboratrice who
> was flitting about watching the packing with keen interest and loving
> care.
>
> It was a strange and interesting sight to see him sitting very quietly
> and patiently watching with regretful interest the dis-assembling of the
> very instrument which he had made so noble use of, and with which he
> had wrested secrets from nature by the labours and work of nearly half
> a century, but the pathetic side of the picture was presented to me when
> I proceeded to pack the great objective in the box.
>
> Lady Huggins had asked me to let her know when I was ready to close
> the box, and when I intimated that I had safely put it in the case, she
> took Sir William by the hand, and brought him across the room to have a
> last look at their very old friend, the object glass which had for so many
> years fulfilled its mission in bringing rays of light from a far distance to
> a focus, there to be submitted to the keen and searching analysis of the
> great scientist. They gazed long and sadly before I closed the lid. (Quoted
> by Ball 1915)

His last years were spent preparing his *Scientific Papers of Sir William Huggins*, published in 1909 as Volume II of his observatory's publications. He was fortunate enough to enjoy good health until the end of his life. He died on 12 May 1910, following a minor operation for a hernia.

Margaret was to defend William's posthumous reputation with extraordinary ferocity. Hugh Frank Newall (1911), one of his obituarists, had stated that William identified the 'nebulium' line as due to nitrogen and secondly that he had received assistance in predicting the Doppler shift from Clerk Maxwell. She took great exception to these statements and poured out her heart to Sir Joseph Larmor, the Irish bachelor physicist who, as Secretary of the Royal Society during William's Presidency, had become an intimate friend:

> I was dismayed when I read the article—at the serious omissions; and not only dismayed but most deeply pained by a statement in connection with a highly important piece of work—a statement which is absolutely untrue, and which I cannot allow to pass unchallenged. It is unpardonable that a man in Professor Newall's position should have made such a blunder. I am sorry for him but it is my duty to see to it that my husband is vindicated. I have the facts before me, as they were before Professor Newall. My position is sad enough without the pain of such an incident as this . . . [15]

This issue resulted in many aggrieved letters being sent to Larmor before the matter was settled. Another source of annoyance was that their telescope after being moved had not been clearly labelled as a memorial to her 'Husband's long and splendid labours'. Such a label was eventually installed.

5.21 . . . and of Margaret

Margaret was made an honorary member of the Royal Astronomical Society at the same time as her friend Agnes Clerke in 1903. Women were still not admitted as regular members, but there had been three previous female honorary members; Caroline Herschel had been one of them (see Section 4.21). She stated:

> From first to last I have striven to qualify myself for the post of first assistant to my husband. [16]

She was granted a pension of £100 per year by the Royal Society after William's death. She wrote to Larmor,

> No doubt you know about my pension . . . 'for my services to Science by collaborating with my Dearest. This I *could* accept without *any* reflection on the memory of my Dearest—& with *honour* to myself as well as to *him*. I do regard the pension as an honour to *him* though it is honourable also to me, & I humbly hope,—*really earned* for the last 35 years of *very hard*

---

[15]Larmor correspondence, Royal Society; quoted by McKenna-Lawlor (1998, p. 98).

[16]'First Assistant' was a term used for the second-in-command at the Royal Observatories. Quoted by Newall (1916).

work. None of you know *how* hard we worked here just our two unaided selves[17].

In fact, her Trust still owned four houses from which she probably received rental income. Her will indicates that she was worth several thousand pounds when she died.

Edwin B. Frost of Dartmouth College and later Director of Yerkes Observatory told how he and his family visited Lady Huggins shortly after William's death:

> We had the pleasure of spending an afternoon with Lady Huggins at their home in Brixton [sic]. The house was quite a museum, many of its doors and tapestries being of Oriental or Italian origin. At one time Sir William must have been a man of a good deal of means, for these collections were costly. We had tea with Lady Huggins, who afterward allowed our daughter Katherine to play on an original Stradivarius. We spent some time looking over the wonderful garden which Lady Huggins had created. Artistic attention had been given to every detail. The iron hinges of the doors of the little summer house and all the other fixtures had been specially designed by her. The garden had a great variety of the most interesting flowers and plants. As she was planning to give up this home within a short time and go to live near the British Museum, she most generously offered to give me anything that I cared to take for my own garden in America. I could not see how this could be done ... (Frost 1933)

At one point, a few years after their marriage, Margaret had written to a friend

> I do wish sometimes William had been a painter, not an astronomer. He'd have been a happier man I think as a landscape painter. And I—aah! the artist is strong in me. But let what will be, be, God has willed things otherwise. Nobody knows how wearing science is: and it does not take faith to be as happy in straining one's eyes to see little patches of light or of darkness ... as in feasting them on the beauties of the fields and skies and woods ... (Quoted by Mills and Brooke 1936, p. 36)

Margaret's last years became lonelier with the gradual disappearance of her friends. In 1906 Madre died, soon after a visit they had made together to the island of Iona. Other dear friends, Ellen Mary Clerke and her sister Agnes Mary Clerke died in 1906 and 1907, respectively. Margaret was moved to write a short and delightful book in their memory, printed privately for their friends (Huggins 1907, see also Brück 1991).

Around 1913 she moved away from Tulse Hill, which 'meant death to her', to More's Garden House, Cheyne Road, Chelsea, overlooking the Thames. She was thrilled to think that this was the very place where Sir Thomas More (1478–1535), once Lord Chancellor of England, had lived and been visited by Holbein

---

[17]Larmor correspondence, Royal Society, quoted by McKenna-Lawlor (1998 pp. 97–98).

and Erasmus. Annie Cannon of Harvard visited her there in 1913, 'surrounded by her work, her favourite pictures, and beloved books, as courageous a spirit as ever lived' (Cannon 1915).

Margaret spent the years after William's death in sorting out the equipment and records of their observatory. Many of their instruments were presented to Cambridge University and Wellesley College, as well as some of their scientific notebooks and pictures (Whiting 1916). Her last letter to Prof Whiting of Wellesley was written in January 1915, giving background information about all the 'precious things' that she was sending over.

Following several operations, she died 'of a long illness' on 24 March 1915, aged 66. She was cremated and her ashes were laid next to William's at Golders Green in London. A memorial by Henry Pegram, in the form of medallions of each partner, was afterwards erected in the crypt of St Paul's Cathedral in the City of London (Brück and Elliott 1992).

The executor of her will was Julia Montefiore, a lifelong friend, who arranged the shipment of numerous items left to Wellesley College. Margaret had intended to write a full life of her husband, but this was forestalled by her own death. When she realised that her end was approaching, she made arrangements for the work to be done by John Montefiore, the brother of Julia. However, he died only two years later and the work was not completed. A short book, *A Sketch of the Life of Sir William Huggins*, based on material that no longer seems to be extant, was eventually written by Charles E. Mills, a solicitor who was the executor of Julia Montefiore's will, and C.F. Brooke, in 1936.

References

André C. and Rayet, G., 1874. *L'Astronomie Practique et Les Observatoires en Europe et en Amerique*, Part 1, Gauthier-Villars, Paris.

Ball, R.V., 1915. *Reminiscences and Letters of Sir Robert Ball*, Cassell & Co., London, p. 112.

Becker, Barbara J., 1993. *Eclecticism, Opportunism, and the Evolution of a New Research Agenda: William and Margaret Huggins and the Origins of Astrophysics*, Ph.D. Thesis, 2 Vols., Johns Hopkins University, Baltimore, MD.

Becker, Barbara J., 2001. Visionary Memories: William Huggins and the Origins of Astrophysics, *J. Hist. Astr.*, **32**, 43–62.

Brück, M.T., 1991. Companions in Astronomy: Margaret Lindsay Huggins and Agnes Mary Clerke, *Irish Astr. J.*, **20**, 70–77.

Brück, M.T., 2002. *Agnes Mary Clerke and the Rise of Astrophysics*, Cambridge University Press, Cambridge.

Brück, M.T., 2003. An astronomer calls: extracts from the diaries of Charles Piazzi Smyth, *J. Astr. Hist. and Heritage*, **6**, 37–45.

Brück, M.T. and Elliott, I., 1992. The Family Background of Lady Huggins (Margaret Lindsay Murray), *Irish Astr. J.*, **20**, 210–211.

Cannon, A.J., 1915. Lady Huggins, *The Observatory*, **38**, 323–324.

Clerke, A.M., 1893. *A Popular History of Astronomy During the Nineteenth Century*, 3rd edn., Adam and Charles Black, Edinburgh.

Crookes, Sir William, 1870. Experimental Investigation of a New Force, *Quart. J. of Sci.*, (I), 339–349.

Darwin, C., 1873. Inherited Instinct, *Nature*, **7**, 281–282, (Letter).

Donkin, Alice E., 1915. Margaret Lindsay Huggins, *The Englishwoman*, May 1915, 152–159.

Fowler, A. and Strutt, R.J., 1917. Absorption Bands of Atmospheric Ozone in the Spectra of Sun and Stars, *Proc. Roy. Soc.*, **93**, 577–586 and Plate 3.

Frost, E.B., 1933. *An Astronomer's Life*, Houghton Mifflin Co., Boston and New York.

Glass, I.S., 1997. *Victorian Telescope Makers, the Lives and Letters of Thomas and Howard Grubb*, Institute of Physics Publishing, Bristol and Philadelphia.

Hale, G.E., 1913. The Work of Sir William Huggins, *Astrophys. J.*, **37**, 145–153.

Huggins, M.L., 1892. *Gio Paolo Maggini, his Life and Work*, Hill & Sons, London.

Huggins, M.L., 1894. The Astrolabe. A Summary, *Astronomy and Astrophysics*, **3**, 793–801 (with plates).

Huggins, M.L., 1907. *Agnes Mary Clerke and Ellen Mary Clerke, an Appreciation*, printed for private circulation (Copy in library of S.A. Astronomical Observatory).

Huggins, W., 1856. Description of an Observatory erected at Upper Tulse Hill, *Monthly Notices of the Roy. Astr. Soc.*, **16**, 175–176.

Huggins, W., 1866. *On the Results of Spectrum Analysis Applied to the Heavenly Bodies*, W. Ladd, London (Armagh Observatory Archives, M176).

Huggins, W., 1879. On the Photographic Spectra of Stars, *Phil. Trans. Roy. Soc.*, **171**, 669–690 + plate.

Huggins, W., 1897. The New Astronomy, a Personal Retrospect, *The Nineteenth Century*, **41**, 907–929.

Huggins, Sir W., 1906. *The Royal Society or, Science in the State and in the Schools*, Methuen & Co. London.

Huggins, Sir W. and Huggins, Lady, 1899. *An Atlas of Representative Stellar Spectra*, London, William Wesley and Son and Hazel Watson and Viney.

Huggins, Sir W. and Huggins, Lady (eds), 1909. *The Scientific Papers of Sir William Huggins*, London, William Wesley and Son.

Kayser, H., 1901. Scientific Worthies XXXIII.–Sir William Huggins, K.C.B., *Nature*, **64**, 225–226.

Lockyer, J.N., 1874, *Solar Physics*, Macmillan, London.

McKenna-Lawlor, S., 1998. *Whatever Shines Should be Observed*, Samton Ltd, Dublin.

Mills, C.E. and Brooke, C.F., 1936, *A Sketch of the Life of Sir William Huggins K.C.B., O.M.*, privately printed, Times Printing Works, Richmond, Surrey.

Newall, H.F., 1911. Obituary of William Huggins, *Monthly Notices R. Astr. Soc.*, **71**, 261–270.

Newall, H.F., 1916. Obituary of Dame Margaret Lindsay Huggins, *Monthly Notices Roy. Astr. Soc.*, **76**, 278–282.

Pearsall, Ronald, 1972. *The Table-Rappers*, Michael Joseph, London.

Perrotin, M.J., 1881, *Visite a Divers Observatoires d'Europe*, Gauthier-Villars, Paris.

Schuster, Sir Arthur, 1932. *Biographical Fragments*. London, Macmillan and Co., p. 219.

See, T.J.J., 1910. Tribute to the Memory of Sir William Huggins, *Popular Astronomy,* **18**, 387–398.

Whiting, S.F., 1915a. Lady Huggins (obituary), *Science*, **51**, 853–855.

Whiting, S.F., 1915b. The Lady Huggins Bequest, *Wellesley College News*, 4 November, p. 4.

Whiting, S.F., 1916. *A Catalog of the Pictures of the Lady Huggins Bequest to Wellesley College*, Wellesley College, Wellesley, MA, 20pp.

6

GEORGE ELLERY HALE: PROVIDING THE TOOLS

Anyone meeting George Ellery Hale must, I think, have felt at once that he was somewhat out of the ordinary run of scientific men. We knew he was a great figure in science, but felt that he could have been equally great at almost anything else. For Nature had not only endowed him with those qualities that make for success in science—a powerful and acute intellect, a reflective mind, imagination, patience and perseverance—but also in ample measure with qualities which make for success in other walks of life—a capacity for forming rapid and accurate judgments of men, of situations, and of plans of action; a habit of looking to the future, and thinking always in terms of improvements and extensions; a driving-power which was given no rest until it had brought his plans and schemes to fruition; eagerness, enthusiasm, and above all a sympathetic personality of great charm. (Jeans, Sir J.H. 1938).

This panegyric by Sir James Jeans was perhaps a little exaggerated, in the manner of obituaries, but not unduly so. Hale did possess the characteristics that he mentioned. His achievements were all the more remarkable when one considers that for much of his life he was debilitated by illness.

George Ellery Hale entered the world in Chicago on 29 June 1868, as the son of William Ellery Hale, a businessman, and his wife Mary Scranton Hale, born Browne. He was their third child but the first to survive.

William Ellery Hale, George's father, was born in Bradford, Massachusetts in 1836, the son of a clergyman. Little is known of his early career. At the time of George's birth in the Chicago boarding house where the family then lived, he was the apparently impecunious representative of a paper manufacturer. A few years thereafter he prospered mightily, having purchased a patent for 'hydraulic' elevators and formed a company in Chicago to exploit it. Following the great fire that swept through the city in October 1871, destroying about a third of its buildings, his business boomed as he supplied elevators for the skyscrapers that began to go up at that time. Later he became a real-estate developer. For two decades, as their wealth grew, the Hales lived in a relatively ordinary middle-class house. Nevertheless, they never spared money on encouraging the hobbies and interests of their children.

William Hale was an 'upright citizen', a deacon of the Congregational Church and a member of the American Board of Foreign Missions. A Congregational minister and his wife formed part of his household for several years and helped to tutor his children as well as to interest them in literary pursuits.

Hale's mother was born in New Hampshire in 1836 to parents who later divorced—then something quite unusual. Her father had also been a Congregational clergyman. She was brought up by a grandfather who had adopted her own mother as a child. She was a strict Calvinist, a delicate woman, always in frail health. Her son inherited her weak constitution, suffering from stomach troubles, backache and occasional fainting spells. Nevertheless, when in pursuit of his goals he was bursting with energy.

As a youngster, George was precocious, preferring to play indoors and pursue relatively serious hobbies like collecting insects, minerals, and fossils. At one point, his father gave him a box of tools. Later he demanded a lathe, which was duly provided. At 13, he wanted a laboratory, so his mother had to give up her dressing-room. His brother Will and sister Martha were drawn into his schemes and each had a private workplace equipped with a Bunsen burner and various electrical gadgets. Soon their laboratory proved too small and they decided to build a workshop in the garden. The necessary plans were drawn up and together they constructed a building of wood and second-hand materials. When it was finished, they built a steam engine to drive the lathe, using a second-hand boiler. Just as he did many times later in his life, George arranged for a grand opening ceremony to which he invited his family and friends.

George was sent first to the local Oakland Public School, which he found dull and where he caught typhoid. Following his recovery, he was sent to a private establishment known as the Allen Academy. Mr Allen was a Harvard graduate who had seen a little of the world. More importantly, he had taught physics and astronomy. Although considered rather brutal by the other pupils, he treated George kindly enough and made him into an assistant for his lectures. George also took some courses in machining at a technical school. As a schoolboy, he was fond of cycling and canoeing, the latter in a canoe he had built himself. He later recalled that he had constructed an ice yacht in the attic, making the classic mistake of forgetting how he was going to get it out of the room. It had to be left behind when the family moved.

Summers were frequently spent at his maternal grandmother's house in Madison, Connecticut, near the shore of Long Island Sound. Here again he experienced the outdoors, swimming, fishing, and watching activities at a nearby shipyard. While still a teenager, he met in Madison a girl named Evelina Conklin to whom he gradually became more and more attached.

Hale's interest in astronomy was aroused by a visit to the private observatory of Sherburne Wesley Burnham, a famous amateur double-star observer who worked as a court reporter by day. His enthusiasm and dynamism quickly won over the older man. Burnham knew of a second-hand 4-in. (10 cm) telescope made by Alvan Clark that was for sale. George immediately importuned his father for the money to buy it but was kept waiting until it was clear that he really had some idea as to why he wanted it. On 6 December 1882 there was to be a 'Transit of Venus'—an occasion when the planet Venus could be seen as a small dark spot crossing the face of the sun, a rare event that caused much excitement among astronomers worldwide because it offered an opportunity to find the scale of the solar system by trigonometry. His father capitulated in good time and bought the instrument.

By 1883, Hale had started on photography and, with the help of his willing slaves—his sister and brother—the telescope was rigidly mounted on the roof of their house. With it, he succeeded in taking some photographs of the moon and other bright objects. He was introduced by Burnham to Washington Hough, a professional astronomer of the Dearborn Observatory. Listening to Hough he learned about the Lick Observatory on Mount Hamilton in California and the much superior astronomical weather encountered there. Later, he was to be lured west himself.

In *Cassell's Book of Sports and Pastimes*, published in 1881, he found a design for a simple spectroscope with a carbon disulphide liquid prism, not the most pleasant of substances to deal with, since it stinks of rotten cabbage. Fifty years later he could still remember the smell. He became entranced by the technique of spectral analysis as practised by Huggins and others and his fate was soon sealed. He soon bought a small diffraction grating[1] from the instrument maker John A. Brashear of Allegheny, near Pittsburg. (A well-made grating can disperse light into a spectrum more efficiently than a prism.) With its aid, he made lists of the lines in the spectrum of the sun. Brashear was well-known as a supplier of optical equipment to many American and foreign astronomers and the seventeen-year-old Hale decided he had to make his acquaintance. The Alleghenian had expected to meet a 'man of about forty-five and certainly ... up in astronomical physics. If some fellow had taken a baseball club and hit me very, very hard, I could not have been any more surprised'.[2] Through Brashear, Hale met Samuel Pierpoint Langley, the Director of the Allegheny Observatory. Langley was the inventor of the 'bolometer', a sensitive instrument for detecting infrared radiation, and was later one of the pioneers of aviation. He had written a book called *The New Astronomy*, largely about the surface of the sun, which Hale proceeded avidly to read. He continued to devour all he could get his hands on about spectroscopy, especially the books by Norman Lockyer, Huggins's great rival.

The following year, 1886, when George was 18, the Hales travelled to Europe with the family of a school friend, Burton Holmes. One place of pilgrimage was the shop of John Browning in the Strand, London, a maker of spectroscopes, who had built instruments for Lockyer and others. There, to the great disgust of Holmes, Hale 'squandered' £40 on optical goods! Holmes spent his pocket money sensibly at a conjurer's supply shop. In Paris Hale visited the observatory at Meudon and was spoken to kindly by Jules Janssen who, independently of, and simultaneously with, Norman Lockyer, had been the first to see the solar

---

[1] A diffraction grating is usually a flat mirror ruled with fine parallel lines. Light reflected by a grating is split up into its constituent colours, as when passed through a prism. An effect of this kind can be seen by looking at a light bulb reflected by a CD–ROM.

[2] Speech of Brashear, quoted by Wright (1994, p. 43).

prominences[3] outside an eclipse, by using a spectroscope.

After his return from Europe in August, he proposed to Evelina and was accepted. Neither set of parents felt an early marriage was a good idea and they persuaded the youngsters to wait four years before marrying, that is, until after Hale had graduated.

6.1   MIT student

William Hale had chosen MIT—the Massachusetts Institute of Technology— for his son, on the recommendation of his expansive Chicago architect friend, Daniel H. Burnham, with whom he was by then closely associated. Burnham's philosophy 'Make no little plans; they have no magic to stir mans' blood' was one that appealed to the Hale family.[4] MIT had been founded in Boston in 1865 and was at first rather narrowly centred on producing engineers as fodder for the growing industries of the United States. Hale was a serious student but did not enjoy the practical engineering orientation of nineteenth century MIT. Already, he knew that his interests lay in the direction of research. Certainly, he did not hang about with the more frivolous students. In fact, as his brother Will once remarked, 'The trouble with George is that he never had any bar-room experience'.[5]

Almost immediately on arrival, Hale became a close friend of Harry Goodwin, a friendship that was to last his whole life. Harry came from a much more liberal background and their late-night student discussions soon led George to doubt the fundamentalist religion of his father. He became interested, for example, in Darwin's theory of evolution by natural selection. Harry was a very bright student, but less goal-directed than George. Nevertheless, he was later to become a Physics professor and Dean of the MIT Graduate School. It was Hale who persuaded him, at the start of their second (sophomore) year, to major in physics.

By the end of Hale's first year at MIT, his parents had moved to a palatial mansion designed by Daniel H. Burnham on Drexel Boulevard in Chicago, a vivid manifestation of their increasing prosperity. This ornate house was approached via a porte-cochère and contained reception rooms decorated in a heavy and ornate Victorian style, right down to a Venus de Milo statue. More importantly, on taking the elevator to the attic floor, George found a new laboratory which had been constructed to his own specifications. Outside, a special mounting pier had been built for his telescope. Aside from a family holiday to the Yellowstone

---

[3]'Prominences' are red flame-like structures seen rising far above the edge of the sun during eclipses, principally radiating in the Hydrogen alpha line. Outside eclipses, they are over-whelmed by the brightness of the nearby disc. Janssen and Lockyers' idea had been to cut out most of the sun's light by restricting the wavelength range to admit only the Hα line of hydrogen (see Fig 5.6).

[4]Burnham was responsible for the design of many famous American buildings and much of the plan of present-day Chicago, especially the development of the lakefront.

[5]W.B. Hale, quoted by Wright (1994, p. 51).

National Park, he spent his annual vacation on observing the sun, learning how to take photographs and observe spectra.

During Hale's second year at MIT, he took a chemistry course from Arthur Noyes, then a young assistant, and became friendly with him. Noyes, whom he nicknamed 'Arturo', shared with him a passion for research and later in life they worked together to build up the California Institute of Technology, which Hale tried to ensure would not suffer from the defects he perceived in MIT. In this year also he wrote, offering his services as a volunteer, to Edward Pickering, the Director of Harvard College Observatory, in Cambridge, just a few kilometres away. Pickering had formerly taught at MIT and welcomed him to the Observatory. Harvard College Observatory was a research institution and was then engaged in surveying the spectra and brightnesses of stars, using photographic techniques. Pickering succumbed to Hale's enthusiasm and arranged that he could assist on Saturday afternoons and evenings. Sometimes he was invited to dine with the family. Mrs Pickering was an upper-class Bostonian—one of the so-called 'Boston Brahmins'—and, no doubt, helped George to polish his manners.

MIT was then situated near the Boston Public Library, where Hale spent a lot of time reading original papers by contemporary physicists and astronomers. In this way he learned 'what was written in the very heat of progress' and longed to make his own contribution to research.

In 1888 he got his father to build a small experimental building in their garden. It contained a darkroom, a laboratory, a workshop, and a room for spectroscopy. During his summer vacation he managed to produce a first-class spectrum of the sun from which he could confirm the presence of carbon. He realised that he was now able to obtain data as good as any that was available. In his following third (junior) year he ploughed ahead with his courses, often fearing that he would not pass. Physics practicals he particularly disliked: 'That is the bitter pill I dread . . . How shall I ever finish the miserable work?'[6]

6.2 Meeting with Rowland

The source of the best available diffraction gratings, the essential element of a successful spectrograph, was Henry Augustus Rowland of Johns Hopkins University in Baltimore, Maryland. Rowland was one of the leading American physicists and had succeeded in making the best gratings in the world, using ruling engines of his own design. With one of his gratings he had made the most detailed photographs of the solar spectrum then available. Hale was determined to get hold of a Rowland grating. While still at school he had written to the notoriously gruff professor and had received a brush-off. Brashear, who supplied the speculum-metal flats on which Rowland ruled his gratings, had in fact warned him of the professor's difficult nature.

As an example, there was a well-known story that on a certain occasion Rowland had to testify as an expert in a law case that hinged around a fact in

---

[6]Hale to Goodwin, undated ca. 1889, quoted by Wright (1994, p. 60).

physics. The judge demanded of him: 'Who is the greatest authority in the field of physics in the United States?' to which he replied without hesitation 'I am'. When asked afterwards how he could have made such a statement, he replied that he was under oath and had had to tell the truth![7]

Hale's interview started in typical fashion.[8] He knocked at Rowland's open door and was ignored. He knocked again.

'What can I do for you, young man?

Hale introduced himself and presented a letter from Brashear. Then he mentioned a possible defect in his 4-in. (10 cm) Rowland grating:

'That grating is good enough for any infant'.

Hale persevered and explained what he was trying to do. Slowly Rowland became more amenable and, after a couple of hours, found himself utterly charmed and giving a complete tour of his constant-temperature vault and ruling machines. Hale had accomplished his mission and learned a lot about gratings besides. In later life he enjoyed telling about this first meeting with Rowland.

He also paid a visit to the pioneering American astronomical spectroscopist, Charles Young of Princeton, who talked to him about Lockyer and Janssen's method of viewing prominences. With his own small spectroscope he had been unable to see them. He afterwards placed an order with Brashear for a large spectroscope that could be attached to the Harvard 15-in. telescope.

On 29 June 1889, Hale was twenty-one years old and his father saw fit to give him shares in a new building he was putting up in Chicago. He again broached the question of getting married as soon as he graduated. Although his father's reply was favourable, the tension of the interview affected him so much that he suffered afterwards from headaches and nervous indigestion, a problem that recurred at similar critical moments throughout his life.

6.3 Invention of the spectroheliograph

The main event of that summer for Hale was his invention of the spectroheliograph, one of his most important contributions to astrophysics. He remembered later that he was riding along in a cable car—in those days common in other places besides San Franscisco—when the idea occurred to him. Until then, the best that anyone could do to view prominences outside the fleeting moments of a total solar eclipse was to look at a small region—outlined by the entrance slit of a spectroscope—at the edge of the sun, as Lockyer and Janssen had done. The prominences were bright in the light of hydrogen atoms and stood out above the continuous background. Hale's idea was to eliminate the background from

---

[7]According to the website of Eugenii Katz, Hebrew University of Jerusalem, the story is at least in part apocryphal. The court had, in fact, noted that he was 'the highest known authority in this country upon the subject of the laws and principles of electricity . . .', and this was how the legend began.

[8]From a letter of Hale to H.M. Goodwin, 6 June 1889, quoted by Wright (1994, pp. 60–61).

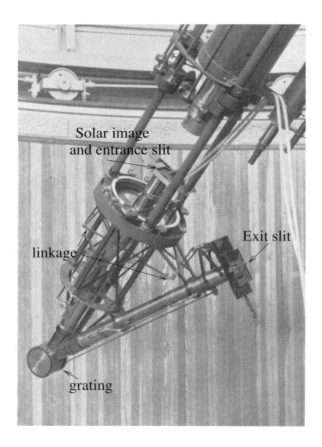

Fig. 6.1. First successful spectroheliograph attached to the telescope at Hale's
    Kenwood Observatory. Only light from a narrow part of the image that fell
    on the entrance slit could enter the spectrograph. The grating at the bottom
    reflected a specific wavelength range to the exit slit, beneath the photographic
    plate at the output of the instrument. By moving the entrance and exit
    slits together through a linkage, a complete picture of the sun at a single
    wavelength could be built up. (From Hale 1908.)

every other part of the spectrum by having a second narrow slit at the exit of
a spectrograph, tuned to the wavelength of hydrogen light. Then, if the image
of the sun was moved across the input slit and a photographic plate was moved
at a similar rate behind the exit slit, a picture of the whole solar surface could
be built up in the light of a particular kind of atom (see Fig 6.1). In fact others,
unknown to Hale, had thought of the same principle, but he was the first person
to construct a usable instrument.

On returning to MIT for his final (senior) year he arranged that his 'senior thesis', a requirement for MIT physics degrees, would be on 'The Photography of Solar Prominences'. He spent much of his time experimenting at Harvard College Observatory with his new Brashear spectrograph and spectroheliograph system, but found it was only marginally capable of producing the images he wanted. However, the results were promising enough to get him his degree and to show that he was on the right track.

6.4 Marriage and honeymoon

A couple of days after graduating, at the start of June, 1890, he was married at last to Evelina from her parents' home in Brooklyn, New York. The honeymoon was an opportunity to see the great tourist sites, such as Niagara Falls, Yosemite, and San Francisco, but also a chance to visit a number of observatories. One suspects the itinerary may have been arranged with astronomy more than anything else in mind. The highlight of the trip was a visit to Lick Observatory on Mount Hamilton, in the hinterland of San Franscisco, near San Jose. At the summit he was greeted by his old friend S.W. Burnham and James Keeler, a young spectroscopist, who was acquiring a reputation as America's most important astronomical physicist. Lick was famous among astronomers for its 36-in. (0.9 m) telescope, the largest lens telescope in the world, and the excellent climate in which it was situated. The telescope had been the gift of James Lick, an eccentric millionaire, who had wished to have himself commemorated by a spectacular monument. At first he had thought of erecting a pyramid occupying a city block of San Franscisco, but he was persuaded that to construct the world's largest telescope would be more fitting. He even specified that his coffin was to be placed inside the telescope pier. He refused to be cremated beforehand, preferring to 'rot like a gentleman' (Osterbrock et al. 1988). At Keeler's disposal was a first-class spectrograph. In fact, everything needed to impress Hale was in place.

6.5 The Kenwood Observatory

As Hale became known in the world of astronomy, he received several offers both of employment and of the right to use the largest telescopes for his experiments. However, he came to the conclusion that he had to do things his own way. He talked his father into setting up a private observatory with a 12-in. (30 cm) telescope, all for him, next to his laboratory in Chicago. This he called the 'Kenwood Observatory' after the district in which they lived.

The construction of the observatory (see Fig 6.2) and telescope took until about April of the following year (1891). In the meantime, Hale worked in his laboratory, leaving his unfortunate wife to keep herself occupied as best she could. They lived in his parents' house, which was hardly a cheerful place as his father worked long hours and his mother spent most of the time in bed, suffering from migraines. She may indeed have been subject to the manic-depressive type of illness that later caused her son so much trouble. They wanted to move out,

FIG. 6.2. Hale's private observatory, Kenwood, in the garden of his parents' house in Chicago (Hale 1908).

to an apartment of their own, but Mrs Hale senior would not hear of it, fearing 'what the neighbours would think'.

Just at this time, the 'new' University of Chicago was started with a large grant from the oil millionaire John D. Rockefeller. Its first President, William Rainey Harper, had been a child prodigy. He was a Hebrew scholar and educationalist who had high ambitions for the new institution. He set about hiring the best academics that he could find. One of his aims was to emphasise research, in the German university tradition. He had heard of the activities of Hale and made enquiries about him. At first, he clearly hoped to tap into the Hale fortune. He proposed quite shamelessly that the Kenwood Observatory should be transferred to the University as the 'Hale Observatory', and suggested to Hale that he should be advanced 'as rapidly as your work and age might seem to you and us to call for advancement', and that the observatory should be endowed with enough money to support two professors. Hale replied rather tartly:

> The possibility of securing a position for me by gift or lease of this observatory and additions cannot be considered favourably by us. If I am not competent to obtain a place on my own merits at present, it will probably be best for me to wait until I shall have gained experience by future study. Possibly I shall not be able to fit myself for an important position; if not, I certainly should not desire to obtain one.[9]

[9]Hale to Harper, 30 May 1891, quoted by Wright (1994, p. 76).

6.6 Discoveries at Kenwood

By May 1891, the Kenwood Observatory was in full swing and Hale soon had his spectroheliograph working properly. He was able to photograph prominences in the light of hydrogen and calcium (Fig 6.3). Writing to his MIT friend Harry Goodwin he said 'Yesterday I got a prominence photograph good enough to prove the success of the method, and the result is that I am just now feeling pretty *neat*'.[10] He quickly published an article in the *Sidereal Messenger*, an astronomical periodical named after Galileo's *Sidereus Nuncius*. He started to write Volume 1, Part 1 of the *Publications of the Kenwood Physical Observatory*. He was also elected a member (i.e. a Fellow) of the Royal Astronomical Society in London, where his article from the *Sidereal Messenger* had been read. On 15 June he held an 'Official Opening' ceremony. His father sponsored Prof Charles Young as the main speaker. Other well known attendees were Brashear and Charles Hastings of Yale, an optical expert. The proceedings were duly reported in *Science*. He very rapidly became well-known as a solar observer.

Hale also busied himself with founding a 'Chicago Section' of the Astronomical Society of the Pacific, the main astronomical society west of the Rocky Mountains. However, he lost interest in this after only about a year and it duly faded away, only to be supplanted by another of his creations, the 'Section of Mathematics and Astronomy' of the Chicago Academy of Sciences. In each case he was elected secretary. He used the second society as a base from which to organise an international conference—the 'World Congress of Mathematics, Astronomy and Astro-Physics'—in connection with a World Fair[11] to be held in Chicago in 1893.

The following month Hale and Evelina sailed for England where he was invited to meet the Hugginses. He spent some time in their observatory and got invitations to all the main private and public astronomical establishments. One of the English amateurs, Andrew Ainslie Common, was a builder of large reflecting telescopes, and had just completed a 60-in. which, though not very successful, must have sown a fertile seed in Hale's mind. This was the year (1891) of Huggins's Presidency of the British Association for the Advancement of Science and the Hales were invited to accompany him to the annual meeting, held that year in Cardiff. Here George presented a paper on 'The ultraviolet spectrum in prominences'. He and his wife were lionised. They were introduced to many of the members of the British astronomical community. Continuing to the continent, Hale was annoyed to find that the French astronomer Henri Alexandre Deslandres was claiming priority in the invention of the spectroheliograph. They subsequently met and attempted to settle the question of priority, but Deslandres later persisted with his claim and was to become a tedious rival.[12]

---

[10]Hale to Goodwin, 15 May 1891, quoted by Wright (1994, p. 77).

[11]The Columbian Exposition.

[12]Today they are regarded as independent inventors of the spectroheliograph.

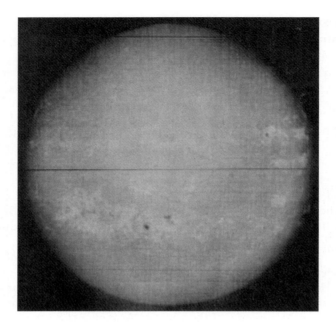

FIG. 6.3. Early spectroheliogram from Kenwood Observatory, dated 23 June
    1892. It shows dark sunspots and white cloudy faculae. For this photograph
    the instrument was tuned to the wavelength of calcium, which is very hot in
    the faculae and therefore shows up as bright patches. (The horizontal dark
    lines were caused by the imperfectly coordinated movements of the entrance
    and exit slits.) (SAAO).

On his return to Chicago, Hale made various improvements to his spectroheliograph and was soon observing the eruptions associated with solar flares
wherever they occurred on the surface of the sun and not merely at the edge,
which was all that could be done before. This was a significant step, because the
development of these 'faculae'[13] could now be observed continuously, whenever
the sun was visible. He could deduce that they formed well above the surface of
the sun. Since the times of their occurrence could not be predicted, long-term
observations had to be made to catch them. Never a person much interested in
routine work, he hired an amateur photographer, Ferdinand Ellerman, to observe
at Kenwood. Ellerman was skilled at technical matters and continued to work
with Hale at each of the establishments he subsequently founded. Hale's father
undertook to pay Ellerman's salary.

Another scheme of Hale's was to found a journal to be devoted to astrophysics. At first he worked with the editor of the *Sidereal Messenger* but later

[13]'Faculae' (from a Latin word meaning 'little flames') are clouds of gas above and 300°
hotter than the surface of the sun, often associated with sunspot regions.

they agreed to combine forces under a new title, namely *Astronomy and Astrophysics*, which duly appeared for the first time in 1892.

6.7 University of Chicago

One of President Harper's most successful appointments was that of Albert Abraham Michelson to head the physics department. Michelson, who was born in Prussia, emigrated with his father to the United States and studied at the US Naval Academy in Annapolis. He had built up a reputation as an optical physicist, especially for his ever-improving determinations of the speed of light, on which he was the world expert. He realised that Hale was a unique phenomenon and persuaded Harper to approach him again. The result was that in July 1892 Hale was made an Associate Professor of 'Astral Physics' [sic][14] and the University was to acquire the movable property of the Kenwood Observatory, but on condition that they should raise $250,000 within two years to build a much bigger establishment.

As a member of the University of Chicago staff Hale observed with approval the methods used by Harper to build up the new university. Later in life he applied them himself, when he became involved in the formation of the California Institute of Technology.

6.8 Yerkes—the largest refracting telescope ever made

Within a few months, Hale's genius as a promoter of observatories was to become evident. At a meeting of the American Association for the Advancement of Science in the summer of 1892 he heard of two 40-in. (1 m) lens blanks, one each of crown and flint glass, necessary for making the objective lens of a telescope, that had been ordered by the leading makers Alvan Clark and Sons of Cambridge, Massachusetts. The project for which they had been intended had fallen apart in one of the financial crashes that plagued the United States. At about that time, President Harper had been trying to persuade Charles T. Yerkes, a streetcar promoter, to donate money to the University of Chicago and thereby earn himself the respect of the citizens of Chicago. This had been sadly lacking before his benefaction because of his reputation for unscrupulous business methods, which had on one occasion landed him in jail. Hale wrote to him in glowing terms of the exceptional opportunity that had now presented itself and how much better a 40-in. would be than the present largest lens telescope, the Lick 36-in. (90 cm). Harper and he duly received an invitation to meet Yerkes, who was completely sold on the idea that he would become the 'owner' of the largest telescope in the world. He told them 'Yes, I'll do it. Build the observatory, gentlemen. Let it be the largest and the best in the world and send the bill to me'.[15] Once they reached the street again after the interview, Hale and Harper

---

[14]Goodspeed to Hale, 26 July 1892, quoted by Wright (1994, p. 92).

[15]Reconstruction of events by Wright (1994, p. 98).

looked at each other. They burst out laughing and danced a jig. They were birds of a feather when it came to the game of raising money.

Yerkes was quite not such a willing giver when he found out that the total cost of the project was much more than he had originally agreed to pay. Hale's response was to discreetly leak the story of his 'princely donation' to the press, making it difficult for him to avoid giving more money to finance the building without seeming mean. Observing that the University's main benefactor, the oilman John D. Rockefeller, was not exactly saintly in his business dealings either, an editorial in the *Chicago Times* suggested that 'Harper's success as a money raiser was due to his having shrewdly represented the Chicago university to divers men of wealth as a sort of conscience fund (Franch 1997)'. This was, of course, right on the mark. Yerkes was ultimately induced to part with $349,000 to pay for the lens, the telescope and the observatory buildings. Hale, aged 23, was to be the director of the new institution.

By the time of the congress that Hale had arranged in conjunction with the Chicago World's Fair of 1893, he had met George W. Ritchey, a teacher of mechanical workshop practice at a local technical school. Although the 40-in. telescope was then well under construction, Ritchey soon persuaded Hale that reflecting telescopes were in reality a better proposition than refractors. Even though no really successful large reflector had been constructed before that time, several quite interesting attempts had been made, such as that of the Irish aristocrat Lord Rosse, whose cumbersome 72-in. (1.83 m) reflector was known as the 'Leviathan of Parsonstown'. Rosse's telescope antedated photography but had yielded the intruiging result, thanks to its unprecedented light-gathering power, that some of the faint nebulae are of spiral form. The seeds of Hale's next big project had thus been sown and Ritchey was to play a large part in Hale's schemes until they fell out over the mirror for the 100-in. (2.54 m) telescope more than twenty years later.

Following the congress, there was little to do while the 40-in. lens was undergoing the slow process of grinding and polishing. Hale and his wife went to Berlin, where he hoped to improve his knowledge of physics and obtain a doctorate, following a path trodden by many young scientists of the time. He sat at the feet of the physicist Hermann von Helmholtz and the spectroscopist Hermann Vogel. Unfortunately, he found that he could not easily learn the language and that German cuisine was not at all to his taste. As the winter wore on, he became depressed by the cold central European weather and was unable to concentrate, a harbinger of later problems. Instead of studying, he preferred the 'displacement activity' of working on the plans for the buildings of the Yerkes Observatory. In the spring of 1894, he abandoned all pretence and headed south for the more benign climate of Italy. He brought a special solar telescope with him in the hope that the clear Mediterranean skies would allow the corona[16] to be seen around the sun. He ascended the volcano Etna in Sicily to make observations but soon

---

[16]The faint, outermost extension of the sun's atmosphere, from the Latin word for 'crown'.

found that conditions there were no better than they had been on Pike's Peak in Colorado, where the year before he had been defeated by the smoke from forest fires on a similar expedition.

6.9 The Astrophysical Journal

Once Hale got back to Chicago late in 1894 he decided that *Astronomy and Astrophysics* had served its purpose. It had been divided into an astronomy part and an astrophysics part and he felt that the time had come for a purely *Astrophysical Journal*. Thus, after only three years, the dual-purpose journal came to an end and the new one began, in January 1895. Hale had carefully lined up an editorial board that included Michelson, Rowland, Hastings, Young, Pickering, and Keeler, the main US scientists with interests in astrophysics. It was to become the primary journal of astrophysics, a place it still occupies today. The astronomy side was re-started as a popular journal, *Popular Astronomy*, which had a long life, succumbing only in 1951. The clever manner in which Hale effected this divorce tempted a colleague to say 'Hale had the ability of putting a knife in your side, and of making you think he was doing you a favor'.[17]

The finances of the *Astrophysical Journal* were precarious for several years and Hale, although supported by the astrophysical community, found himself begging from his friends and relations for funds to support it. There were also delicate problems such as how to reject papers by the wealthy Percival Lowell, who had convinced himself (and many others), quite erroneously, that the surface of Mars was covered by canals.

In 1895 Hale found time once again to do some interesting science. The British chemist William Ramsay had just discovered an element on earth whose spectrum matched that of a substance in the sun found by Norman Lockyer and given the name 'helium'. However, the laboratory spectrum showed a close double line whose counterpart had not been seen in the sun. With his solar spectrograph at Kenwood Hale was able to resolve this line, finding it to be double and making the identification more secure.

6.10 Yerkes Observatory completed

By October 1895, the 40-in. lens had been finished and was ready for inspection at the factory. Its two components weighed 500 pounds (227 kg). The cumbersome tube that held it had to be about 19 m long and the dome would have to be 23 m in diameter. The site fixed on for the Yerkes telescope and observatory was Williams Bay, Wisconsin, about 130 km north of Chicago. The building itself was designed to accommodate two smaller telescopes besides the 40-in. and contained offices, laboratories, workshops, and dark rooms for the effective pursuit of the latest ideas in astrophysics. One of Hale's great strengths was his belief in using the latest technology to obtain unique results, as a statement of his dating from 1915 shows:

[17]Quoted by Wright (1994, p. 115)

In departing from accepted standards and in preparing to overcome difficulties, the initiator of new methods almost necessarily become an instru-
ment-maker, and hence a machine shop may be his first requirement. He
can not afford to intrust construction to instrument-makers thousands
of miles away, with whom he is unable to discuss details of the design,
necessarily subject to frequent modification in the light of newly-acquired
ideas. To be most efficient, he must be his own designer and builder, ready
to take immediate advantage of those new points of view and new possibilities of attack which his investigations are certain to disclose. (Hale
1915)

The Hales rented a cottage in Williams Bay. At last Evelina and he had a
valid excuse for leaving his parents' home. It was a pleasant place, located on the
shore of Lake Geneva and much favoured by the wealthy of Chicago for summer
homes, though exceedingly cold and subject to being snowed in in winter. Their
first child was born a year later. Evelina immediately asked the attending doctor
whether the new baby would become an astronomer, like his father. The reply
was 'Well, yes, but of the Maria Mitchell kind'—the latter being a pioneering
American woman astronomer.[18] The baby was given the name Margaret.

As the observatory reached completion, Kenwood was dismantled and Hale's
telescope and dome were moved to Williams Bay. The Hales moved into a house
of their own and they were the proud possessors of a carriage and four horses.
The staff of the new observatory began to arrive: Ellerman and Ritchey were
followed by Edward Emerson Barnard, an acute visual observer rather than an
astrophysicist.

The completion of the telescope, marked by the installation of the 40-in. lens,
was in April 1896. The following day, the University's President, the trustees
and other important 'friends' were invited for a look at Jupiter. Barnard found a
faint companion to the bright star Vega and his discovery was speedily exploited
'...this new star would in all probability have remained undiscovered with the
Lick telescope' he reported to Yerkes, who was, of course, suitably impressed.[19]

On one night during the following month, Ellerman and Barnard had been
observing and had closed the dome at 3:15 in the morning and gone to bed. The
contractor for the building arrived a few hours later to do some work and, just
before entering, heard a loud crash. It turned out that the rising observing floor,
installed to avoid the necessity of using long ladders, had given way, fallen from a
good height, and disintegrated. It was sobering to contemplate what might have
happened had it been the night of the Trustees' visit!

The repairs took several months but the official opening on 21 October 1897
was a great triumph for Hale and Yerkes, whose gift was successful in providing
this disreputable character with the wished-for key to entering Chicago society.
Everybody who was anybody in American astronomy was there. William Huggins
had been invited from England, but felt too frail in the end to attend. This was

---

[18]E.C. Hale, quoted by Wright (1994, p. 121).
[19]Hale to Yerkes, 31 May 1897, quoted by Wright (1994, p. 129).

the occasion for which the older astronomer had written his famous article on
the *New Astronomy*. Hale, though only twenty-nine-years old, found himself in
control of the largest professional telescope in the world.

Besides Hale, the early staff of Yerkes included Barnard, Burnham, Ellerman, and F.L.O. Wadsworth, the last an instrumentalist who had worked with
Michelson. Hale expected a high degree of enthusiasm; those who showed a lack
of interest were soon got rid of. In 1898, they were joined by Edwin B. Frost,
who had been educated at Dartmouth and Potsdam, with Vogel and Scheiner,
two of the leading German astrophysicists.

The Kenwood spectroheliograph was the first to be attached to the 40-in. and
with it Hale studied carbon in the sun's atmosphere. However, in 1899, a much
better spectroheliograph with higher wavelength resolution was ready, known as
the Rumford, from the source of funds for its construction. The budget for it
was so small that one of its lenses had been bought from a pawnbroker. Using
the Rumford spectroheliograph Hale found patches of incandescent material just
above the sun's surface, but not high enough to be seen, like the faculae and
prominences, at the edge. These clouds he called 'flocculi', because of their cloud-
like appearance.

One of the first graduate students at Yerkes was Walter S. Adams, who
was to become Hale's closest and most loyal astronomical associate and who
later took over many of his administrative functions during his periods of severe
depression. Adams had been born near Antioch, in the Middle East, and was the
son of missionaries. His family returned to the United States when he was eight
years old and he eventually attended Dartmouth College, followed by graduate
study at Chicago. He was encouraged to spend a year in Munich and given an
assistantship in Yerkes when he returned. Attempts to attract other astronomers
failed because of the low salaries that could be offered, but a number of famous
foreigners were happy to spend short periods at Yerkes to take advantage of the
large telescope.

6.11   The 60-in. mirror

Even as the 40-in. refractor was being constructed, Hale's thoughts were dwelling
on his next big project, i.e. a large reflector. Large lens telescopes were cumbersome and required large, expensive, domes. Worse, large lenses had to be thick in
order to support themselves and so absorbed some of the light. They were also
unsuited to studying infrared and ultraviolet radiation. The Yerkes telescope,
at the limit of practicality, remains to this day the largest refractor ever constructed. The mirror of Rosse's early 72-in. reflector (1845) had been made of
speculum metal, the reflectivity of which was rather poor. The much superior
technology of silver-on-glass mirrors was developed seventeen years later, by the
French physicist J.B.L. Foucault. Since then, there had been various attempts at
large reflectors with glass mirrors, mainly by amateurs such as A.A. Common,
and often with results that were rather poor, through no fault of the actual principle. However, one of Common's telescopes, a 36-in. (0.9 m) had found its way

to Lick Observatory as a gift from Edward Crossley, a wealthy English textile manufacturer whose taste had migrated from astronomy to theology, rendering the possession of a telescope superfluous. After much improvement it had been used successfully by Keeler and had even become his favourite instrument. Also, an Irish amateur, W.E. Wilson, had a 24-in. (61 cm) reflector made by Grubb of Dublin in 1881 and with it had produced some very fine photographs.

Ritchey built another very successful 24-in. reflector for the Yerkes Observatory between the years 1898 and 1901. In the process he invented methods for figuring and testing large mirrors which enabled him to achieve new levels of perfection. With his new instrument he took many excellent photographs, superior to those made with the Crossley reflector and by W.E. Wilson. These led to widespread recognition of his skills. Not only did he produce the best mirrors and mountings, he was a very meticulous observer and had developed techniques for guiding telescopes with great accuracy during long exposures. He was able to choose the moments of best 'seeing' or atmospheric steadiness for his exposures. His pictures of nebulae and of the moon were in great demand. For this work Ritchey was made an Associate of the Royal Astronomical Society in 1904.

Hale sensed that:

> The investigation of these nebulae, with the great reflecting telescopes of the future, should lead to results of fundamental importance. (Hale 1908, p. 45)

In pursuit of his vision, in 1896 he ordered a 60-in. (1.52 m) diameter glass mirror blank from the French firm St Gobain, following technical advice from Ritchey, who became his full-time employee with a salary paid by his father. The glass cost $2000. When it arrived, Ritchey went to work on the rough grinding, using specially constructed machines of his own design. Hale was very careful to emphasise that the disk was his personal property rather than the Observatory's. As Osterbrock (1993) remarks, 'Hale was one of the first observatory directors to grasp the importance of buying a glass blank and starting to work it into a mirror, even if there was no money to build the rest of the telescope, as the best way to attract the remainder of the necessary funds'!

6.12   First attempt on Carnegie

Hale tried over several years to raise funds for an observatory in southern California, but it was only in 1902 that the opportunity he was waiting for at last presented itself. That was when the immensely wealthy steel manufacturer Andrew Carnegie set up the Carnegie Institution of Washington with a gift of $10,000,000. Carnegie (1889) had written an essay on *Wealth*, in which he expressed his strong views on how a rich man should dispose of his fortune for the greater good of mankind, rather than leave it to his family members, who might be corrupted in character by large legacies. He felt that a charitable legacy after a donor's death showed a lack of social responsibility, whereas the careful support of useful causes within one's own lifetime was praiseworthy. The foundation of educational institutions, such as universities, were among the things he approved of. However, he had a special regard for astronomy, one which accorded with Hale's own philosophy:

> We cannot think of the Pacific coast without recalling another important work [besides Stanford University] of a different character which has recently been established there—the Lick Observatory. If any millionaire be interested in the ennobling study of astronomy,—and there should be and would be such if they but gave the subject the slightest attention,— here is an example which could well be followed, for the progress made in astronomical instruments and appliances is so great and continuous that every few years a new telescope might be judiciously given to one of the observatories upon this continent, the last being always the largest and best, and certain to carry further and further the knowledge of the universe and of our relation to it here upon the earth. (Carnegie 1889)

Hale immediately sent the trustees of the new Institution a proposal, complete with some of Ritchey's photographs, for the erection of an observatory 'at high elevation in southern California or Arizona'. The result was that he was invited to become a member of an 'Advisory Committee on Astronomy', composed of well-known astronomers. Unfortunately, the other members were old and traditionally-minded, so that Hale's propaganda in favour of dramatic innovation fell on deaf ears. The Secretary to the Trustees, Charles D. Walcott, an eminent geologist, was personally keen on the project and discussed it privately with Hale and Carnegie himself. The main impediment was that Advisory Committee. They at last agreed to provide some funds for the investigation of possible sites for a new southern and solar observatory—Hale having emphasised the 'solar' in his proposal. This work was carried out by W.J. Hussey of Lick Observatory, who duly suggested that Mount Wilson, overlooking the Los Angeles basin from an elevation of 5900 ft (1830 m), would be ideal both for its weather and accessibility. In fact, Harvard had had a station there several years before, but had abandoned it for logistical reasons. One of these was that the trail that led to the mountain top was only two feet wide (about 60 cm) in places.

6.13 Early days on Mount Wilson

In June 1903, Hale and W.W. Campbell, another Lick astronomer, trekked up Mount Wilson, where Hussey was still at work, to see it for themselves. Hale appeared to be completely convinced that it was the place to go for, but the owners of the land, the Mount Wilson Toll Road Company, were at first uncooperative. He became uncertain and dithered for a while, but ultimately decided firmly in favour of the site. The Advisory Committee on Astronomy was persuaded to recommend the establishment of a high-altitude solar observatory as well as a Southern-Hemisphere observatory. However, there were many other calls on Carnegie's funds and little cash was forthcoming in the short term. Meanwhile, Hale's friends among the Trustees continued to argue in his favour. They pointed

out to Carnegie that Hale was 'an exceptional man', just the sort that he claimed
should be supported.

Even though no positive decision had been made, Hale decided to go to California at his own expense with his family, which now included a son, William
Ellery Hale (born in 1900). On arrival in December he set about carrying out further tests with a solar telescope and some weather instruments that he borrowed
from Yerkes. He was enchanted by the mild climate of Pasadena after the rigours
of a Chicago winter. Further, his children rapidly overcame their bronchitis and
other problems. He enlisted the help of a local teenager and together they carried
a small telescope and its mounting up the mountain trail. Hale climbed up a tall
pine tree with the telescope and made seeing tests at 32 (9.8 m) and 68 ft (20.7
m) above ground level. They found a ramshackle building known as the 'Casino'
in which to sleep. This expedition convinced him to set up a proper experimental
station. In early 1904 he engaged some workmen to make the Casino habitable.

Meanwhile, he propagandised among the citizens of Pasadena, then still a
beautiful village much favoured by the rich and set among orange groves. He
gave a lecture on astronomy at Throop Polytechnic Institute, a small college
which had been founded in 1891 by the philanthropist Amos G. Throop. There
he was introduced to a local hardware millionaire, John D. Hooker, who had
helped to form the Southern California Academy of Sciences and was interested
enough in astronomy to possess a small telescope. Hooker very soon found himself agreeing to finance the travel of Barnard and the transport of a small photographic telescope from Yerkes to the temporary observatory. Hale was rather
taken by Hooker's sophisticated wife Katherine, who read Italian poems to him
in their beautiful garden. She was then about fifty-four and he about thirty-
seven years old. Although he did not fully understand the words, he afterwards
enjoyed reciting them from memory on the Mount Wilson trail. In that same
agreeable household, he met, in early 1905, another lively lady, Alicia Mosgrove,
nicknamed Ellie. She was a friend and companion of Katherine and often travelled with the family. As a group, they were familiar with many of the California
intellectuals of the period, such as the naturalist John Muir, who often stayed
with the Hookers. Some believe that Hale may have had an affair with Ellie.
Certainly he found in her at least a soul-mate. She appealed to the whimsical
side of his nature. He introduced her as the 'gal' who sold drinks in a bar on the
Embarcadero in San Franscisco, or as the keeper of a seraglio in Constantinople!
She was an adventurous lady who founded parks and schools and did unladylike
things such as climbing mountains, sailing to Tahiti and floating down the Nile in
a native boat. Ellie and her friend Maude Thomas would sometimes accompany
Hale on his trips up the mountain.

Yerkes Observatory at this time lost the support of Yerkes himself and Rockefeller was not willing to put more money into the University of Chicago. One
consequence was that Walter Adams was going to be without a salary though
still keen to keep working with Hale for 'Omar's loaf of bread and jug of wine'.
Hale found the money for Ellerman to come out and told Adams not to worry—a

solution would be found. He informed President Harper that his child's health was such that he would have to remain in California, 'where he was establishing a small station'. Ellerman and he soon set up a modest fixed solar telescope on Mount Wilson, housed in a simple building of paper and canvas.

In the meantime, Hale had been awarded the Gold Medal of the Royal Astronomical Society for his invention of the spectroheliograph and his discovery of flocculi. In April 1904, he travelled to Washington to receive the Draper Medal from the National Academy of Sciences, of which he had become a member two years previously. In his hotel, he found the chairman of the Carnegie Trustees and Mrs Henry Draper, a benefactress who had made gifts to several observatories, waiting for him. After his medal presentation, the Carnegie Executive Committee met. They awarded Hale an immediate $20,000 and promised that before the end of the year they would pay $30,000 more for the Yerkes expedition to Mount Wilson. His brother Will, now a lawyer, and his uncle George lent him further sums and he decided to go for broke in developing Mount Wilson. He carefully negotiated an understanding with Harper that, should Carnegie come up with sufficient money, a separate institution would be founded.

On returning to Mount Wilson, Hale made an agreement, in his own name, to lease a part of the mountain at no cost for 100 years from the Mount Wilson Toll Road Company. His romantic imagination had previously been stirred by a famous travel book *Visits to Monasteries in the Levant* by Robert Curzon (1849). He chose a spectacular spur, with the ground falling off on three sides, on which to build a 'Monastery' for the (male) astronomers to stay in while working on the mountain. By December, the building was ready for occupation and Hale, Ellerman, and Adams celebrated the occasion by moving ceremoniously down from the Casino carrying lighted candles.

The living room of the monastery was the scene of much *camaraderie* in these early years. On cloudy nights, the young astronomers would 'shoot the breeze', whether discussing future plans or listening to Hale tell stories of the interesting people he had met or even reciting poems by Keats and Shelley. The lifestyle was a healthy one, with plenty of exercise going up and down the mountain or helping with the construction work that was always going on.

6.14   Success with Carnegie

Hale did not spend all that year on Mount Wilson. When he had become a member of the National Academy of Sciences in 1902, he had found that body to be moribund, little more than a mutual admiration society. He set about making it more active in national affairs and turning it into a policy-making organisation. He got them to create a Committee on Solar Research, with himself as chairman, which then formulated plans for an 'International Union for Cooperation in Solar Research', scheduled to have its first meeting at the St Louis World's Fair in September, 1904. To this event he succeeded in attracting astronomers from almost all the countries active in research. A by-product was the enhancement of his own reputation within the United States. Hale's International Solar Research Union was ultimately to develop into the International Astronomical Union.

While in St Louis he heard that a big application to the Carnegie Institution for the study of coral atolls had been withdrawn and, in spite of a debilitating bout of depression, he went to see John Billings, a member of the Executive Committee, in New York. Billings seemed pessimistic and uninterested, but listened carefully to what Hale was saying. In Washington, Charles Walcott, its secretary, encouraged him to present his 60-in. project to the whole committee. They showed greater enthusiasm and Hale became hopeful. He soon heard that his proposal had been recommended to the Trustees. At a meeting of the latter on 13 December 1904, Billings and Walcott put forward his plan and defended it against some vocal opposition. The matter was referred back to the Executive Committee. Then finally, on 20 December, it was announced that Hale's Solar Observatory would receive $150,000 a year for two years! Hale himself had been so certain that he would win in the end that he had already bought himself a house in Pasadena.

He lost little time in resigning from Yerkes and set about getting the Mount Wilson Solar Observatory running. Ritchey, Ellerman, and Adams were now given secure salaries. He removed his personal scientific equipment from Yerkes and his furniture from his house in Williams Bay. Frost took over the directorship of Yerkes, but the fact was that most of the glory had departed with Hale. Harper died soon after of cancer. Fortunately for the Observatory that Hale had founded and abandoned, Yerkes, who died about the same time as Harper, left $100,000 to endow it.

On Mount Wilson the 60-in. telescope project was now in full swing. Hale had to settle with the University of Chicago for some of the costs involved in grinding and figuring his mirror, but it was soon shipped to Pasadena, where a workshop for the observatory was under construction on Santa Barbara Street, an address to become famous later as the headquarters of the Mount Wilson Observatory.

The year 1905, though busy for Hale, was a difficult one for Evelina, with the children sick and her husband spending much of his time away. On his return from one of his Solar Union conferences in Oxford in November, she had a 'nervous collapse' and had to spend some time in a 'sanitarium' (mental hospital).

6.15   The Snow Telescope

To make further progress in observing sunspots, Hale required a telescope that could give a larger and more detailed image than the Yerkes refractor. The spectrograph to match it would have to be very long and heavy and therefore quite unsuitable for attaching to the end of a moving telescope. He decided to use a fixed horizontal telescope onto which the sun's image could be reflected by a 'coelostat', consisting of two flat mirrors, one of which could be driven to counteract the rotation of the earth. Before leaving Yerkes, he and Ritchey had set up such an instrument, but it was housed in a wooden building and was destroyed

FIG. 6.4. George Ellery Hale at his desk on Mount Wilson in 1908 (Carnegie Institution of Washington).

by fire even before it could be used. Undaunted, Hale arranged for a second one to be constructed, using a grant from Miss Helen Snow of Chicago. He borrowed this instrument from the Yerkes observatory and set it up on Mount Wilson, even before persuading Miss Snow to lend it to him. Once in position in April 1905 it was, in fact, the first major installation of the new observatory. It was placed in a special shielded building whose inside remained cool in spite of the hot sun outside. It could feed a large fixed spectrograph and a spectroheliograph. Adams especially looked forward to using the new telescope for astrophysical work with Hale and was glad to have got away from the more pedestrian atmosphere at Yerkes. Hale, Adams, and Henry G. Gale (a physicist from the University of Chicago) studied the spectra of sunspots and were able to show that they differed from the rest of the solar surface by being lower in temperature (For many years, the *Astrophysical Journal* was edited by Hale, Gale, and Frost!). Comparison spectra made under controlled conditions in Pasadena enabled them to determine the temperature of the spots. Thus they had found the physical explanation of sunspots which had been an enigma since their discovery by Galileo and his contemporaries. It was a major feather in the cap of the new observatory. The Snow telescope was eventually bought outright from the University of Chicago.

The sun, taken as a whole, has the spectrum of a G-type star, according to the classification scheme developed at Harvard (see Section 7.6), but the sunspot spectra resemble a K or M type. This led Hale and Adams to realise that the Harvard spectral types were probably a temperature classification. A fuller understanding came only about twenty years later, after the development of quantum mechanics.

6.16   The first solar tower

Almost immediately, it became obvious that the horizontal Snow telescope was not the ideal solution in spite of all the precautions that had been taken in its construction. Heating of the mirrors still distorted the images of the sun as the day advanced. Hale, hardly a year after the instrument had come into service, now decided to construct a new kind of solar telescope in the form of a 65-ft (19.8 m) high tower, so that the objective lens and coelostat could be placed at the top of the structure, well away from the hot ground. The image was to be formed at ground level and the spectrograph would be situated directly beneath, in a 'well' 30 ft (9.1 m) deep. There the temperature would be kept stable by the surrounding rock.

The tower telescope was ready towards the end of 1907. The most important discovery associated with it was made in the middle of the following year (1908)— the detection by Hale and Adams of split spectral lines, which were caused by the magnetic fields present in sunspots—a celestial manifestation of the Zeeman effect discovered in a laboratory by the Dutch physicist Peter Zeeman twelve years before. By placing light polarisers at the entrance of the spectrograph, the relative intensities of the split lines could be studied and an estimate made of the strength of the magnetic field, which turned out to be near 3000 times that of the earth's.

Hardly was the 60-ft tower ready for use when Hale designed a 150-ft (46 m) version that offered an image of the sun with twice the size. It included more special features to avoid wind shake and some remaining problems with the heating of the optics. The associated spectrograph would now be placed in a well of 75 ft (23 m) depth. The Carnegie Trustees, by then properly appreciative of Hales 'indefatigable labors', almost immediately granted the necessary money and the telescope was completed by 1912. Hale's main aim with this instrument was to find if the sun had an overall magnetic field. Unfortunately, as is now known, this field is too weak to have been detected using the techniques of the time.

6.17 The 60-in. reflector

The 60-in. reflector was a much larger and heavier instrument than the Snow and required a much wider trail to transport it up the mountain, as well as a special truck that could negotiate the sharp corners. The heavy parts were cast at a foundry in San Franscisco and fortunately escaped damage in the earthquake of April 1906. In August 1907, the mirror was ready and by the spring of the following year the whole telescope had been pre-assembled and tested in Pasadena (Fig 6.5). The building and dome were finished around September and the telescope was finally placed inside, ready for service, in December.

The technology available for supporting heavy telescope parts was then very primitive. The delicacy of the motion that was required meant that the bearings of the main axis could not be allowed to take the whole of its weight. Most of the load, which is driven by precise clockwork to counteract the rotation of the earth, was therefore supported by a cylindrical steel tank, 10 ft (3 m) in diameter, floating in a giant mercury bath. This method had first been employed by Common, the English amateur telescope builder.

The telescope could be used for photography and spectroscopy. For the first of these, plateholders were installed at two of the focal stations and for the second, the light from a star could be directed by means of additional mirrors to a fixed spectrograph, with which Hale hoped to take detailed spectra of bright stars for comparison with the sun.

In the midst of the frenetic activity of 1907, Hale wrote a copiously illustrated book on his achievements, called *The Study of Stellar Evolution* (Hale 1908). In it, he expressed some of his private philosophy:

> But though dangerous when unrestrained, the imagination, when rightly exercised, is the best guide of the astronomer. His dreams run far ahead of his accomplishments, and his work of today is part of the development of a plan projected years ago. He perceives that only a few generations hence many of the instruments and methods of his time are to be replaced by better ones, and he strains his vision to obtain some glimpse, imperfect though it be, into the obscurities of the future. As he sits in his laboratory, surrounded by lenses and prisms, gratings and mirrors, and the other elementary apparatus of a science that subsists on light, he cannot fail to entertain the alluring thought that the intelligent recognition of some well-known principle of optics might suffice to construct, from these very elements, new instruments of enormous power ... (Hale 1908, p. 13)

6.18 Hale and the development of Caltech

In 1906 Hale, at the age of 36, and a potential victim of his own fame, had been offered prestigious scientific administrative posts such as the Secretaryship of the Smithsonian Institution (a government body) and the Presidency of MIT.

FIG. 6.5. The 60-in. telescope under construction in Pasadena. The large circular structure near the telescope tube contained a mercury flotation system to relieve the weight on the bearings (Hale 1908).

Both these he turned down, preferring to continue with astrophysical research. He turned his attention instead towards the local Throop Polytechnic Institute, then a modest establishment which he decided should be upgraded to become the MIT of the west. Within a decade, he had persuaded its trustees to phase out the basic technical and trade training that it originally offered and instead to enrol only exceptionally meritorious students who would study science and engineering at the feet of the best teachers they could afford.

> In developing such a school, we must provide the best of instruction and the most perfect equipment that modern engineering offers. But in laying stress upon the practical aspects of the problem, we must not forget that the greatest engineer is not the man who is trained merely

to understand machines and apply formulae, but is the man, who, while knowing these things, has not failed to develop his breadth of view and the highest qualities of his imagination. No great creative work, whether in engineering or in art, in literature or in science, has been the work of a man devoid of the imaginative faculty.[20]

In 1907, he was given the task of finding a new President for Throop. He started by asking some of his MIT friends if they were interested, but met with no success. The problem had to be laid aside for the moment. Shortly afterwards he took a ship to Europe for a meeting of the Solar Union in Paris. He was fortunate enough to find the Carnegies aboard the vessel he had booked on. Carnegie was interested to discover more about this energetic young scientist and invited Hale and his wife to join them at table. For Hale it was a heaven-sent opportunity to impress the normally inaccessible philanthropist with his plans. Also on board was James A.B. Scherer, the President of a small college called Newberry, in North Carolina. Scherer had heard that Carnegie would be making the trip and had deliberately booked on the same boat in the hope of meeting him. Unfortunately, Newberry was a Lutheran college and Carnegie was decidedly secular in outlook. Scherer and he argued every time they met! Carnegie evidently appreciated a good fight, because at the end of the voyage he asked both the now despairing Scherer and the Hales to stay with him at his Scottish castle, Skibo, and even arranged for a private railway carriage to take them there. Scherer's determination also impressed Hale who eventually persuaded him to take on the Presidency of Throop. Their relationship became a familial one in 1918 when Scherer's son Paul married Margaret Hale.

Towards the end of the First World War, Hale talked his friends, the chemist Arthur Noyes of MIT and the physicist Robert Millikan of Chicago, into spending part of each year at Throop. Ultimately he was able to persuade them to leave their positions and become the academic founding fathers of the reborn institution.

Throop narrowly escaped becoming a state institution as part of the University of California. Had it done so, interference by state politicians would undoubtedly have limited the freedom it needed if it was to accomplish Hale's purpose. Only in 1920 was it officially re-named the California Institute of Technology, known widely today as Caltech. The fund-raising skills of Hale and the other trustees soon saw to it that Caltech was well endowed through donations from the Rockefeller, Carnegie and Guggenheim foundations as well as from many California patriots.

6.19 The 100-in. Hooker telescope

Always thinking of his next project, even before the 60-in. mirror was finished Hale's thoughts had turned to a larger reflector still. He judged that the time was ripe for an attack on the fortune of J.D. Hooker:

---

[20]Address of Hale to Throop Trustees, 1907, quoted by Wright (1994, p. 247).

> Mr. Hooker is getting to be an old man, his oil wells happened to be giving an extraordinary flow, and there were grounds for believing that the psychological moment had arrived ... Mr. Hooker was fully informed as to all the difficulties in the way, but it was he who proposed that we should increase the size, first from 7 feet, the size I suggested to him, to 8 feet, and ultimately ... to 100 inches (2.54 m).[21]

Carnegie joined the chase. He wrote to Hooker, welcoming him 'to the number of those who see in wealth only a sacred trust to be administered in the service of man'.[22] Ultimately, Hooker agreed to finance the cost of the mirror. Though the telescope was named after him, much of the funding was in the end provided by the Carnegie Foundation.

To cast the glass mirror blank weighing 4.5 tons for such a telescope was no easy matter. A pouring on this scale had never before been attempted. Three separate melting pots, each containing 1.5 tons of molten glass, had to be poured into the mould and the red-hot liquid stirred to rid it of air bubbles and make it absolutely homogeneous. Even the smallest variation in the composition of the disc would cause strains that would make it crack as the glass cooled down. The cooling process, which had to be very slow on account of the high expansion coefficient of the plate glass then in use, lasted several months. The first attempt at a casting was made at St Gobain in September 1907 but after cooling the glass was found to be full of small cracks. The second pouring, made in 1908, was much better, but the quality of the result was still very poor compared to the 60-in. blank. It was nevertheless forwarded to Pasadena where Hale and Ritchey made an inspection and rapidly decided that it was not acceptable.

Trials and further attempts went on for some years, but met with no success. The St Gobain company lost a lot of money on the project and asked for a more favourable contract if they were to continue with expensive experiments; further, Hooker was getting balky about supporting it. Worse still, he felt that Hale was becoming too familiar with his wife and forbade her to have male guests when he was not present. In Pasadena, Hale, egged on by Adams, fell out with Ritchey, who had made the mistake of approaching Hooker for separate funding of some experiments he was doing on photographic plates. In Hale's eyes, benefactors were important people who should not be messed with. Ritchey also chose this time to start his own telescope business and negotiated a rather disadvantageous revised employment contract with Hale which eventually allowed for his dismissal.

6.20   Mental illness

It soon became obvious that Hale was working too hard. He began to suffer from nervousness and an inability to sleep or to concentrate. His doctor advised him

---

[21] Hale to Frost, 25 September 1906, Hale papers, Caltech, quoted by Osterbrock (1993, p. 93).

[22] Quoted by Osterbrock, ibid.

to take a long break. Even before this time he had occasionally found it necessary to go on long holidays to recover from stress. Throughout his life, he had shown a tendency to become depressed, especially during bad wintry weather, but a few months of relaxation had usually been adequate to restore his equilibrium.

In 1909, as he approached his 41st birthday, his mental health showed more serious signs of failure under the pressure of the problems with Hooker and the 100-in. mirror. Reluctantly, he took his doctor's advice and gave up work almost completely. At first he tried his usual remedy of visiting Europe, where he felt well enough to give a public lecture at the Royal Institution in London in May. He travelled through the continent that spring, attending receptions, participating in meetings and giving lectures. He received an honorary doctorate from Oxford.

Back again on Mount Wilson in the summer, he found himself suffering from 'terribly hard dreams' and, in a state of half-sleep, he would try to climb the picture frames on the wall. Following some observational work on Mars in the autumn, he could no longer concentrate and became increasingly nervous. The fourth meeting of his International Union for Cooperation in Solar Research took place on Mount Wilson in January 1910 but he himself was so ill that he could only manage to attend a few social functions. He was fortunately able to accompany Carnegie on a visit to Mount Wilson with his wife and daughter in March 1910 (Fig 6.6). A scientific friend who was present on this occasion saw that he was at the end of his tether and urged him to take a break. Finally he did agree to go on a fishing expedition, accompanied by James Scherer, the President of Throop. The latter, in a letter to a mutual friend, wrote 'The whole thing is a tragedy of which I can speak to you more freely when you get home ... He is badly broken.[23]

Much of the rest of that year he spent in Europe. In December, Sir David Gill, the doyen of British observational astronomers, found him 'looking exceedingly ill and suffering from severe nervous pains and noises in his head, symptoms which demanded rest from all excitement'. An examination by one of the most famous physicians in London, Sir William Osler, could turn up no physical cause for his problems. In 1911, he fulfilled a long-time desire by touring ancient Egypt. On the way back, via the south of France, he wrote to Goodwin on 25 March

> Until I got back from Egypt I was able to read, with pleasure, a great variety of books. But now I can't keep my mind on the subject, as a little demon stands by my side, and every few minutes prods me with the suggestion that, after all, the book is not interesting, and that all my attention belongs to him ... If I could only do a little of my regular work there would be no difficulty. But work excites me and sets the back of my head to aching, and so appears to be out of the question.[24]

He did tell Goodwin in a postscript that 'I can get rid of the 'demon' by having some regular kind of work to do each day—something light and easy,

[23]Scherer to A. Fleming, 27 September 1910, quoted by Wright (1994, p. 260).
[24]Hale to Goodwin, 25 March 1911, quoted by Sheehan and Osterbrock (2000).

FIG. 6.6. Hale with Carnegie in front of the 60-in. telescope on Mount Wilson
in 1910. At this time, he was suffering from a severe depression (Carnegie
Institution of Washington).

such as tracing drawings, writing library cards, or even proof reading ... The
work must keep my mind busy, without taxing it too much'.

In Chicago, where he stayed with his brother, his troubles soon recurred.
Instead of returning to Pasadena, he went back to England, where he found
repose with Hugh Frank Newall, a wealthy astronomer whose house, Madingley
Rise, lay in the countryside next to the Cambridge Observatories. For Hale, it
was a place 'where I could sit quietly in your garden on heavenly English days.
From that time the *demons* slipped into hiding, and have never emerged since

in their original fury' as he wrote to his erstwhile host many years later[25].

When he got back to the United States in June 1911 his own doctor, James H. McBride, told him that his symptoms were the result of twenty-five years of over-intensive work and that he would probably never be able to work so hard again. 'As simple as it may seem to you, as well as you seem to be, I can tell you that your improvement is only on the surface, and that you are now, and will be for some time, living on the edge of a precipice'.[26] He was sufficiently terrified by McBride's words to check into a sanitarium, where he stayed for six months. He was sane enough to be sceptical of the treatment that was meted out to him by Dr John G. Gehring, the 'wizard' who controlled the institution and attempted various quackish cures. On his return home he enjoyed relating his experiences to Ellie Mosgrove:

> 'Close your eyes and lie entirely limp. Your body is quite limp. Your body is quite relaxed and inert. If I lift your hand it falls back dead and motionless. Your legs begin to feel as heavy as lead.' (But they don't, said the patient to himself.) 'Your extremities become somewhat numb and seem to be asleep. You involuntarily take long, deep breaths.' (Not I, said the patient.) And so on and on the voice droned, while Hale lay, convulsed with hilarity, unable to take the whole business seriously. Later he delighted in describing these periods of 'floating on the bosom of the great sub-conscious.'[27]

In spite of the nonsensical side of his treatment, he had recovered enough to work in the mornings on his return to Pasadena in December 1911, after an absence of nearly two years. Even working less hard he had a short relapse in March 1913 and was sent back to Dr Gehring.

In their paper on Hale's illnesses, Sheehan and Osterbrock (2000) diagnose him as a 'Bipolar II' manic-depressive, one of several classes now recognised in this type of illness. Persons with this syndrome suffer from hypomanic and depressed periods. Most of those associated with Hale in his early years could clearly see the manic side of his personality, commenting on his elevated and expansive moods, his unusual energy and restlessness, his decreased need for sleep, and the sharpened and unusual creativity associated with the condition.

6.21   Life and work on Mount Wilson

During his intervals of good health, Hale was as active as ever. The 150-ft tower had been completed in 1912 and he set to work studying the magnetic fields of sunspots with it. It had been known since the discovery in 1843 by the German amateur astronomer Heinrich Schwabe that sunspots appear and disappear over an eleven-year cycle. Usually they occur in pairs. Hale noticed that on the northern hemisphere of the sun, between 1908 and 1911, the first spot of each pair

---

[25]Hale to Newall, 11 November 1931, Hale papers, Caltech, quoted by Sheehan and Osterbrock (2000).

[26]McBride to Hale, 1911, quoted by Wright (1994, p. 270).

[27]Hale to Goodwin, 1911, quoted by Wright (1994, p. 271).

to appear as the sun rotated was a south magnetic pole and its follower was a
north one. In the new cycle, which started in 1912, the opposite behaviour was
seen. This showed that the true solar magnetic cycle is of about twenty-two years
duration, double the length of the basic spot cycle.

To make the most of its facilities, Hale did his best to attract active young
astronomers to Mount Wilson. Walter Adams, already well established, was one
of the first of those interested in spectroscopic astrophysics but others were soon
recruited. Spectra for comparison with the sun's atmospheres were generated in
the observatory's physics laboratory in Pasadena under the direction of Arthur
S. King, who worked there from 1908. The advent of the 60-in. telescope brought
about a change in emphasis from solar to general stellar work and Hale began
to hire astronomers with expertise in this area. In 1914, Harlow Shapley, who
had taken his PhD at Princeton, was among the first of the stellar specialists
to arrive. Using the 60-in. telescope, he was soon to make a second Copernican
revolution by discovering that the sun is not the centre of the Milky Way galaxy.
He was able to find the direction and estimate a distance to the true centre.
Though he found Hale a slightly remote figure, he was impressed by his concern
for others. In his autobiography he mentions the Director's generosity and how
he had been asked to look out for fellow staff members who might be in need
and could be helped anonymously.

Adams was Hale's right-hand man and often, when the Director was ill,
acted on his behalf. He struck Shapley as tough and competent, good at sports,
whether tennis, mountaineering or even billiards. He had a way of encouraging
people without being a slave driver, though he pushed himself hard and expected
others to do the same. Using the 60-in. telescope with a spectrograph, Adams
discovered in 1915 that the binary companion of Sirius, called Sirius B, has a
density that seemed incredibly high compared to other stars. Ralph H. Fowler,
a theoretician of Cambridge, England, showed in 1926, using the newly developed quantum theory, that it is composed of 'degenerate' matter whose atoms
had collapsed under extreme pressure. Later, a famous controversy would erupt
between Eddington and Chandrasekhar over the stability of stars like this one
(see Chapter 7).

One of the most remarkable careers in astronomy was that of Milton Humason
who around 1909 dropped out of high school to became a mule driver on the
Mount Wilson trail. After a year or two, he found other work but in 1917 he
joined the Observatory as a janitor. Soon afterwards he became a 'night assistant'
and began to help the astronomers, such as Shapley, in the taking of photographic
plates. By 1920 he was employed as a staff member. Hale had been reluctant at
first to appoint him because of his lack of a formal education but afterwards came
to appreciate his merits. Humason went on to work closely with Edwin Hubble
with whom he established a good rapport and published several papers. In 1950
his valuable contribution to the redshift-distance question was recognised by the
award of an honorary degree by the University of Lund (Mayall 1973).

In 1915, Hale wrote a small book *Ten Years' Work of a Mountain Observatory*, in which he summed up the accomplishments of Mount Wilson and his ideas as to how an observatory should be run. In line with this thinking, he had set up a well-equipped laboratory and a workshop in Pasadena to support the telescope operations on the mountain. The Santa Barbara Street laboratory contained specialised equipment such as furnaces, electric arcs, giant magnets, and pressure vessels by means of which atoms and molecular substances could be made to emit their characteristic spectra under differing physical conditions. By comparing what was found in the laboratory with observations from the telescopes, the physical environment that gave rise to the spectra of the sun and stars could be deduced. The instrument workshop that he had also set up made precision parts for the telescopes and special machines needed for measuring the photographic plates on which the observations were recorded. In addition, Hale decided that Mount Wilson should have a ruling engine of its own to make the large diffraction gratings needed for state-of-the-art spectrographs.

His own solar discoveries he could justly boast of; but he also cast around for other topics that could be addressed with the equipment now available on Mount Wilson. He had been impressed by the work of J.C. Kapteyn of Groningen on the streaming motions of the stars.[28] Mount Wilson lacked astronomers like him with the theoretical knowledge to interpret the rapidly accumulating results being obtained by its telescopes. Kapteyn, whose Astronomical Laboratory in Groningen had no telescopes of its own, was easily persuaded to visit the Observatory each summer to help guide its work. There was only one small problem: Kapteyn did not like to spend long periods away from his wife and she could not stay with him on the mountain because women were not permitted to reside in the Monastery. In the summer of 1908, Kapteyn's first on Mount Wilson, they lived together in a tent (Hertzsprung-Kapteyn 1928). But Hale so valued his presence that the following year they found a cottage waiting for them (later called the 'Kapteyn Cottage'). Much telescope time was devoted to Kapteyn's programme for investigating the star streaming to greater and greater distances. Hale believed that the streams were of fundamental importance and merited a large investment of telescope time. After the nature of the Milky Way galaxy became better understood through the discoveries of Shapley and many others, the cause of the streaming motions became clearer. In essence, the younger stars rotate smoothly around the centre of the galaxy whereas old ones swarm about it less systematically.

6.22   World War

With the outbreak of the First World War, Hale sought a useful role in promoting America's participation. Shapley and van Maanen were pleased to find that he was not blinded by patriotism but could see that there might be two sides to some of the issues. Strangely enough, during the years 1916–1917, which he spent

---

[28]See also Section 7.2.

in Washington, Hale was thoroughly happy and suffered little from depression, almost as though it was intense cogitation about astronomy that brought about his collapses.

Working through the National Academy of Sciences, of which he was the foreign secretary from 1910 to 1921, he had access to President Wilson and played a big part in mobilising the scientific community to help with the war effort. One of his great ideas was the formation of the National Research Council, into which he roped many leading scientists. For example, Millikan, the Chicago physicist and later Nobel Prize winner, was put to work on anti-submarine warfare.

At the end of the War, Hale proposed setting up an 'Inter-Allied Research Council'. Then he helped to found the International Research Council, which ultimately became the International Council of Scientific Unions (ICSU), which remains today the chief player in promoting international scientific contacts and preserving the freedom of scientific exchange. He also persuaded the astronomer Henry Norris Russell of Princeton to take a leading part in the National Research Council. Russell was a pure scientist in his outlook and usually tried to avoid entanglement in administrative matters, but this exercise increased his power and influence. He was later very useful to Hale in his efforts to raise money for the 200-in. telescope.

6.23   Completion of the 100-in.

Although most of the parts of the 100-in. telescope were well under way soon after the funding had been secured, a perfect primary mirror blank was still lacking. An attempt would clearly have to be made to use the one that had been delivered, however imperfect it was. Ritchey was ordered to get working on it, which he did with great reluctance. Not only had he decided that it would probably not be successful, but he felt that a new wide-field optical design that he had invented in collaboration with the French astronomer, Henri Chrétien, should be used instead of the conventional Cassegrain[29] arrangement that had been standard up to then. Neither Hale nor Adams could be persuaded to take an interest in this scheme, although many late twentieth century instruments have used it. With all the difficulties involved, the grinding and figuring of the giant mirror was to take six years.

Ritchey made no secret of his belief that the 100-in. would be a failure, an attitude which Hale and Adams regarded as disloyalty. The first tests of the 100-in. mirror in the laboratory seemed to confirm Ritchey's gloomy prognostications concerning the quality of the glass but, after experiments that included input from Adams and Hale, it was found that the testing arrangement itself had been faulty. In an atmosphere of increasing paranoia, Hale actually went so far as to prevent Ritchey from being nominated for a medal of the Royal Astronomical Society, by pointing out he was subject to epileptic fits, and was therefore mentally unstable!

[29]In a conventional Cassegrain telescope, the primary mirror is a paraboloid and the secondary a hyperboloid (see Fig 3.3). In the Ritchey–Chrétien, both mirrors are hyperboloidal.

FIG. 6.7. Early photograph of the 100-in. telescope on Mount Wilson. The long
stepladder for access to the bent Cassegrain focus was later replaced by a
motorised seat attached to the telescope pier (SAAO).

The first test of the 100-in. telescope (Fig 6.7) was carried out on the evening
of 1 November 1917. Only the Newtonian arrangement was ready. To the hor-
ror of Hale and Adams, Jupiter appeared as a number of overlapping images.
However, the dome had been open and the mirror had heated up during the
day. They decided to go to bed and return at 3 a.m. when, it was hoped, the
mirror would have regained its equilibrium shape. Neither could sleep and they
each returned to the telescope before the appointed hour. To everybody's intense
relief, the image was then perfect!

As soon as the telescope was complete, Hale and Adams decided together
that Ritchey would have to go. At the age of 55, the great optician found himself
on the street. Totally incompatible with Hale, but an essential component of his

advances in telescope technology and an excellent observer, he was effectively frozen out of the American astronomical scene (Osterbrock 1993).

Edwin Hubble, then a Ph.D. student at Yerkes, was offered a position at Mount Wilson in 1916, before he had completed his thesis. However, the United States entered the first World War in 1917 and Hubble requested Hale to grant him a postponement so that he could volunteer for war service. When he eventually took up his post in 1919, he commenced a wide ranging programme on nebulae with the 100-in. telescope. Within a few years he was able to conclude that the spiral galaxies lie far beyond the Milky Way, and thus to settle a question that had been around for decades. Later, his work on the distances and radial velocities of the galaxies was accomplished with the same telescope. These observations, which he made during the 1920s and 1930s, were to furnish the evidence that the universe is expanding. As a result, he became the most famous of all the Mount Wilson astronomers.

The 100-in. telescope also provided a spectacular confirmation that the ideas of stellar structure then being put forward by Eddington and Russell were on the right track. Michelson had designed a 'stellar interferometer' as an accessory to the 100-in., an optical bench 20 ft long mounted across the top of the telescope, that increased its resolving power. Following a prediction by Eddington that the nearby red supergiant star Betelgeuse should be within its range, Michelson used it to determine the diameter of a star for the first time. The result was well-publicised (DeVorkin 2000) as a demonstration of the power of the new instrument.

Continuing his policy of encouraging theoreticians to visit Mount Wilson, after the World War Hale invited summer visitors such as Russell from Princeton as well as James Hopwood Jeans and others from England. Russell was one of the few theorists in the United States who could be compared with people such as Eddington or Jeans in England and Hale ensured that he came often. His presence was immensely stimulating, if intimidating. Some of the astronomers found him rather overwhelming as he brought forth one new idea after another! Hale, mindful of his own problems concerning over-excitement, felt an obligation to keep him calm and relaxed by providing novels for him to read (DeVorkin 2000)!

For thirty years the 100-in. Hooker telescope remained the most powerful instrument available to astronomy. During this time, Mount Wilson became the Mecca of the astronomical world. The many discoveries made there more than justified the efforts of Hale and the magnificent support that had been provided by private donors and the Carnegie foundation.

6.24   The years as a recluse

In 1922, active as ever with schemes such as planning a building for the National Academy of Sciences and a spectroscopic laboratory for Caltech, Hale had a severe recurrence of his illness. He sought relief by setting off for Europe again. In March 1923 the inevitable had to be faced, and he wrote from Rome to the

Carnegie Institution to resign his directorship of Mount Wilson. In his letter he estimated that, although he was only 55, he had not enjoyed one-third working capacity during the previous sixteen years. He gave details of the many operations he had undergone, which included appendectomy and gall-bladder removal,

> Add to this the daily evidences of congestion of the head, with frequent acute phases; many attacks of hemorrhoids, several of which have kept me in bed for weeks, while one involved a severe operation; and repeated cases of lumbago or similar trouble, and you have a catalog fit to rejoice the soul of a pathologist! He would be still more delighted if it were extended back to include the typhoid fever, repeated dysentery, colitis, and other difficulties of earlier days ...
>
> Congestion of my head is caused chiefly by worry, excitement, responsibility, discussion of any scientific subject, attendance at scientific meetings, lecturing (mainly because of the defective memory of faces and inability to recall names—even when well-known—when needed), and continued mental work.[30]

He nevertheless took the opportunity to visit Arcetri observatory outside Florence, where he persuaded the Director, Giorgio Abetti, a former visitor to Mount Wilson, to mount one of Galileo's original telescopes so that he could observe Jupiter through it.

Following his resignation, the Carnegie Institution found a new and less taxing role for him. He was made 'Honorary Director' of Mount Wilson while Adams became the actual Director, both at salaries of $8000 per year. Meeting people had become too much for him and thereafter he became a semi-recluse. Though during the following year (1924) he attended the dedication of a new building that he had done much to realise for the National Academy of Sciences in Washington, he was not able to take an active part in the proceedings. The Academy passed a unanimous resolution asking him to sit for a portrait 'as a permanent memorial and an adornment of the walls of the fine building which it owes in large measure to his unselfish and untiring efforts in furthering the material and intellectual interests of the Academy these many years ...'[31]

6.25 The Hale Solar Observatory

Using for the most part his own funds, he had a small, tastefully designed, solar observatory built near the Huntington Library, (see below) where he could work quietly on his own, away from all disturbances, including apparently his wife. His main preoccupation during his later years was to try to detect the general magnetic field of the sun which, as mentioned, he never succeeded in doing. The Observatory contained a laboratory and a machine shop, where he could make mechanical items. He evidently thought of himself as an amateur

---

[30] Hale to Merriam, 29 Mar 1923; CIW, Mt Wilson Papers, f. General, 1902–30, no. 2, quoted by Christianson (1995, p. 169).

[31] Michelson to Hale, 27 April 1927, quoted by Wright (1994, p. 316).

FIG. 6.8. Hale operating a spectroheliograph in his private observatory, the 'Hale Solar Observatory', in San Marino, California (Carnegie Institute of Washington).

once again, following the tradition of Huggins and others. In many ways, the 'Hale Solar Observatory' was a reincarnation of Kenwood. Its main instrument was a spectroheliograph (Fig 6.8), partly buried underground, as in his tower telescopes.[32] It was later used by the Mount Wilson astronomers Harold D. and Horace W. Babcock to develop the solar magnetograph, with which the general magnetic field of the sun was finally detected in 1952.

[32]The observatory was donated, on completion, to the Carnegie Institution.

The Hale Solar Observatory contained a large, comfortable library where Hale now spent most of his time quietly reading. Throughout his life, he had enjoyed novels, biographies and scientific histories. As an adult, he had never been religious and had developed a dislike of dogmatism. He once told his wife 'Of course you must see that it is hard—really impossible—for me to reason one way through the week, and another way on Sunday. My creed is truth, wherever it may lead, and I believe that no creed is finer than this'. As for the children: 'Probably the best way is to have them learn that there is a fine underlying idea which they should value, but which does not require them to believe the many absurd doctrines of the church'.[33] He expressed the deist view 'Every new fact observed and every underlying law formulated enlarges the known scope of the Creator's powers'.[34]

In his later years, he was continually frustrated by his inability to concentrate. To Goodwin he wrote (1926) 'My bursted old head is the bane of my existence, always preventing me from doing what I want ... after a period of apparent improvement, the very sight of my new spectrograph being assembled would set it boiling for the day'.[35]

He wrote a number of popular books during this period. Research ideas had never come easily. In a letter to Edwin Frost (Yerkes) he explained how the urgency of administrative matters had made it difficult for him to think coherently about scientific matters. This was why he had rejected the offers of administrative posts in his earlier career.

> I am convinced that I can accomplish comparatively little in research without devoting my *entire* thought to it. New ideas come to me very slowly, and only as the result of continual thinking in and out of working hours. If I have other things on my mind, especially journal work, which must be completed by a certain date, I am so much disturbed that I make no headway. As I look back upon my record, I find that I accomplished nothing whatever in solar research during the entire period of the organisation of the Yerkes Observatory which was not altogether completed when I came to California.[36]

Evelina did not find it easy to share her husbands quiet and contemplative way of life. Her children were by now out of the house but she still liked being surrounded by activity and meeting other people. Increasingly left to her own devices, she started to build a new life for herself by developing new interests. She busied herself with civic affairs and worked with institutions such as the Pasadena Hospital. She helped to support cultural groups such as local musicians and the Pasadena Playhouse. As time went on, Hale and she seem to have increasingly gone their own ways.

[33] Hale to E.C. Hale, 19 April 1909, quoted by Wright (1994, p. 363).
[34] Hale, unpublished, quoted by Wright (1994, p. 363).
[35] Hale to Goodwin, 7 January 1925, quoted by Wright (1994, p. 364).
[36] Hale to Frost, 3 December 1926, quoted by Wright (1994, p. 243).

6.26   Continued public service

In spite of his semi-retirement, Hale was still putting to use his talent for fund-raising and persuading millionaires to part with their riches. To complete the set of cultural institutions that he had established in Pasadena, he talked Henry Huntington Jr into leaving his art collection and library to the public as his monument, with a suitable endowment of $10,000,000. The Huntington Library, Gallery and Gardens which resulted are located in San Marino, next to Pasadena. They contain priceless collections of paintings, incunabula and manuscripts. Together with Mount Wilson Observatory and the California Institute of Technology, they form a trio of great California institutions that owe their existence to Hale.

He also had a scheme for a 'National Science Endowment', which though it ultimately failed, gave him the opportunity to badger yet another millionaire. The future US President, Herbert Hoover, then Secretary of Commerce, was Chairman of the intended fund. Julius Rosenwald, the President of the chain store Sears Roebuck, was one of those approached for a contribution. In a letter to Goodwin, Hale gleefully recounted his experience. Rosenwald started off by saying that he did not believe in endowments *at all* and admitted that he was a well-known crank on the subject. However, Hale knew that he had just given $3,000,000 for an industrial museum in Chicago, and steered the conversation towards that subject. Soon they were on the best of terms and Rosenwald gave him a book which Hale told Goodwin he would treasure 'as a memento of an uphill struggle'. He eventually presented Rosenwald with a letter from Hoover that suggested a $10,000 annual contribution. This was more in line with Rosenwald's ideas. 'I saw my finish and sparred for wind'. Hale built up an imposing picture of the fund and made 'ten thousand dollars a year look like thirty cents'. Summoning up his courage he decided to ask Rosenwald for $100,000 per year. At first he demurred, but Hale pressed him to set a good example so as to encourage other donors. He feared he had gone too far. Then Rosenwald suddenly said he might agree to be one of five to give such a gift. Hale promptly took him up on it and 'without losing a second got away before anything else could happen'.[37]

6.27   The 200-in. Palomar reflector

Hale still enjoyed writing popular articles. In 1928, he wrote for *Harper's Magazine* about 'Possibilities for Large Telescopes'. He showed how 'Each expedition into remoter space [i.e. by looking through larger and larger telescopes] has made new discoveries and brought back permanent additions to our knowledge of the heavens ... ' He mentioned the great work then being done by Hubble and others on 'Island Universes', stressing how these studies were only just beginning. He praised the great benefactors of American astronomy—Lick, Yerkes, Hooker, and

[37]Hale to Goodwin, 31 January 1926; quoted by Wright (1994, pp. 366–367).

Carnegie—and hoped that a successor could be found to carry on the tradition by financing even larger instruments.

As soon as he had a galley proof of his article [February 1928], he sent a copy to Wickliffe Rose of the Rockefeller Foundation, with a request for a grant to investigate how large a mirror could, in fact, be made. Rose responded favourably and Hale went East to see him. He took up Hale's suggestion of a 200-in. telescope, but it was clear that it could not be given to the Carnegie Institution, a rival organisation. Instead, it was proposed that Caltech should be the recipient. Still, the cooperation of the Carnegie scientists was necessary if the project was to be a success. John C. Merriam, the President of the Carnegie Institution, was the chief stumbling block. The now aged and ailing Elihu Root, Chairman of the Carnegie Trustees, was favourable, saying that the Carnegie Institution could furnish the brains if the Rockefellers could provide the funds! Hale and Root used all their persuasive powers, and Merriam was eventually won around.

According to Woodbury (1940), the crucial meeting with Rose went like this:

> "How large a telescope do you contemplate?" Rose asked.
> "Two hundred inches."
> "And its cost?"
> "Six million dollars."
> Rose did not even blink. "Dr. Hale," he said, "I think you can count on Rockefeller support for the whole sum."
> *"For the whole thing?"*
> "For the whole thing."
> Hale was not a demonstrative man ordinarily, but he burst out now with his pet quotation from Jules Verne's *Journey to the Moon.*
> " 'A frightful cry was heard,' " he shouted gleefully, " 'and the unfortunate man disappeared into the telescope!' "
> Then he seized the startled Dr Rose's hand.

The Rockefeller Foundation had agreed to provide the funds ($6,000,000)—but only for building the instrument, and on condition that Caltech would endow the operating costs. Hale's powers of persuasion were required once again. With extreme generosity, Henry M. Robinson, one of Caltech's Trustees, came forward with the necessary money. Hale privately thanked Root for his help, comparing him to his own father who had in his time done so much to keep his plans moving along.

There was considerable opposition from Shapley and other observatory directors. 'Shapley described the channeling of so much money into "the small autocracy in Pasadena" as "criminal" and said that Hale's success was "very little short of embezzlement!"' (DeVorkin 2000). Shapley's jaundiced remarks were a reflection of the fact that, at the time, he was trying to persuade the Rockefeller foundation to support his own expansion plans in South Africa and was only induced to keep quiet by a promise from Russell that he would promote his (Shapley's) own application for funds.

Hale was now sixty and his health was more precarious than ever. The previous year he had been recommended to Dr Riggs' sanitarium in Stockbridge, Massachusetts, where he found the regime more scientific. As the 200-in. project got under way, his energy revived and he became the chairman of an 'Observatory Council', founded to coordinate the planning for the new instrument, which was to be located at Mount Palomar. This was a darker site than Mount Wilson, which by then suffered seriously from light pollution generated by the nearby city of Los Angeles. At first, it was hoped that the mirrors could be made from fused quartz, which promised to be much less sensitive to changes in the air temperature than the traditional plate glass that the main mirror of the 100-in. had been made of, but the necessary experiments at the General Electic Company dragged on, using up too much money, and the idea had to be abandoned. The contract was given instead to the Corning Glass Works, and the mirror was to be made of 'Pyrex', a glass with better properties than the ordinary plate that had been used for previous reflectors, but not as desirable as fused quartz.

In 1929, Hale published details of a new instrument—his 'spectrohelioscope', a version of the spectroheliograph that enabled an image of the sun in a single spectral line to be built up many times each second, so that it could be viewed by the eye directly, taking advantage of its slow reaction time ('persistence of vision') to hide the piecewise construction of the image. This instrument could also be used to make continuous movies of the sun's surface.

6.28   Slow decline

In 1931, Hale was elected President of the International Council of Scientific Unions, and was still well enough to attend its meetings in Europe, though he found them very tiring. However, his blood pressure became too high in 1932 and he started to suffer from nosebleeds and dizziness. He fell a victim to nightmares and again had to stay at the Riggs sanitarium. In 1932, he experienced a severe attack of depression and in 1933 he went to stay with Newall in England, which seemed to be a certain, if temporary, curative method. Honours, such as the Copley Medal of the Royal Society (1932) continued to arrive. He made a quick trip to Corning, in upstate New York, to see how the 200-in. mirror project was getting along. The first attempt at casting had been a failure but success was achieved on 2 December 1934. The finished blank arrived in Pasadena for grinding at Easter, 1936. Hale was just well enough to go to look at it: he marvelled that the Cassegrain hole in the centre of the disc was the same diameter as the Yerkes lens had been—40 in.

That year, Harlow Shapley organised a special symposium in his honour, but he was too weak to attend. On this occasion many of Hale's friends and admirers spoke of him and his work. The *New York Times*, in an editorial on the event, referred to Hale as a 'Priest of the Sun' and a 'Zoroaster of our time'. In the summer he set off for another trip to England, but on his way he suffered a stroke from which he recuperated with his brother in Chicago. On his return to Pasadena, he spent some time in a sanitarium. His wife had in the meantime moved into an apartment, where Hale was able to join her for a short time. Unfortunately, his condition soon deteriorated and he had to spend the last year and a half of his life in a nursing home, where he died on 21 February 1938.

The 200-in. telescope was dedicated on 3 June 1948. Watched by a bronze bust of Hale, Lee DuBridge, the President of Caltech, asked Mrs Hale to step forward as he named it in her husband's honour. In a report of the occasion[38] she was described as 'white-haired and wan'. She managed a barely audible thank you before resuming her seat, where she touched a handkerchief to her eyes. In spite of her apparently delicate constitution Evelina Hale lived on in Pasadena to the age of 99, dying only in July 1967.

References

Carnegie, Andrew, 1889. Wealth, *N. Am. Rev.*, **148**, 653–684.

Curzon, Robert, 1849. *Visits to Monasteries in the Levant*, John Murray, London.

Christianson, Gale E., 1995. *Edwin Hubble, Mariner of the Nebulae*, University of Chicago Press, Chicago, IL.

DeVorkin, David H., 2000. *Henry Norris Russell, Dean of American Astronomers*, Princeton University Press, Princeton, NJ.

Franch, John, 1997. Charles Tyson Yerkes 1837–1905, *University of Chicago Alumni Magazine*, February 1997.

Hale, G.E., 1908. *The Study of Stellar Evolution, An Account of Some Recent Methods of Astrophysical Research*, University of Chicago Press, Chicago, IL.

Hale, G.E., 1915. *Ten Years' Work of a Mountain Observatory*, Carnegie Institution, Washington DC.

Hertzsprung-Kapteyn, H., 1928. *J.C. Kapteyn, Zijn Leven en Werken*, P. Noordhoff, Groningen.

Jeans, Sir J.H., 1938. *Nature*, **141**, 502. ©*Nature* (1938).

Mayall, N.U., 1973. Milton L. Humason—some personal recollections, *Mercury*, **2**, (1), 3–8.

Osterbrock, Donald E., 1993. *Pauper and Prince: Ritchey, Hale, and Big American Telescopes*, University of Arizona Press, Tucson, AZ.

Osterbrock, D.E., Gustafson, J. and Unruh, W.S. 1988. *Eye on the Sky - Lick Observatory's First Century*, University of California Press, Berkeley, CA.

Sheehan W. and Osterbrock, D.E., 2000. Hale's 'Little Elf'; The mental breakdowns of George Ellery Hale, *J. Hist Astr.*, **31**, 93–114.

Woodbury, David O., 1940. *The Glass Giant of Palomar*, William Heinemann Ltd., London.

Wright, Helen, 1994. *Explorer of the Universe: A Biography of George Ellery Hale*, American Institute of Physics, New York. (History of Modern Physics and Astronomy, Vol 14; reprint of original edition of 1966.)

[38]Quoted by Christianson (1995, p. 314)

7

ARTHUR EDDINGTON: INSIDE THE STARS

In the application of mathematics to the study of natural phenomena, it is necessary to treat, not the actual objects of nature, but idealised systems with a few well-defined properties. It is a matter for the judgement of the investigator, which of the natural properties shall be retained in his ideal problem, and which shall be cast aside as unimportant details; he is seldom able to give a strict proof that the things he neglects are unessential, but by a kind of instinct or by gradual experience he decides (sometimes erroneously it may be) how far his representation is sufficient.[1]

Arthur Stanley Eddington was born on 28 December 1882 in Kendal, in the Lake District, where his father was the headmaster of a Quaker school called Stramongate. Although this school no longer exists, it had an interesting history, being associated with another great Quaker scientist, John Dalton, the originator of the atomic theory of matter, who had also been headmaster there in the late eighteenth century. Arthur Henry Eddington, father of A.S., died at the age of 34 during a typhoid epidemic only two years after his son's birth. His mother, born Sarah Ann Shout, found herself with Arthur and his sister, Winifred, to support. They migrated to Weston-super-Mare in Somerset, to the home of Mrs Rachel Eddington, A.H.'s mother. Fortunately, they had sufficient income to support a simple but comfortable existence.

Eddington showed precocity by learning the 24×24 table before he could read. He liked counting up to large numbers. Once he started to count all the letters in the Bible and got to the end of Genesis before giving up. It is also said that he tried to count the stars from the seashore. His fascination with large numbers remained with him all his life. Although short-sighted, he only began to wear glasses at twelve years of age. He told his student R.O. Redman that until then a tree had just been an amorphous mass but now took on an entirely new significance with its structure of twigs and leaves.

His early education was at home and then at a small preparatory school. He was lent a 3-in. telescope by the headmaster. From 1893 to 1898 he attended Brynmelyn school in Weston. Later, he wrote that his teachers had stimulated his interest in literature, natural history, mathematics, and physics. 'it was the personality and enthusiasm of the master that illumined the field of English literature for us. He not merely opened the door; he swept us through with him'. (Douglas 1956, p. 3). His eyesight could not have been so very bad, as he played cricket and soccer for the school teams and had already begun to

---

[1] Eddington, *Stellar Movements and the Structure of the Universe*, 1914, p. 201.

show enthusiasm for cycling. He met Sir Robert Ball, Professor of Astronomy at Cambridge and popular lecturer, when he visited the school. He received a typical warning—most astronomers have heard it—that the astronomical life is not an easy one!

At the age of 15 he won a scholarship from the Somersetshire County Council that enabled him to go to Owen's College in Manchester, part of the Victoria University. He stayed there at Dalton Hall, a university residence, and even played football as a member of its second team! He took the honours course in physics and had lectures from Professors Arthur Schuster, who had worked as a young man with Bunsen and Kirchhoff, and Horace Lamb, famous for his textbooks on mechanics and hydrodynamics. As top of his classes in mathematical and some humanities subjects he won further scholarships. In 1901 he completed his degree, obtaining first place. At the close of that year he won a scholarship to study Natural Science at Trinity College, Cambridge, to which he moved in 1902.

7.1 Trinity College, Cambridge

Trinity College had, since the days of Newton, become the college of choice for students of science. Eddington's main subject was, in fact, mathematics. It was then the custom at Cambridge to have private lessons on problem-solving with a specialist coach, in this case R.A. Herman, who helped candidates to prepare for the highly competitive 'Tripos', effectively the honors degree in the subject. Herman not only taught Eddington how to answer the kind of questions likely to be asked in the Tripos, but stressed mathematical style and elegance. The individual results in the examinations were then listed in order, and it was a great honour to become 'Senior Wrangler', i.e. to obtain first place, or to reach one of the other high places on the list. Among the lecturers whose courses Eddington attended were A.N. Whitehead, author with the philosopher Bertrand Russell of *Principia Mathematica*, and E.T. Whittaker, a mathematical physicist and later the writer of a history of electricity and magnetism. Whittaker, who outlived Eddington, was to champion some of the wilder speculations that he made towards the end of his life.

He was a quiet and reserved student, though he took part in tennis, bowls, and cycling. In Cambridge he also met C.J.A. Trimble, who was to be his closest friend. They had

> an undemanding, happy, almost casual relationship which had yet a rather rare quality of mutual understanding ...In general outlook they had much in common and since both enjoyed long strenuous walks and climbs, many holidays were spent together in such pursuits. With this one friend Eddington could throw off all the hesitant diffidence which formed an almost impenetrable barrier to intimacy with others. With this friend he was often gay, light-hearted and full of fun, revealing a side of his nature which his other associates rarely or never saw; and this was as true in later years as in these student days. (Douglas 1956, p. 7)

Trimble was a mathematics student and a Foundation Scholar. He also received a scholarship for sacred music! He was fourth Wrangler in 1905. On graduation, he became a civil servant for a few years and then, in 1910, turned to teaching mathematics at his old school Christ's Hospital. He continued as a mathematics teacher until he retired. Eddington and he remained good friends for life, often going on cycling or hiking holidays together.

Eddington took the Tripos examination in 1904 and emerged as Senior Wrangler, an unprecedented achievement for a second-year student. He was given ovations at Brynmelyn school and Dalton Hall. He paid tribute to his former Manchester mechanics lecturer in typical Eddington fashion: 'While he now knew what it was to be treated as something of a lion his ambition was to become something of a Lamb!' (Douglas 1956, p. 11). In the following year he carried on his studies—a minimum period of three years was necessary before he could receive his B.A.

In the summer of 1905, he took a job as tutor to the son of W.E. Wilson, an Irish landowner and amateur astronomer of great ability who owned a 24-in. reflecting telescope and had taken some of the best early photographs of spiral nebulae. He played tennis, fished, and made expeditions to the nearby 'beauty spots'. On his return to Cambridge that autumn, Eddington started to prepare for the Fellowship examinations and took some private pupils as well as lecturing to first-year engineering students on spherical trigonometry—lectures which one participant described as 'dull'. He also spent some time in the Cavendish Laboratory in trying to determine the velocity with which electrons left the surface of incandescent metals, but was soon disheartened by the problems of experimental research.

7.2   Royal Greenwich Observatory; Kapteyn's 'Star Drifts'

In January 1906, Eddington received a letter from the Astronomer Royal, Sir William Christie, offering to nominate him as Chief Assistant at the Royal Greenwich Observatory. Encouraged by Whittaker, he accepted the post and by mid-February he was working there. In compliance with the staff rules of the Observatory, he had to live within a mile of his work and he took lodgings locally. He was elected a Fellow of the Royal Astronomical Society in April of the same year. In 1907, he became a member of the R.A.S. Club, a kind of 'Inner Party' of the Society that dines together after each meeting. He was soon put to work on routine programmes such as 'checking the places of 12,000 stars for the new catalogue', which can hardly have been the most stimulating of tasks.

One of his first research programmes concerned the 'star drifts' discovered by the Dutch astronomer Jacobus Kapteyn. Kapteyn, the Professor of Astronomy at Groningen, had made his name largely through analysing plates taken by Sir David Gill for one of the first photographic sky surveys, the *Cape Photographic Durchmusterung*. From the small lateral motions of the nearby stars against the more distant background and the increasing availability of measurements of their velocities along the line-of-sight, he had shown that those in our celestial neighbourhood seem to be streaming towards two particular directions in the sky. Eddington was able to confirm these results from Greenwich observations. This work was the basis of his first contribution to the *Monthly Notices of the Royal Astronomical Society*, called *The Systematic Motions of the Stars*, dated 1906. He visited Kapteyn near his home in Groningen to discuss this work, which came to dominate his research for the next eight or nine years.

Meanwhile, the star-drift phenomenon had also attracted the attention of the German astronomer, Karl Schwarzschild. He, in fact, was able to derive a better mathematical representation of it. Schwarzschild, perhaps the greatest German astronomer of his generation, shared some of Eddington's later interests also, such as general relativity and stellar structure, and his early death from an illness contracted while serving in the First World War was felt by the latter as a personal loss.

Some years were to elapse before the explanation of star streaming was found, following the work by Shapley (see Chapter 8) and others who showed that the Milky Way, of which we form a part, is a rotating spiral galaxy. Nevertheless, in his studies of the phenomenon, Eddington had been led to think deeply about the movements ('kinematics') of the stars and what caused them. This work led him into the field of 'stellar dynamics', whose foundations he may be said to have laid.

In 1909, Eddington lectured at the Royal Institution on *Some Recent Results of Astronomical Research*. By then it had been the main venue for popular science lectures in London for over a hundred years and it was without doubt an honour, as well as a sign of growing recognition, for the twenty-six-year old Eddington to have been asked to speak there. Nevertheless, one has the impression that he was just 'jogging along', enjoying the opportunities for travel—Dublin, Malta, Winnipeg—that the job brought. He had not yet made much use of his ability.

His Director, Sir William Christie, was succeeded as Astronomer Royal in 1910 by Frank Watson Dyson, a person with whom Eddington had a good relationship. Dyson had also been a scholarship boy from a non-conformist family, in his case a Baptist one. In 1912 he asked Eddington to lead an expedition to observe a solar eclipse in Brazil which, though unsuccessful on account of the weather, gave him some valuable experience which he was able to draw upon seven years later when seeking an observational confirmation of the General Theory of Relativity.

7.3 Professor at Cambridge

In 1913, the Plumian Professorship of Astronomy at Cambridge, one of the prestigious ancient chairs of the University, fell vacant. The electors decided to appoint Eddington, a theoretician, rather than the 'First Assistant' at the Cambridge Observatory, Arthur R. Hinks, who was an observational astronomer. Soon afterwards, Sir Robert Ball, Director of the Cambridge Observatory, died and Eddington succeeded also to his post. With his mother and sister he moved into

the Director's wing of the neo-classical Observatory building, where he continued
to live until the end of his life.

Although as a university lecturer, Eddington was not inspiring, he made a
tremendous impression on those with whom he had close contact. R.O. Redman,
later himself a Professor of Astronomy at Cambridge, attended his courses in
1925:

> Already acquainted with his early books, I was a trifle disappointed at
> his rather hesitant manner when lecturing, but I found little to criticise
> in his matter, at a time when I put most Cambridge mathematicians into
> one of two classes, (a) the lucid but incredibly dull, (b) the enthusias-
> tic but incomprehensible. Eddington appeared never entirely at his ease
> when addressing an audience, but seemed most at home when speaking in
> the familiar and friendly atmosphere of the R.A.S. [Royal Astronomical
> Society] ... On the biggest occasions his speech became more deliberate
> and nearer his style of writing, probably as a result of very painstaking
> preparation. (Redman 1945)

Fred Hoyle had a similar experience a decade later and described him as
being paradoxically 'among the world's worst lecturers and yet ... one of the
best ... As far as I could tell, he began in mid-sentence and stopped at the end
of the hour, without any full stops in between. He drifted along from one subject
to another, never finishing anything. Nor did he write any too clearly on the
blackboard'. The paradoxical thing was that 'You remembered and thought a
lot about the big issues he raised, long after you'd forgotten apparently much
better presented lectures from others' (Hoyle 1994).

7.4 'Stellar Movements and the Structure of the Universe'

As mentioned, in his last year at Greenwich, Eddington had devoted himself to
stellar dynamics. After his arrival in Cambridge his first book, *Stellar Movements
and the Structure of the Universe*, was published (Eddington 1914), summarising
recent work by himself and others such as Kapteyn and Schwarzschild. In was
his first great achievement, the perceptive summary of a new field.

Besides the discovery of star drifts by Kapteyn, two other major observational
advances had occurred as the nineteenth century ended and the twentieth began.
The first was the large-scale classification of stellar spectra by a group of ladies
(Williamina Fleming, Antonia Maury, and Annie Jump Cannon) at Harvard
under the direction of Edward C. Pickering. They assigned 'spectral types' to
stars according to their dominant lines and bands. Although a scheme invented by
Father Secchi had for some time been widely accepted, the new Harvard classes
O–B–A–F–G–K–M–R–N gave a better description and soon became standard.[2]

---

[2]One remembers the order of these classes by the mnemonic 'Oh Be A Fine Girl, Kiss Me
Right Now'. R and N are carbon star classes and an extra S class (Smack!) was added later.
The sequence O to M, in the (revised) order given, is now known to be one of decreasing
temperature.

FIG. 7.1. Group photographed at the Royal Observatory, Cape of Good Hope,
in July 1914. Sitting, left to right: Sir F.W. Dyson (Astronomer Royal), S.S.
Hough (HM Astronomer at the Cape), and A.S. Eddington. In the back row
are members of the Observatory staff, with J.K.E. Halm in the middle. The
black man sitting on the ground was a Krooman, a seaman from the west
African coast around Sierra Leone (SAAO).

By 1911, Kapteyn had concluded that a star's spectral type strongly influenced which of the two star drifts it belonged to. As one proceeds along the
Harvard O to M classes the average velocities of the stars increase, as do their
proper motions. What could be the cause of this trend? A theory had been put
forward by J.K.E. Halm, Chief Assistant at the Royal Observatory, Cape of Good
Hope (Fig 7.1), that stars might behave like the frequently colliding particles of

a gas and obey the law of 'equipartition of energy', which would require heavy ones to move slowly and light ones to move quickly. In his book, Eddington was able to demonstrate that this could not be the case. He showed that there is no real resemblance between a system of stars and the particles of a gas. In essence, their density is far too low for collisions: '*the stars describe paths under the general attraction of the stellar system without interfering with one another*' (Eddington's italics). This was one of the first major results in the new field of 'Stellar Dynamics'. The explanation of Kapteyn's observation had clearly to be sought elsewhere.

The second important observational development in the new century besides star drifts was that the distances to many nearby stars were becoming available. This was the result of laborious trigonometric parallax surveys made with the aid of photographic plates and precision measuring machines. As a result, the *absolute magnitudes* or *luminosities* i.e. the power in watts[3], of a large sample of stars became known.

7.5   The Hertzsprung–Russell diagram

A paradox soon appeared: though the velocities increased in a systematic way from O- to M-type stars, their average distances did not. Going from O-types, they at first increased but by the middle of the sequence they turned around and decreased again. The clue to the resolution of this strange behaviour was to come through a new way of presenting the data: the Hertzsprung–Russell diagram.

In 1911 the Dane Ejnar Hertzsprung, then working in Potsdam with Schwarzschild, and in 1913 the American Henry Norris Russell of Princeton, independently used the newly emerging data to plot diagrams showing the absolute magnitudes of the nearby stars versus their spectral types.[4] This type of plot ultimately became known as the Hertzsprung–Russell or HR diagram. Crude though the first diagrams were, they revealed that stars could be further divided into 'giant' and 'dwarf' classes even though they might have similar spectra (Nielsen 1963). Furthermore, the differences between the two new classes were seen to be much greater for K and M stars than for O and B types.

It now became evident that the paradox described at the start of this section had arisen from the way the K and M stars had been selected. The radial velocity sample was dominated by the dwarfs, while the average distance sample had been affected by a strong admixture of giants.

In his book, Eddington reproduced Russell's diagram (Fig 7.2), which was then quite new. It was now quite obvious that stars were limited to certain ranges of temperature and luminosity. The diagonal region occupied by most of them seemed to define a track along which they might be evolving. His interest in the Hertzsprung–Russell diagram was destined to play a highly important role in his

---

[3]The absolute magnitude of a star is a way of expressing its power output or wattage. Formally, it is $-2.5 \times \log(\text{power output}) + \text{constant}$.

[4]Strictly speaking, Hertzsprung used a related quantity, 'effective wavelength'.

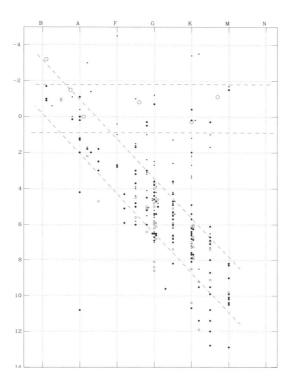

Fig. 7.2. Russell's (1914) version of the Hertzsprung–Russell diagram for stars
   near the sun (re-plotted for clarity). Their absolute magnitudes (luminosities)
   were plotted vertically and their spectral types horizontally. The horizontal
   dashed lines (added by Eddington) enclose the giants and the diagonal lines
   the dwarfs. The large open symbols are averages over several stars; the large
   and small solid points relate to high- and low-accuracy distances and the
   open points are those with only a single observation.

research on stellar structure. It is rather typical of his perspicuity that he was
one of the first people to have recognised its importance.

7.6   Eddington's 'physical intuition'

Eddington was to become famous for his 'physical intuition', the faculty that
enabled Galileo to cast aside ancient ways of thought and see what factors were
the truly important ones when making theories about the behaviour of matter.
Besides his comments on the Hertzsprung–Russell diagram there are some other
examples in *Stellar Movements* ... of his remarkable insight, concerning what
were then called 'spiral nebulae' and are now known to be galaxies.

It must be admitted that direct evidence is entirely lacking as to whether these bodies are within or without the stellar system ... the [second possibility] is that, lying altogether outside our system, those that happen to be at low galactic latitudes are blotted out by great tracts of absorbing matter similar to those which form the dark spaces of the Milky Way.

If the spiral nebulae are within the stellar system, we have no notion of what their nature may be. That hypothesis leads to a full stop. ...

If, however, it is assumed that these nebulae are external to the stellar system, that they are in fact systems coequal with our own, we have at least an hypothesis which can be followed up ... For this reason the 'island universe' theory is much to be preferred as a working hypothesis; and its consequences are so helpful as to suggest a distinct probability of its truth.

If each spiral nebula is a stellar system, it follows that our own system is a spiral nebula ... There is one nebula seen edgewise ... which makes an excellent model of our system ... (Eddington 1914)

Curiously, he also emphasizes a fact about globular clusters[5], first noted by Hinks (1911), who had been the other candidate for the Plumian chair in 1913:

The distribution of these globular clusters in the sky is very remarkable; they are to be found almost exclusively in one hemisphere of the sky, the pole of which is in the galactic plane in galactic longitude 300°. This result ... is clearly of great significance; but it does not seem possible at present to attempt any explanation of it.

It was this asymmetry in the distribution of globulars that, a few years later, led Shapley to realise that the sun is far displaced from the centre of our galaxy. Although Eddington recognised that a general rotation of the Milky Way was necessary if it was not to collapse under gravity, the general picture that he drew of the overall structure of the universe was an antiquated one, resembling that of William Herschel, and fated to be completely overthrown by Shapley only a few years later. In Eddington's view, the Sun was still the centre of a universe which consisted mainly of the Milky Way (see Fig 7.3). Yet the book obviously contained the germs of many important later developments.

7.7   Prophet of relativity

Einstein invented his special theory of relativity in 1905 and worked on its generalisation for many years afterwards. Galileo had dealt with the observations made by two observers moving with constant speed relative to one other but Einstein broadened his result into a more general one. He showed that time appeared to slow down and the length of bodies appeared to get smaller as the two observers' relative speed approached that of light.

In a paper *On the influence of gravity on the propagation of light* (Einstein 1911), he derived a formula for the deflection by gravity of the light rays from

[5]Globular clusters are very dense clusters of old stars with spherical symmetry. There are only about 150 in the whole Milky Way galaxy.

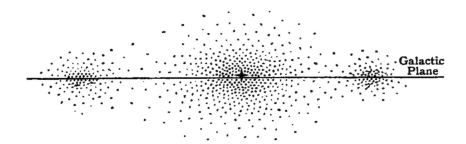

FIG. 7.3. Eddington's idea of the cross-section of the Universe given in his 1914
book, *Stellar Movements and the Structure of the Universe*. The sun is at the
centre of the system and is marked by a cross symbol in the diagram. A few
years later, Shapley's discovery completely altered this simple picture.

a star near the edge of the sun.[6] Effectively, Einstein treated light quanta as
though they were a stream of Newtonian corpuscles possessing mass, according
to his famous equation $E = mc^2$. Using the data available at the time he found
that the angle of deflection should amount to about 0.83 arcsec.[7] He made an
urgent appeal for astronomers to check his theory by making observations during
a solar eclipse (the only circumstance under which stars could be seen near the
sun).

In response, an application for funds to the Joint Eclipse Committee of the
Royal Society and the Royal Astronomical Society was made in 1912 by Erwin
Findlay-Freundlich, a friend and future colleague of Einstein's, for an observation to be made during an eclipse that was to occur in eastern Europe in 1914
(McCrea 1991). Dyson was chairman and Eddington a member of this committee
which, however, turned down the request for lack of funds, though the idea was
clearly not forgotten by them. Findlay-Freundlich did eventually succeed in raising money from the industrialist Krupp von Bohlen und Halbach and the chemist
Emil Fischer. Unfortunately, while he was in the Crimea in 1914 preparing for
the event, the First World War broke out and he was arrested as an enemy alien.
He was unable to make the measurement and was later released in an exchange
of prisoners (Clark 1971). It is indeed ironic that, had he been able to carry out
his observation, he would have found that Einstein's first prediction was wrong.

In 1915 Einstein published his General Theory of Relativity in the form of
papers read before the Berlin Academy of Sciences. One of the consequences
of his new theory was that the deflection of the position of a star by the sun
would be *twice* the amount of his previous prediction based on special relativity

---

[6]Interestingly enough, this effect had been predicted, though largely forgotten about, as
early as 1804, by Johann Georg von Soldner, based on the corpuscular theory of light (see
Soldner 1921).

[7]One arcsec is one part in 3600 of a degree.

(Einstein 1915).

Because of the war, scientific journals from Germany were not reaching England. However, the astronomer Willem de Sitter of Leiden, by virtue of Holland's neutrality, had access to Einstein's work. He recognised its fundamental importance and passed on copies of Einstein's papers with a summary of his own to the Royal Astronomical Society, where they fell into the hands of Eddington, then its Secretary. Eddington immediately comprehended the new theory. His mathematical background enabled him to understand Einstein's calculations, which made use of the new and unfamiliar Tensor Calculus of the Italian mathematicians Ricci and Levi-Civita. He prepared de Sitter's papers for publication.

Even though ordinary, or 'classical' mechanics, derived from Newton's laws, is a satisfactory theory for nearly all normal observations, it was realised that there are three situations in which extreme conditions lead to effects in which general relativity gives answers that are slightly different from the classical ones:

1. The fact that the axis of the elliptical orbit of Mercury rotates slightly faster than can be explained by the gravitational influence of the other planets. The new theory could explain the observations, but it was feared that there might be some other influence causing such a small effect.
2. General relativity predicts the displacement redwards of the spectral lines formed in the intense gravitational field of a very dense and compact star. This effect is also hard to verify since such stars have very broad spectral lines that overwhelm any subtle effects.
3. The general theory predicts that a ray of light should be deflected by a predictable amount as it passes through a strong gravitational field.

The last one of these, the bending of a ray of light in a gravitational field, is related to the effect that Einstein had suggested should be observable during a solar eclipse. The general theory now predicted that the apparent position of a star should be altered by twice the original amount, or 1.75[8] arcseconds, when its light grazed the sun's surface. Unfortunately, it was clear that, even during an eclipse, scattered light and the radiation from the solar corona would make it hard to measure the change of position of a star close to the sun's surface with the necessary degree of accuracy.

Eddington set about publicising Einstein's work within the scientific community by writing a *Report on the Relativity Theory of Gravitation*, for the Physical Society in 1918. Probably fired by Eddington's enthusiasm, the Astronomer Royal (Dyson) pointed out that an eclipse predicted for 29 May 1919 offered a unique opportunity for verifying the theory by the third method listed. The occasion was particularly suitable because the sun was due to pass in front of a cluster of bright stars, the Hyades, which should be easily visible in spite of the bright background.

---

[8]The precise deflection depends on the speed of light, the constant of gravity and the angular size of the sun as seen from the earth; the last-named varies slightly according to the time of year.

S. Chandrasekhar (1983) relates the following piece of bravado on Eddington's part:

> I once expressed to Eddington my admiration of his scientific sensibility in planning the expeditions under circumstances when the future must have appeared very bleak. To my surprise, Eddington disclaimed any credit on that account and told me that, had he been left to himself, he would not have planned the expeditions since he was fully convinced of the truth of the general theory of relativity! (Chandrasekhar 1983, pp. 24–5)

7.8   Conscientious objector

Eddington, as a Quaker, was a pacifist. During the First World War he found it hard to accept the extremely anti-German sentiments expressed, for example, by Herbert Hall Turner (1916) in his 'Oxford Notebook' article of May 1916 in *The Observatory*. Turner suggested that normal scientific relations could not be resumed with so barbarous a nation as the Germans when the war ended. Eddington urged a calmer approach in a letter to the editors, which appeared in the following issue. He said that

> we should not attach undue importance to judgments made in the heat of conflict ... Fortunately, most of us know fairly intimately some of the men with whom, it is suggested, we can no longer associate. Think, not of a symbolic German, but of your former friend Prof. X., for instance—call him Hun, pirate, baby-killer, and try to work up a little fury. The attempt breaks down ludicrously. No doubt, he is a most ardent supporter of his fatherland, passionately convinced of the righteousness of his cause. Call this wrong-headed, if you will, but surely not morally debased. Far be it from me to deny his individual responsibility for his country's share in the evil that has befallen. The worship of force, love of empire, a narrow patriotism, and the perversion of science have brought the world to disaster. (Eddington 1916)

When Britain decided to introduce conscription, Eddington was liable to be called up even though he was thirty-four years old and short-sighted. Because of the odium attached to conscientious objectors at the time, some of the senior physicists in Cambridge tried to have his enlistment deferred on the grounds that he was a distinguished scientist and that he would be of more value to his country if he continued in that capacity. This manoeuvre succeeded and the Home Office sent Eddington a letter which he merely had to sign and return. However, he added a postscript that if he was not given deferral for his science he would claim it as a conscientious objector anyway. He would be quite content to peel potatoes in the north of England like some of his Quaker friends.

By 1918, the Ministry of National Service was looking desperately for new recruits and it then appealed against Eddington's exemption. At a hearing before the Appeal Tribunal for Cambridge, it was suggested that he should be called up for some employment in connection with the war effort, though not necessarily as a common soldier. Eddington stated 'I am a conscientious objector' but, since he had not been excused on those grounds, his exemption was to be revoked. There was an appeal a few weeks later and Eddington stated

> My objection to war is based on religious grounds. I cannot believe that God is calling me to go out to slaughter men, many of whom are animated by the same motives of patriotism and supposed religious duty that have sent my countrymen into the field. To assert that it is our religious duty to cast off the moral progress of centuries and take part in the passions and barbarity of war is to contradict my whole conception of what the Christian religion means ... [9]

Following a procedural delay, his case came up again in July and this time was supported by a strong letter from the Astronomer Royal, Dyson, who pointed out

> I should like to bring to the notice of the Tribunal the great value of Professor Eddington's researches in astronomy ... They maintain the high tradition of British science at a time when it is very desirable that it should be upheld, particularly in view of the widely spread but erroneous notion that the most important scientific researches are carried out in Germany ... The Joint Permanent Eclipse Committee, of which I am Chairman, has received a grant of £1000 for the observation of a total eclipse of the sun in May of next year, on account of its exceptional importance ... Professor Eddington is peculiarly qualified to make these observations and I hope the Tribunal will give him permission to undertake this task. [10]

Eddington was questioned by the chairman and he explained how the eclipse of 1919 would be the best opportunity for an observational test of Einstein's theory that was likely to arise for several centuries. Although he would go to make the observations if circumstances permitted, he was willing to work for the Friends' Ambulance Service or become a labourer for the harvest if they should so decide. The judgement of the Tribunal was that Eddington was, and had been, a conscientious objector. Their decision was to grant him twelve months further exemption on condition that he continued in the work he was doing, more especially in connection with the coming eclipse!

7.9   The solar eclipse of 1919

Expeditions were organised as the war progressed, but with great difficulty as instrument makers had no spare capacity for civilian work. However, the armistice at last came about, the War ended, and the plans went ahead. Eddington described what followed as 'the most exciting event, I recall, in my connection with astronomy' (Chandrasekhar 1983, p. 24).

To reduce the chance of being wiped out by bad weather, two separate observing sites were chosen. Andrew Crommelin and C.R. Davidson of the Royal

---

[9] Quoted by Douglas (1956, p. 93).
[10] Quoted by Douglas (1956, p. 94).

Greenwich Observatory went to Sobral (Fig 7.4) in Brazil with two telescopes and Edwin Cottingham (also of Royal Greenwich Observatory) went with Eddington and one telescope to the island of Principe, off the West African coast.

On the night before sailing, the participants were discussing the expected result. Some believed that light quanta might not be influenced by gravity at all and that there might be no deflection. If general relativity was wrong a deflection of 0.87 arcsec might be expected from ordinary Newtonian mechanics and special relativity. If right, the change in apparent position of a star should be 1.74. Worried, no doubt, by the factor of two between Einstein's first and second estimates, Cottingham asked 'What will it mean if we get double the Einstein [1.74 arcsec] deflection?', to which Dyson replied 'Then Eddington will go mad and you will have to come home alone!'

In the event, the weather at Eddington and Cottingham's site in Principe was rather poor, but they did succeed in getting some usable material. Eddington made a preliminary measurement of the photographs they had taken and assured Cottingham that he would not have to go home alone. The conditions were better at Sobral, but the result seemed at first, based on the results with the larger of the two telescopes, to be more in accord with the classical value. However, when the plates taken with the smaller instrument were developed and measured on the observers' return to England, it was found that the quality of its data was superior to the others and the Einstein value was confirmed. The earlier spurious result was probably due to distortion of the optics in the coelostats feeding light to the larger telescopes in the hot sun—the comparison plates needed to show the undeflected positions of the relevant stars had to be taken at night when the temperature was much lower.

When the news got to Einstein, via a telegram from the Dutch physicist H.A. Lorentz, he showed it to a student, Ilsa Rosenthal-Schneider, saying 'Here, this will perhaps interest you'. While she was expressing her joy that the results coincided with his calculations, he said, quite unmoved 'But I knew that the theory is correct'. When she asked what he would have said if his prediction had not been confirmed, he replied 'Then I would have been sorry for the dear Lord— the theory *is* correct' (Clark 1971). In spite of his deadpan remark, there is plenty of evidence that Einstein was very pleased by the observational confirmation of his theory (Pais 1982).

The joint meeting of the Royal and Royal Astronomical Societies on 6 November 1919, when the formal results were presented by Dyson, on behalf of the participants, was a historic occasion. The philosopher Albert North Whitehead was there:

> The whole atmosphere of tense interest was exactly that of the Greek drama: we were the chorus commenting on the decree of destiny as disclosed in the development of a supreme incident. There was dramatic quality in the very staging:–the traditional ceremonial, and in the background the picture of Newton to remind us that the greatest of scientific generalisations was now, after more than two centuries, to receive its first

FIG. 7.4. The two telescopes of the British expedition at the Sobral eclipse site in Brazil. The tilted mirror coelostats, driven to follow the sun by clockwork motors, reflected its image into the telescope tubes. The small instrument, on the right, gave the better results. The large coelostat mirror was found to have been distorted by the heat of the sun and did not perform as well. From Eddington's book *Space, Time and Gravitation* (1920).

> modification. Nor was the personal interest wanting: a great adventure in thought had at length come safe to shore. (Whitehead 1932)

At the end of the meeting, Ludwik Silberstein, a Polish physicist with an interest in relativity, came up to Eddington and said

> Professor Eddington, you must be one of three persons in the world who understands general relativity'. On Eddington's demurring to this statement, Silberstein responded 'Don't be modest, Eddington' and Eddington replied that 'On the contrary, I am trying to think who the third person is.' (Chandrasekhar 1983, p. 30)

The following day, newspapers worldwide carried headlines about the results. The success of his prediction brought enormous acclaim to Einstein whose name, up till then, although well known among physicists, meant nothing to the general public. He became an instant celebrity. In an explanatory article he wrote for *The Times* of London, he expressed his amusement at being called a 'German man of science' in the German press and a Swiss jew in *The Times*. 'If I come to be a *bête noir* the description will be reversed and I shall become a Swiss jew for the Germans and a German man of science for the English.' A *Times* editorial remarked in a coy, almost Eddingtonian, manner 'We conceded him his little jest, but we note that, in accordance with the general tenor of his theory, Dr Einstein does not supply an absolute description of himself' (Clark 1971).

Cecilia Payne, later to become a famous astronomer herself, entered Cambridge in 1919. She was present when Eddington presented a lecture in the Great Hall of Trinity College on the results of the eclipse expedition (2 December 1919). A quarter-hour beforehand a queue of students and dons stretched halfway across the Great Court of the College. She was so moved by his brilliant presentation that she was afterwards able to write down the lecture word for word. The following day she decided to change her subject from biology to physics. Sometime later she managed to meet the great man by attending an open night at the Observatory. She told him that she would like to be an astronomer. His reply was 'I can see no *insuperable* objection!', the recollection of which, she claimed, was to sustain her through many later rebuffs. Later she attended his lectures, which included practical work on numerical computations. In those days, the tedious method of logarithms was used rather than calculators. After each session, she and the other two or three students present would be asked to tea with the family. Quite the bluestocking, she was shocked to discover that Eddington's favourite composer was Humperdinck,[11] and that the music he liked best included Harry Lauder's[12] songs, especially *Roamin' in the gloaming*! (Payne-Gaposchkin 1984, p. 120).

In a recent discussion of the 1919 eclipse expedition and Eddington's part in it, Stanley (2003) emphasises the place of Eddington's Quaker background and pacifism in his motivation and his courage in standing up against the jingoism that mushroomed even in scientific circles during the First World War. He knew and, of course, approved of Einstein's anti-war position as early as October 1916. Undoubtedly, the international character of the effort concerning General Relativity made a strong impression on the scientific community and the thinking public.

7.10 Aftermath of the eclipse

In an after-dinner conversation in Trinity College, Cambridge, some fourteen years after the excitement had died down, Rutherford, the discoverer of the atomic nucleus, Eddington, Chandrasekhar, and some others were chatting about public fame. One of them said to Rutherford:

> I do not see why Einstein is accorded a greater public acclaim than you. After all, you invented the nuclear model of the atom; and that model provides the basis for all of physical science today and it is even more universal in its applications than Newton's laws of gravitation. Also, Einstein's predictions refer to such minute departures from the Newtonian theory that I do not see what all the fuss is about.
>
> Rutherford, in response, turned to Eddington and said 'You are responsible for Einstein's fame'. And more seriously, he continued: 'The war

---

[11] Composer of the opera *Hansel und Gretel*.

[12] Lauder was a popular comedian and singer of Scottish songs. Payne's recollection must refer to a later encounter, since *Roamin' in the Gloamin'* was written in 1928.

had just ended; and the complacency of the Victorian and the Edwardian times had been shattered. The people felt that all their values and all their ideals had lost their bearings. Now, suddenly, they learnt that an astronomical prediction by a German scientist had been confirmed by expeditions to Brazil and West Africa and, indeed, prepared for already during the war, by British astronomers. Astronomy had always appealed to public imagination; and an astronomical discovery, transcending worldly strife, struck a responsive chord. The meeting of the Royal Society, at which the results of the British expeditions were reported, was headlined in all the British papers: and the typhoon of publicity crossed the Atlantic. From that point on, the American press played Einstein to the maximum'. (Chandrasekhar 1983, p. 28)

In retrospect, Eddington and Dyson obtained the 'right' result, but there was a good deal of criticism of their conclusions at the time, during the joint meeting and afterwards. The maximum amount of the deflection of the star images was about $1/30$ mm at the focus of the telescopes involved and non-astronomers found it hard to believe that this could be measured satisfactorily, especially since the diameters of the images themselves were much larger. However, the problem is a routine one in measuring the positions of stars. The case against the British team has been discussed in detail (e.g. Earman and Glymour 1980). Silberstein was particularly sceptical and, at the famous joint meeting, had pointed his finger at a portrait of Newton that looked down on the assembly and warned that 'we owe it to that great man to proceed very carefully in modifying or retouching his Law of Gravitation' (Coles 2001). In fact, the eclipse method is inherently difficult and the observational errors of later expeditions were little better than the Eddington and Dyson ones (von Klüber 1960). The best modern determinations of the bending of electromagnetic radiation by the sun have been made in the radio region.

Eddington too became quite well-known to the general public as a result of the successful eclipse expedition. In 1920, he wrote the first of his semi-popular books, *Space, Time and Gravitation*, in which he attempted to give a simple explanation of General Relativity and discussed in detail the eclipse observations.[13]

By 1922, a Lick Observatory expedition reported had results which also confirmed the predictions. Eddington, a great admirer of Lewis Carroll, commented at a meeting of the Royal Astronomical Society:

> I think that it was Bellman in 'The Hunting of the Snark' who laid down the rule 'When I say it three times, it is right'. The stars have now said it three times to three separate expeditions, and I am convinced their answer is right. (Eddington, 1923)

---

[13]Eddington's popular writing was very successful, to the extent that at his death he left a fortune of £47,000 (Crowther 1952), the equivalent of several millions today.

FIG. 7.5. Five famous relativists who met in Leiden on 29 September 1923. Front: Eddington, H.A. Lorentz; Back: Einstein, P. Ehrenfest and W. de Sitter (Sterrewacht Leiden).

7.11 Influence on Lemaître

During the academic year 1922–3, the young Belgian cleric and mathematician, Georges Lemaître (1894–1966), a former student of the [Roman Catholic] University of Louvain, worked with Eddington. Many years afterwards he showed G.C. McVittie, who had taken his PhD with Eddington, a letter that his now-famous professor had generously written to a colleague at the Free University of Brussels: 'I found Mr Lemaître a very brilliant student, wonderfully quick and clear-sighted, and of great mathematical ability. He did some excellent work whilst here . . . In case his name is considered for any post in Belgium I would be able to give him my strongest recommendation'. As McVittie read this, Lemaître 'chuckled and remarked how little Eddington realized that any recommendation coming by way of the Free University of Brussels would be complete anathema to the officials at the University of Louvain!' (Douglas 1967).

In 1927 Lemaître was to write a paper of fundamental importance to cosmology (see Section 9.6) in a relatively obscure Belgian journal. This went unnoticed by the astronomical community in spite of a copy having been sent to Eddington. Three years later, when he heard that Eddington had set the very problem he had already solved to McVittie, then his PhD student, he realised that it had never been read and shot off a second copy, which this time hit its mark. Eddington did his best to made amends, by translating the paper and publishing it in the *Monthly Notices of the Royal Astronomical Society*.

7.12   The internal constitution of the stars

Eddington's most important work is generally reckoned to have been his study of the internal structure of the stars, which he made between 1916 and 1926. Apart from his interest in the Hertzsprung–Russell diagram, he seems to have been attracted to the subject of the insides of the stars by the work on the Cepheid variables that Shapley had been doing with Russell at Princeton (see Section 8.6). These stars, whose light was observed to vary in a regular way with time, changing from bright to faint states with periods of a few days to tens of days, had first been thought to be eclipsing binaries. Shapley had shown that pulsation, during which the star changed its size cyclically, was a more probable mechanism.

Little was known about the state of matter in stars or about how they evolved. Some of the pioneers, such as Huggins and Lockyer, had suggested schemes of evolution in which growing older would be accompanied by a change of temperature, but there was no theoretical backing for their points of view.

Any stable star is held together by gravity but the question was: What keeps it from collapsing altogether? The obvious suggestion was gas pressure, but the temperatures required to produce enough pressure were far too high and the loss of heat from the star would be much more than that observed. Karl Schwarzschild, the German astronomer who died during the First World War, had studied the outer atmospheres of stars and realised that pressure due to the outgoing photons of light was in fact an important source of support. Eddington developed Schwarzschild's ideas and found that he could make use of results from a study of gravitating spheres of gas that had been worked out by a Munich-based Swiss mathematician, J.R. Emden. Emden had taken various models of how the pressure in a gas might vary with its density and had published his results in a book *Gaskugeln* (Gas-spheres) in 1907. This work was criticised heavily by James Jeans, who frequently was mathematically correct but lacked physical insight—the gut feeling for reality. What Eddington eventually found was a relationship between the luminosity—the power output of a star, measured in watts—and its mass. The results given in Eddington's first paper did not match real stars and several physicists (including Jeans) pointed out to him that he had overlooked an important fact: atoms in the centre of stars would be at very high temperatures where the (light) electrons would become detached from the (heavy) nuclei and act as independent particles. Thus the *average* mass of a gas particle would not be that of a whole atom as he had assumed, but much less.

James Hopwood Jeans was five years older than Eddington and had also attended Cambridge University. In 1907, he married Charlotte Tiffany Mitchell, an American heiress (Crowther 1952). By 1912 he had quit his academic position and opted instead to work from home. His interests included many of the same topics as Eddington's. He made important and basic contributions to theoretical physics, especially on the behaviour of gases, and his name is attached to the 'Jeans instability', 'Jeans length', and the Rayleigh-Jeans radiation law. He and Eddington grew up in the same competitive atmosphere. Jeans had been joint Second Wrangler in the 1898 Tripos. They carried their rivalry into meetings of the Royal Astronomical Society where, for many years, they took delight in jousting with each other whenever either gave a paper. The meetings were considered to be of high entertainment value—in an intellectual sort of way—and the famous pure mathematician, G.H. Hardy, a contemporary of Jeans, specially joined the Society so that he could have a ringside seat! These occasions have been preserved for posterity by the reports that appeared in the magazine *The Observatory*.

In November 1917, Jeans presented some results arising from Eddington's work. Eddington had assumed that energy was being generated by some unexplained mechanism throughout the interior of a star. Jeans made the condescending remark that Eddington 'obtained some results which seemed to me to be rather astounding, and I thought I would examine whether the mathematics is altogether above suspicion' (Jeans 1917). Then he derived many of Eddington's results, supposedly invalid as originally found, without the energy generation mechanism, and went on to say that 'he may not use his own equations, but I may' (Jeans 1917). Jeans took the traditional view that a star derives its power solely from the energy released as it shrinks under the effect of gravity.

Eddington then commented

> I should like to make a few remarks, chiefly about the first portion of Mr. Jeans's remarks, in which I saw my equations being ruthlessly taken from me and adopted by Mr. Jeans, whilst I was forbidden to use them . . . (Eddington 1917).

He pointed out that his equations worked whether the unexplained energy generation mechanism, such as what we now call nuclear energy, was included or not. Then followed one of his remarkable physical insights:

> In any case it seems to me that, whether the energy is supplied by radioactivity or not, there must be radioactive processes going on—either disintegration or formation of radioactive elements,—so that I do not think that at a given moment the outflow of energy corresponds to the [gravitational] contraction solely (Eddington 1917).[14]

---

[14]The actual process that powers the sun, the conversion of hydrogen into helium, was elucidated by H.A. Bethe in 1938.

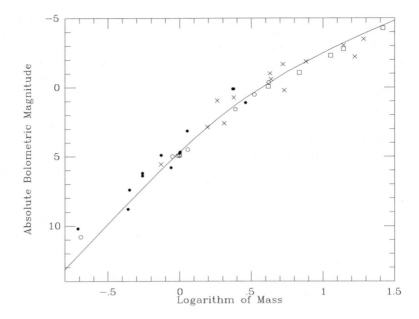

FIG. 7.6. The 'Mass–Luminosity Curve' re-drawn from Eddington (1926). Here
he plotted his theoretical curve (the line on the diagram) against the luminosities and masses (in terms of the mass of the sun) of the few stars for
which these quantities were then known. The vertical axis is related to the
logarithm of the wattage of the star, which increases upwards. 'Bolometric'
means summed over all wavelengths. Circles represent well-studied binary
stars, dots less well-studied binaries, crosses eclipsing binaries and squares
Cepheids.

7.13   The mass–luminosity relation

Eddington's theory of stars showed that their luminosities (output power) are
related to their masses and hardly at all to their sizes. When he plotted the
masses and luminosities of all the stars for which these quantities were known,
he found good agreement with his 'law' (Fig 7.6). This proved as a by-product
that the stars were behaving like gases, i.e. that they were not liquid or solid
inside as many, such as Jeans, believed.

Neither Eddington nor anybody else knew at that time what the source of
energy within the stars could possibly be. Yet their luminosities depended only
on their masses. Jeans and others felt that Eddington's theory could not be right,
that the source of the energy had to be an important consideration also. It was
felt that a gas could not reach the densities required by Eddington's theories –
the atoms would be packed together too tightly to be able to move around freely
and the centres of the stars would have to be solid or at least liquid.

Generally speaking, Eddington's insights turned out to be correct even if he could not really show why, i.e. his methods were not always sound from a mathematical point of view, as Jeans had said. He realised that the material of stars stayed gas-like at the high densities in their centres because the conditions were so hot that there were X-rays everywhere, which drove off the electrons from the outer layers of the atoms and left bare the relatively tiny nuclei. The packing could therefore be much tighter without hindering movement of the now much smaller gas particles.

In *The Internal Constitution of the Stars*, Eddington (1926) summarised his stellar work of the previous decade. His style is a mixture of formal mathematics and very informal descriptive passages, of which the following is one of the most famous; in fact, he liked it so much that he repeated it almost verbatim in his popular book *Stars and Atoms* (Eddington 1927):

The Inside of a Star

The inside of a star is a hurly-burly of atoms, electrons and aether waves. We have to call to aid the most recent discoveries of atomic physics to follow the intricacies of the dance. We started to explore the inside of a star; we soon find ourselves exploring the inside of an atom. Try to picture the tumult! Dishevelled atoms tear along at 50 miles a second with only a few tatters left of their elaborate cloaks of electrons torn from them in the scrimmage. The lost electrons are speeding a hundred times faster to find new resting-places. Look out! there is nearly a collision as an electron approaches an atomic nucleus; but putting on speed it sweeps around it in a sharp curve. A thousand narrow shaves happen to the electron in 10^{-10} of a second; sometimes there is a side-slip at the curve, but the electron still goes on with increased or decreased energy. Then comes a worse slip than usual; the electron is fairly caught and attached to the atom, and its career of freedom is at an end. But only for an instant. Barely has the atom arranged the new scalp on its girdle when a quantum of aether waves runs into it. With a great explosion the electron is off again for further adventures. Elsewhere two of the atoms are meeting full tilt and rebounding, with further disaster to their scanty remains of vesture.

As we watch the scene we ask ourselves, Can this be the stately drama of stellar evolution? It is more like the jolly crockery-smashing turn of a music hall. The knockabout comedy of atomic physics is not very considerate towards our aesthetic ideals; but it is all a question of time-scale. The motions of the electrons are as harmonious as those of the stars but in a different scale of space and time, and the music of the spheres is being played on a keyboard 50 octaves higher. To recover this elegance we must slow down the action, or alternatively accelerate our own wits; just as the slow-motion film resolves the lusty blows of the prize-fighter into movements of extreme grace–and insipidity. (Eddington 1926, pp. 19–20)

7.14   Recreations

At this point we may pause and mention something of Eddington's private life. Like many Cambridge professors of the time, he had no office and worked from his home, in a wing of the Observatory building, where he had a study. He was looked after by his mother and sister. He dined in his College (Trinity) several times a week and was one of the most regular supporters of the Dining Club, which met after the meetings of the Royal Astronomical Society in London. He was a 'Double Centurion', i.e., had attended over 200 dinners, with a maximum of eight per year (see Fig 7.7).

He did not neglect physical exercise. In his younger days, at the Royal Greenwich Observatory, he had been an enthusiastic member of the Observatory hockey team. In Cambridge

> as Plumian Professor he played golf fairly regularly on the Gog Magog course, not however—it must be confessed—with very conspicuous success; also, he was a very good swimmer, and on most summer afternoons he was to be seen disporting himself in the waters of the Cam. (Smart 1945)

He continued to enjoy watching soccer, which he had played as a student in Manchester. Traditionally, it was regarded in England as a working-class game and quite inappropriate for a Cambridge professor to be associated with. On one occasion he was away from home when he was being looked for. His visitor said 'I suppose you were at the [cricket] Test Match?'. 'No', he replied sadly, 'I'm afraid it was a less interesting engagement; I lunched at 10 Downing Street'! [residence of the Prime Minister] (Redman 1945).

Redman related further that 'He had a liking for Scandanavia and in Copenhagen visited the Tivoli amusement part as regularly as Bohr's famous institute of theoretical physics. But often, if he wanted a holiday, he just put a map in his pocket, mounted his bicycle, and set off without further ado, occasionally with a companion (usually his friend C.J.A. Trimble), but often alone.

On a more mental plane, he enjoyed difficult crossword puzzles which, according to Chandrasekhar, he usually solved within five minutes. The absurd and paradoxical appealed to him. He liked to tell how, when he got an odometer for the first time on his bicycle he, the relativist, found himself checking its readings against the milestones and putting on speed to beat it to the next stone. On a large map of England he marked the routes he had taken on his bicycle holidays. He told Chandrasekhar that his first such map had been chewed up by his dog and that he had had to transcribe them onto a new one. From the records of his rides he derived a number n which was the number of days on which he had cycled n miles or more. His final value for n was 77 (Chandrasekhar 1983, p. 6). The Second World War put a stop to his long-distance cycling because he could not rely on finding places to stay the night.

He was usually quite hopeless at small talk. At one Cambridge lunch

> It proved almost impossible to get Eddington to say a word ... every possibility was given for Eddington to join in, but he remained completely silent. At last about half way through lunch my father asked him his opinion of the work of Conan Doyle. Then Eddington spoke, naturally, clearly and at some length ... (Quoted by Douglas 1956, p. 114).

During 1924 he visited Canada for a meeting of the British Association for the Advancement of Science in Toronto and he took the opportunity to visit the US West Coast observatories. Another story of his lack of small talk was related by Harold Babcock of his visit to Mount Wilson:

> In those days, trips up the mountain used to be far less easy than they have since become; and the visitor had to be accompanied *en route* by a local member of the staff. One of the technicians was detailed to accompany Eddington; and, since the trip lasted several hours, tried to engage the visitor in some talk. This was, it seemed, mostly unilateral, the guide pointing out the place where there had recently been some forest fire, where some of the Observatory's people had seen a mountain lion, etc – to all of which Eddington apparently responded only with non-committal grunts.
>
> Some days later, the same technician met in Pasadena a colleague who, in the meantime, had been detailed to return Eddington back to civilisation; and the two happened to exchange some comments. 'Wasn't that English professor you took down a stuffshirt?' asked the first one. 'What do you mean?', retorted the other incomprehendingly. 'Well, when I took him up the mountain, he said hardly a word for the whole trip.' 'Impossible', said the other; 'when we came down he was a very pleasant companion.' 'What did he tell you?' 'Well', replied the second guide, 'as we were coming down, he pointed out to me where we recently had a forest fire, or where one of our chaps had seen a mountain lion—and he had it all right'—not realising that what he had listened to was only a replay of what Eddington had learned going uphill. (Kopal 1986, pp. 142–3)

7.15 Sureness or cocksureness?

Eddington attended Quaker meetings regularly but was rarely moved to speak. In 1929, he gave a lecture at Swarthmore College, a Quaker foundation, on *Science and the Unseen World* (Eddington 1929). He talked of the obstacles imposed by religious creeds and dogma. He himself had a kind of non-specific religious belief, but he went on:

> Religious creeds are a great obstacle to any full sympathy between the outlook of the scientist and the outlook which religion is so often supposed to require ... The spirit of seeking which animates us refuses to regard any kind of creed as its goal. It would be a shock to come across a university where it was the practice of the students to recite adherence to Newton's laws of motion, to Maxwell's equations and to the electromagnetic theory of light ... Science may fall short of its ideal, and although the peril scarcely takes this extreme form, it is not always easy, particularly in popular science, to maintain our stand against creed and dogma ...
>
> Rejection of creed is not inconsistent with being possessed by a living belief. We have no creed in science, but we are not lukewarm in our beliefs. The belief is not that all the knowledge of the universe that we hold so enthusiastically will survive in the letter; but a sureness that we are on

the road. If our so-called facts are changing shadows, they are shadows
cast by the light of constant truth ...
　　There is a kind of sureness which is very different from cocksureness.
(Eddington 1929, pp. 55–56)

Would that Eddington himself had avoided cocksureness! He could be very
harsh towards other scientists. Some, like Jeans, could give as well as he got,
but others he put down in public without reflecting how his own increasingly
God-like reputation lent damaging weight to his criticisms. Jeans accused him
of failure to reference his prior work and of allowing the impression to grow that
it was Eddington rather than he who had developed certain key ideas about the
behaviour of matter inside stars. He was sufficiently unhappy at Eddington's
cavalier approach that he wrote twice to *The Observatory* to complain

> So much work has been done on isothermal equilibrium that it is difficult
> to understand how Prof. Eddington can harbour the illusion that he is
> doing pioneer work in unexplored territory, yet his complete absence of
> reference to other theoretical workers ... suggests that such is actually
> the case. (Jeans 1926a)
> 　　May I conclude by assuring Prof. Eddington it would give me great
> pleasure if he could remove a long-standing source of friction between
> us by abstaining from making wild attacks on my work which he can-
> not substantiate, and by making the usual acknowledgments whenever
> he finds that my previous work is of use to him? I attach all the more
> importance to the second part of the request, because I find that some of
> the most fruitful ideas which I have introduced into astronomical physics
> – *e.g.*, the annihilation of matter as a source of stellar energy, and highly
> dissociated atoms and free electrons as the substance of stars – are by
> now fairly generally attributed to Prof. Eddington. (Jeans 1926b)

This rivalry had, however, its good-natured side. G.H. Hardy, who knew that
Eddington sometimes went with his sister to the races, once asked him if he had
ever bet on a horse.

> Eddington admitted that he had, but 'just once'. Hardy was curious to
> know the occasion; and Eddington explained that a horse named Jeans
> was running and that he could not resist the temptation of betting on it.
> Questioned whether he had won, Eddington responded with his charac-
> teristic smile and a 'NO!' (Chandrasekhar 1983, p. 21)

When presenting the Gold Medal of the Royal Astronomical Society to Jeans
he slyly referred to their years of intellectual sparring:

> The *Monthly Notices* [of the Royal Astronomical Society] records how we
> have hurled at each other mathematical formulae—the most undodgeable
> of missiles, when they are right—and the onlooker will perhaps conclude
> that *someone* was badly annihilated. But it is possible that Jeans and
> I may still have a difference of opinion as to precisely whose corpse lies
> stricken on the field. (Eddington 1922)

FIG. 7.7. Eddington and Jeans in mellow mood after a Royal Astronomical Society Club dinner on 10 May 1935. The person at left is S.C. Roberts, a guest at the time. Eddington was particularly fond of Club dinners (Royal Astronomical Society Club).

7.16 The Expanding Universe

General relativity had appealed to Eddington because of his interest in scientific epistemology—the study of the nature of scientific knowledge. Conventional or Newtonian physics seemed to be something arising from our particular measuring instruments and viewpoint, whereas general relativity was much more universal. He felt that many other truths of nature might be hidden from us because of what he believed to be our blinkered way of viewing things.

Eddington's interest in the geometry of space revived with the publication of Hubble's paper of 1929 on the relation of the speed of recession of the galaxies to their distances. He gave a public lecture on the subject at the meeting of the International Astronomical Union held in Cambridge, Massachusetts, in 1932, and afterwards turned it into a semi-popular book called *The Expanding Universe* (Eddington 1933). In this work he made considerable use of Lemaître's solution to the equations of General Relativity as applied to the large-scale structure of space. He also wrote of his somewhat mystical ideas on the connection between large quantities such as the curvature of space and the total number of fundamental particles in the universe and microscopic ones such as the mass of the electron. Studies of this kind were to become an important part, at least in his own estimation, of his later work.

7.17   The strange case of Chandrasekhar and Sirius B

Sirius, the brightest star in the sky (after the sun!), was long suspected of being a binary or double, mainly because the early nineteenth-century German astronomer Friedrich Wilhelm Bessell had found that its position altered periodically by a very small amount, about two seconds of arc, over a period of fifty years, in a manner indicating that a massive object, not just a giant planet, must be in orbit around it. The companion object, which turned out to be very faint, was seen for the first time by the American telescope maker, Alvan Clark, on 31 January 1862, in Cambridge, Massachusetts, while he was testing the lens for a telescope ordered for the University of Mississippi—one that was never delivered owing to the outbreak of the Civil War between the Northern and Southern states.

The companion received the name of Sirius B. Because Sirius is close enough to have a measurable distance, the properties of the orbit of the two stars and Newton's law of gravitation can be used to show that Sirius A has 2.28 times the mass of the sun but its companion has only 0.96 times this amount. A has the brightness to be expected from Eddington's mass–luminosity relation, but B is much too faint. Its size was first inferred by Walter S. Adams (1915) at Mount Wilson Observatory by estimating its temperature from a spectrum and making use of the physics of radiation. He found it to be tiny, a veritable 'white dwarf', with a diameter only about twice that of the earth. So much mass in such a small volume could only mean that its density had to be extraordinarily high—about 60,000 times that of water.

Eddington's theory, by its failure to predict the luminosity of Sirius B, could clearly not be applied to every star. It was Ralph H. Fowler, a Trinity College colleague, who in 1926 used the newly discovered quantum mechanics to rescue the situation. He pointed out that the material of a star could behave differently from a normal gas under conditions of extreme density. The new state of matter was called 'degenerate'. According to a study of the behaviour of electrons developed by the physicists Enrico Fermi and P.A.M. Dirac, each electron in the star had to be slightly different in position and velocity from every other electron, even in the lowest possible energy state of that star. This results in a kind of pressure that prevents the star from collapsing to such a small size that it becomes a 'black hole' (Laplace, the great eighteenth-century mathematician, had predicted that a massive body, if it was sufficiently small, would have such a strong gravitational pull at its surface that radiation would not be able to escape from it).

It soon became evident that even Fowler had not taken quite everything into account. The fact that many of the electrons in a white dwarf would be travelling near the speed of light meant that relativistic effects would bring back the old problem of collapse that the Fermi-Dirac theory sought to avoid. The young Indian astrophysicist, Subramanyan Chandrasekhar, showed that a white dwarf star simply cannot be stable if its mass exceeds 1.4 times that of the sun.[15] Eddington obstinately refused to accept this result. Unfortunately, the other leading British theoretician in the field, Edward Athur Milne, for once supported Eddington's point of view, because the new idea disagreed with a theory of his own. Chandrasekhar reluctantly concluded that, in the face of so much opposition, he would never be able to get his result published in England. He therefore decided to send it in November 1930 to the *Astrophysical Journal* in the United States, which accepted it.

Early in 1935, Chandrasekhar presented a more precise calculation of his theory of the limiting mass of white dwarfs at the Royal Astronomical Society. His paper was followed by a talk by Eddington who said such things as 'I do not know whether I shall escape from this meeting alive, but the point of my paper is that there is no such thing as relativistic degeneracy!' then 'Dr Chandrasekhar has got this result before, but he has rubbed it in, in his last paper ... I think there should be a law of Nature to prevent a star from behaving in this absurd way' (Chandrasekhar 1983, p. 52).

Later in 1935, at meeting of the International Astronomical Union in Paris, Eddington again stated that there was no such thing as relativistic degeneracy and that Chandrasekhar's work was absurd. Henry Norris Russell was chairman of the session, and, even though he told Chandrasekhar privately that 'Out there, we don't believe in Eddington' (Wali 1984), he refused to let him reply.

Chandrasekhar tried to get some of the most famous physicists of the time to support his views, without success. They either had not thought about the problem enough, or did not want to get into a controversy which they felt was an astronomical rather than a physical one.

Speaking during the Tercentenary of Harvard University in 1936 Eddington wrote about the relativistic degeneracy formula in his most cocksure manner:

> All seemed well until certain researches by Chandrasekhar brought out the fact that the relativistic formula put the stars back in precisely the same difficulty from which Fowler had rescued them. The small stars could cool down all right, and end their days as dark stars in a reasonable way. But above a certain critical mass (two or three times that of the sun) the star could never cool down, but must go on radiating and contracting until heaven knows what becomes of it. This did not worry Chandrasekhar; he seemed to like the stars to behave that way, and believes that is what really happens. But I felt the same objections as 12 years earlier to this stellar buffoonery; at least it was sufficiently strange to rouse my suspicion that there must be something wrong with the physical formula used.
>
> I examined the formula—the so-called relativistic degeneracy formula— and the conclusion I came to was that it was the result of a combination of relativity theory with a non-relativistic quantum theory. I do not regard the offspring of such a union as born in lawful wedlock ... I was not

---

[15]In fact, there are white dwarfs in binary systems that slowly accrete mass from their companions and, when they reach Chandrasekhar's limit, they explode. For a few hundred days they become very luminous and are called 'Type I supernovae'.

FIG. 7.8. Photograph of Eddington taken by Howard Coster, ca. 1936. (SAAO).

surprised to find in announcing these conclusions I had put my foot in a
hornet's nest; and I have had the physicists buzzing about my ears—but
I don't think I have been stung yet. Anyhow, for the purposes of this
lecture I will assume that I haven't dropped a brick. (Eddington 1938)

Not long after he uttered this string of mixed metaphors, Eddington loftily
told Chandrasekhar 'You look at it from the point of view of the star; I look at
it from the point of view of nature' (Chandrasekhar 1983, p. 58). Chandrasekhar
eventually decided not to risk antagonising the profession by continuing to harp
on his views, although he was sure they were correct. He reluctantly decided to
move on to other fields of research.

In spite of these often vindictive professional differences, Eddington maintained a friendly enough attitude to Chandrasekhar in other matters. They attended the Wimbledon tennis championships in 1935 and even went on a bicy-
cling trip together. Eddington wrote to him about quite personal things such as
his own relatively poor childhood and loneliness. On one occasion, he told him
how he 'had to put on a weird costume—knee breeches and silk hose—and get
my order [decoration] from the King' (Chandrasekhar 1983, p. 6).

In a sense, Chandrasekhar had the last word. In 1983, many years after his
work on the subject, he received the Nobel prize in physics.

7.18   Eddington's wilder speculations

Eddington's later work was characterised by the mystical approach to physics that he began to discuss in *The Expanding Universe*. In some ways his outlook resembled that of Johannes Kepler. The latter had had a strong, essentially mystical, feeling that the orbits of the planets ought to obey simple mathematical laws and he spent many years speculating as to what these might be. In the end, he hit upon some important truths which arose from his deep studies of Tycho Brahe's observations. If Brahe's work had been less precise, Kepler's planetary ellipses would have been hard to discern and, if they had been much more so, small deviations from the elliptical form would have confused the issue so much that he would not have discovered his laws! Unfortunately, Eddington's speculations were not, in general, verifiable. Typical of his later years was the statement:

> I believe that there are 15,747,724,136,275,002,577,605,653,961,181,555,
> 468, 044,717,914,527,116,709,366,231,425,076,185,631,031,296 protons in
> the universe, and the same number of electrons.

This quotation forms the opening line of Chapter XI, 'The Physical Universe', of *The Philosophy of Physical Science* (Eddington 1939). The number is 136×2^{256}. Its formal mathematical simplicity appealed to his mystical outlook. 'Bertrand Russell asked Eddington if he had calculated this number himself, or if he had someone else do it for him. Eddington replied that he had done it himself during an Atlantic crossing' (Chandrasekhar 1983, p. 3).

This was one of the quantities in physics that Eddington came to believe could be derived as the solution to certain equations without reference to experiment. Of course, the number of protons in the universe could not be counted, but he predicted other numbers which could be verified. He believed that the ratio of the electrical force to the gravitational force between a proton and an electron (around 10^{39}) was one of these. He 'proved' that the ratio of the mass of a proton to that of an electron should be 1836 and that the fine structure constant, a number that appears in many physical calculations, should be 137 precisely. His methods of arriving at these numbers were not easy to follow. Essentially he sought equations whose roots they were and then tried to justify his choices in physical terms. One relativist who knew him, Sir William McCrea, had this to say:

> He was convinced that there must be mathematical reasons for the numerical constants of physics to possess their particular values, and he proceeded to present them in his fundamental theory.[16] In fact, the values he produced were in good—some in unbelievably good—agreement with experimental evidence. The development involved much clever mathematics, some of it invented for the purpose by Eddington, and it should be said that no one ever detected a serious mathematical error in anything he wrote.

---

[16]Eddington's last manuscript was published under the name of 'Fundamental Theory' by Cambridge University Press in 1946, under the editorship of E.T. Whittaker.

> Yet it is a fact that no particular results of this part of Eddington's work have been accepted. Nor has any physicist ever claimed to be certain of the postulates from which Eddington started or to follow his reasoning all the way through to any of his main conclusions; always there appeared to be some gap—some infuriating gap—in the logic. (McCrea 1991)

At one time Chandrasekhar asked Eddington to what extent his views on relativistic degeneracy were fundamental to his theory of electrons and protons. Eddington replied 'Absolutely fundamental. If my ideas on relativistic degeneracy are wrong, my entire theory of electrons and protons is wrong'. Chandrasekhar did not reply to this and Eddington asked him why. He answered 'I am sorry to hear what you just said' (quoted by Wali 1991, p. 142). In a sense he was genuinely sorry, feeling that the work of Eddington's later years then depended on ideas which were completely wrong.

Einstein, another physicist with a profound respect for intuition, had this to say late in his life, several years after Eddington's death:

> Eddington made many ingenious suggestions, but I have not followed them all up. I find that he was as a rule curiously uncritical towards his own ideas. He had little feeling for the need for a theoretical construction to be logically very simple if it is to have any prospect of being true. (Einstein, quoted by Rosenthal-Schneider 1949)

His speculations about numerical values were parodied by various eminent physicists. For example the philosopher and astronomer Herbert Dingle of Imperial College, London, wrote a rhyme about them:

> He thought he saw electrons swift
> Their charge and mass combine,
> He looked again and saw it was
> The cosmic sounding line.
> The population then, said he,
> Must be 10^{79}.[17]

On a more positive note, Eddington's beliefs have encouraged a number of physicists to speculate on whether there really could be ways in which the constants of physics might be predicted (see, e.g., Barrow and Tipler 1986).

Not all Eddington's later views were unorthodox. He felt that Hubble had not discussed the accuracy of his determination of the recession constant of the galaxies in an adequate way, and did not like his reluctance to interpret the phenomenon as a consequence of general relativity. In a letter to A.V. Douglas, a former student and later his biographer in 1943, he wrote:

> I just don't understand this eagerness to find some other theory than the expanding universe ... If you do away [with Einstein's theory], you throw back relativity theory into the infantile diseases of 25 years ago. And why the fact that the solution then found has received remarkable

[17]Pronounce 10^{79} as 'ten to the seventy-nine'. Quoted by Douglas 1956, p. 154.

confirmation by observation should lead people to seek desperately for ways to avoid it, I cannot imagine. They do not seem to have the same urge to find some explanation of light which avoids identifying it with electromagnetic waves[!] (Douglas 1956, p. 113)

7.19   Recollections of Eddington's students

Eddington had few research students but all of these seem to have admired him thoroughly. They included Richard Woolley (afterwards Astronomer Royal), Roderic Redman (afterwards Professor of Astronomy at Cambridge), G.C. McVittie (a cosmologist), and Richard Stoy (afterwards His Majesty's Astronomer at the Cape of Good Hope). They probably knew Eddington better than anyone else and we are fortunate that several of them wrote down their impressions of him.

David Evans, University of Texas, was in the late 1930s a student in the Cambridge Observatories and saw Eddington on many occasions. In his book *The Eddington Enigma* he says:

> I, perhaps go farther than most, in calling attention to his limitations as well as his achievements, but in spirit I remain the awe-struck lad who knew him, infinitely privileged to approach one of the giant intellects of my time ... (Evans 1998, p. 13)
>
> If one engaged him in conversation, requiring a thoughtful answer, he would cock his head to one side and gaze half upwards into the middle air, finally replying with a diction in which his lower jaw seemed a little thrust forward. His hat, when needed, was a somewhat elderly felt trilby. He wore a well-used business suit with a slight embonpoint, and shoes so down-at-heel as to merit urgent attention from the cobbler. In short, he was not a snappy dresser. (Evans 1998, p. 24)

Redman believed that Eddington's whole outlook on life had been influenced by his myopia. 'He never used his eyes much, even as an adult. It was well-known that he could pass a member of his own family in the Observatory drive without recognition, and I have often seen him slowly cycling up the Madingley Road at Cambridge, with an old felt hat and a flapping raincoat, crouched over the handlebars and lost in thought, with unseeing eyes apparently focused on a point in mid-air about twenty feet ahead'. In his recollections, Redman went on to describe his study around the year 1926 as 'the only untidy spot in an otherwise immaculate house. You sat on an ancient sofa, sufficient space for the purpose being cleared by turning off the Aberdeen terrier, or pushing a pile of reprints aside. Among the masses of books lining the walls was a shelf full of P.G. Woodhouse [sic] and detective stories. He seemed usually to be in the middle of writing a paper or a book, his desk littered with sheets in his own very regular and legible hand (he never used a type writer)' (Redman 1945).

Stoy was his student a few years later. He reminisced to McCrea as follows:

> Redman's description of a PhD tutorial describes exactly how it was for me five years later (1931–33). As far as I can remember, I went up to

Eddington's house (the East Wing of the Cambridge Observatory) at about 6 p.m. on Tuesday evenings two or three times a term. I too sat on the old sofa from which the black Aberdeen terrier now five years older was duly removed. The sofa faced the fire. Eddington sat on an equally old but comfortable arm chair set at right angles to the settee and facing his desk just beyond the other end of the settee. My subject was the Planetary Nebulae of which nobody at that time appeared to have any very clear ideas. Usually I would go up to the Observatory feeling very depressed about the progress I was making. For a short while we seemed to sit looking at each other shyly then somehow we got talking and I tried to tell him what I was trying to do as I spread more and more papers out on the hearth rug. The hour passed very quickly and I went off home feeling very much happier and very much encouraged ...

I do not know if you ever went to The Observatory Club that Eddington used to run on Monday afternoons in term time ... beginning with tea in the Cambridge Observatory Library presided over by Miss Eddington and followed by a paper from one of the students or staff. I first got a neat card from Eddington inviting me to go to the first meeting of the Michaelmas Term of 1929 ... the beginning of my second year as I was starting astronomical lectures for schedule B. As was usual, Eddington gave the first talk of the year but used the meeting to conscribe speakers for the rest of the year. His subject that day was the receding nebulae and Hubble's relation. This was the first time I had heard of it and "H" [18] then stood at about 536 ...

During my last two years I assisted fairly regularly in the Michaelmas and Lent Terms with the Saturday evening viewing sessions at the Northumberland and Thoroughgood Telescopes. Quite often Eddington would come over from his house and mingle inconspicuously with the visitors talking freely to them and answering questions. Very few of them actually knew who he was.[19]

Zdenek Kopal was at Cambridge for a short while in 1938 and relates how Eddington was available to his students at almost any time of day or night in the study of his residence:

When a student ... came to see him, the door of Eddington's study was never closed to him; and the Professor would receive him (usually in front of the fireplace, with a pipe in his mouth), and in his shy manner but seldom looked at the visitor. If the latter did not, however, come at the right time, he would soon become aware of it; for when Eddington happened to be immersed in his own thoughts (which was very often), he probably registered only half of what the visitors were asking; and his responses then usually consisted of monosyllables which left his interlocutors wondering whether or when the discussion had ended.

---

[18]The Hubble Constant, see Section 9.12.

[19]Letter from R.H. Stoy to Sir William McCrea, 9 July 1988, SAAO Archives, Stoy correspondence.

I soon learned that, in order to get the best out of this remarkable man, one had to wait till he came to one on his own. My desk was located in what used then to be the Observatory Library; there I spent most of my time. And, usually in the afternoon, Eddington emerged from his study and came to sit down (sometimes at the corner of my desk) for a chat, lasting sometimes an hour or more. It was then that he wanted to know about my work, present or contemplated, and was prepared to open his own mind on the subject—sometimes even to engage in 'small talk' about astronomy in many of its aspects. In such talks he was always very gentle (often exhibiting a considerable sense of humour); but I never heard him say an unkind word on anyone or anything. (Kopal 1986, pp. 144–5)

Other staff members told Kopal that Eddington was perhaps kinder to him than to most other youngsters. He thought this had to do with the fact that he was not a public school boy[20]. Eddington felt closer to him than to typical Cambridge students of the period, having had a simple non-conformist education himself.

Sir Richard Woolley, Astronomer Royal, who had at one time been Eddington's Chief Assistant in Cambridge, once said to the writer 'I could tell you some amusing stories about Eddington, but I won't'. This, he explained, was because of his respect for the memory of one he regarded as a great man, whose reputation was to be kept sacred. As McCrea remarked privately to Stoy, Woolley usually managed to work in a reference to Eddington in each of his published papers![21]

7.20   Frustrations

In 1943, Eddington gave a course of lectures at the Institute for Advanced Studies in Dublin, where the eminent Austrian physicist, Erwin Schrödinger, one of the founders of quantum theory, had found a wartime home. Although he stated:

> At no time during the past sixteen years have I felt any doubt about the correctness of my theory. (Chandrasekhar 1983, p. 45)

his frustration was expressed in a letter written in 1944:

> I am continually trying to find out why people find the procedure [by which he derived the constants of physics] obscure. But I would point out that even Einstein was considered obscure, and hundreds of people have thought it necessary to explain him. I cannot seriously believe that I ever attain the obscurity that Dirac [P.A.M. Dirac, a famous quantum theorist] does. But in the case of Einstein and Dirac people have thought it worthwhile to penetrate the obscurity. I believe they will understand me

[20]In Britain, the 'Public' schools are a carefully defined group of expensive private schools, some of great antiquity. The term 'public' originally conveyed that they were unlike the private tutoring arrangements found in upper-class households.

[21]Letter from Sir William McCrea to R.H. Stoy, 9 February 1988, SAAO Archives, Stoy correspondence.

all right when they realize they have got to do so—and when it becomes the fashion 'to explain Eddington'. (Chandrasekhar 1983, p. 58)

7.21   Final year, illness, and death

The young Fred Hoyle, noted for his brash manner, found himself disagreeing with Eddington on occasion. In 1943, he presented a controversial paper, originally condemned as unsuitable for publication, at the Royal Astronomical Society. Eddington stood up afterwards and criticised the work, leading to a considerable argument. Hoyle learned afterwards that, in spite of their difference, Eddington had used his influence to get the Society to accept his paper! 'So this was my experience, an experience very far from suggesting him to be a mean and vengeful person'. (Hoyle 1994)

Hoyle noticed at the time of his clash in 1943 that Eddington had dark circles under his eyes. His health deteriorated seriously during 1944, though he did not tell his friends that anything was wrong. He was taken to a nursing home on 7 November and he died on the 22nd. The cause of death was not stated, but was probably cancer.

Eddington was the subject of many touching obituaries, including one by Milne, one of those who had been attacked by him but who, like others who had encountered similar treatment, remained friendly at a personal level. His friends funded in his memory an annual 'Eddington Memorial Lecture' and the Royal Astronomical Society initiated an 'Eddington Medal' which is awarded to this day for outstanding work in theoretical astronomy. His work was well-recognised by the scientific community during his lifetime: he received the main medals of the Royal and Royal Astronomical Societies, was knighted in 1930, and was elected President of the International Astronomical Union in 1938.

McCrea may have the final word:

> 'Eddington had the aspect of one who enjoyed special access to the secrets of the universe, and what he set down came to be regarded by many as verging on the status of holy writ.
>
> Of course, the lasting value of his work does not depend on the almost mystical air he bore at the height of his powers. But his charismatic appeal did attract attention to his ideas. That appeal seems to have come from the quite enormous faith he put in his own intuition, in part because it had so often turned out to be justified ...
>
> It is hard to convey to the present-day reader the widespread respect that was given to Arthur Stanley Eddington in the years between the world wars. He exerted an enormous influence on the development of physical thought. The influence came first from his own contributions to astronomy and astrophysics and second from his insights into the contributions of others; he often seemed to grasp the significance of advances more profoundly than those who made them and to explain the advances more skillfully'. (McCrea 1991)

References

Adams, W.S., 1915. The Spectrum of the Companion of Sirius, *Publ. Astr. Soc. Pacific*, **27**, 236–237.

Barrow, J.D. and Tipler, F.J., 1986. *The Cosmic Anthropological Principle*, Oxford University Press, Oxford.

Chandrasekhar, S., 1983. *Eddington, the most distinguished astrophysicist of his time*, Cambridge University Press, Cambridge.

Clark, R.W., 1971. *Einstein, The Life and Times*, Avon Books, New York.

Coles, P., 2001. Einstein, Eddington and the 1919 Eclipse, in *Historical Development of Modern Cosmology*, ASP Conference Series, Vol. 252, eds. Martinez, V.J., Trimble, V., and Pons-Bordería, M.J., pp. 21–41.

Crowther, J.G., 1952. *British Scientists of the Twentieth Century*, Routledge & Kegan Paul, London.

Douglas, A. Vibert, 1956. *The Life of Arthur Stanley Eddington*, Nelson, London.

Douglas, A. Vibert, 1967. Georges Lemaître, 1894–1966 *J. Roy. Astr. Soc. Canada*, **61**, 77–80.

Earman, John and Glymour, Clark, 1980. Relativity and eclipses: The British eclipse expeditions of 1919 and their predecessors, *Hist. Studies Phys. Sci.*, **11**, 49–85.

Eddington, A.S., 1914. *Stellar Movements and the Structure of the Universe*. MacMillan, London.

Eddington, A.S., 1916. The Future of International Science, *Observatory*, **39** 270–272.

Eddington, A.S., 1917. *Observatory*, **40**, 434.

Eddington, A.S., 1920. *Space, Time and Gravitation: An Outline of the General Relativity Theory*, Cambridge University Press, Cambridge.

Eddington, A.S., 1922. Address Delivered by the President, Prof A.S. Eddington, on the award of the Gold Medal to Dr. James Hopwood Jeans, *Mon. Not. Roy. Astr. Soc.*, **82**, 279–288.

Eddington, A.S., 1923. *Observatory*, **46**, 142.

Eddington, A.S., 1926. *The Internal Constitution of the Stars*, Cambridge University Press, Cambridge.

Eddington, A.S., 1927. *Stars and Atoms*, Clarendon Press, Oxford.

Eddington, A.S., 1929. *Science and the Unseen World*, Allen & Unwin, London.

Eddington, A.S., 1933. *The Expanding Universe*, Cambridge University Press, Cambridge.

Eddington, A.S. 1938. Constitution of the Stars, *Annual Report of the Smithsonian Institution for 1937*, Smithsonian Institution, Washington, pp. 131–144.

Eddington, A.S., 1939. *The Philosophy of Physical Science*, Cambridge University Press, Cambridge.

Einstein, A., 1911. Über den Einfluss der Schwerkraft auf die Ausbreitung des Lichtes, *Annalen der Physik*, **35**, 898–908.

Einstein, A., 1915. Erklärung der Perihelbewegung des Merkur aus der allgemeinen Relativitätstheorie, *K. Preuss. Akad. d. Wiss.*, 1915, 831–839.

Evans, David S., 1998. *The Eddington Enigma, A Personal Memoir*, Xlibris Corp., Princeton, NJ.

Hinks, A.R., 1911. On the Distribution of Gaseous Nebulae and of star Clusters, *Monthly Notices of the Royal Astronomical Society*, **71**, 693–701.

Hoyle, Fred, 1994. *Home is Where the Wind Blows: Chapters from a Cosmologist's Life*, University Science Books, Mill Valley, CA.

Jeans, J.H., 1917. *Observatory*, **40**, 432.

Jeans, J.H., 1926a. Diffuse Matter in Interstellar Space, *Observatory*, **49**, 247–250.

Jeans, J.H., 1926b. Diffuse Matter in Interstellar Space, *Observatory*, **49**, 333–335.

Kopal, Z., 1986. *Of Stars and Men*, Adam Hilger, Bristol.

McCrea, Sir William, 1991. Arthur Stanley Eddington, *Scientific American*, June 1991, 66–71. ©1991 Scientific American Inc. All rights reserved.

Nielsen, Axel V., 1963. Contributions to the History of the Hertzsprung–Russell Diagram, *Meddelelser fra Ole Rømer-Observatoriet i Aarhus*, Nr 30.

Pais, Abraham, 1982. *'Subtle is the Lord . . . '*, the Science and Life of Alfred Einstein, Clarendon Press, Oxford.

Payne-Gaposchkin, C., 1984. *An Autobiography and Other Recollections*, ed. Haramundanis, K., Cambridge University Press, Cambridge.

Redman, R.O., 1945. Sir Arthur Eddington, O.M., *Mon. Notes Astr. Soc. South Africa*, **4**, 26–27.

Rosenthal-Schneider, Ilse, 1949. *Reality and Scientific Truth*, Wayne State University Press, Detroit, MI.

Russell, H.N., 1914. Relations between the Spectra and Other Characteristics of the Stars, *Popular Astronomy*, **22**, 275–294 and 331–351.

Smart, W.M., 1945. Sir Arthur Stanley Eddington, O.M., F.R.S., The Astronomer *The Observatory*, **66**, 1–6.

Soldner, J., 1921. Über die Ablenkung eines Lichtstrahls von seiner geradlinigen Bewegung durch die Attraction eines Weltkörpers, an welchen er nahe vorbeigeht. With an introduction by Lenard, J., *Annalen der Physik*, **65**, 593–604.

Stanley, Matthew, 2003. An Expedition to Heal the Wounds of War, The 1919 Eclipse and Eddington as Quaker Adventurer, *Isis*, **94**, 57–89.

Turner, H.H., 1916. *The Observatory*, **39**, 240–242.

von Klüber, H., 1960. The Determination of Einstein's Light-Deflection in the Gravitational Field of the Sun, *Vistas in Astronomy*, **3**, 47–77.

Wali, K.C., 1991. Chandra, *A Biography of S. Chandrasekhar*, University of Chicago Press, Chicago, IL.

Whitehead, A.N., 1932. *Science and the Modern World*, (cheap edition), Cambridge University Press, Cambridge.

8

HARLOW SHAPLEY: DEFINING OUR GALAXY

Shapley did for the Milky Way system what Copernicus had done for the solar system: He placed our sun and earth in the outskirts of the Milky Way system. He proved conclusively that our sun and earth are definitely not located close to the center of our galaxy. (Bok 1978)

Harlow Shapley was born on 2 November 1885 in Missouri. His birthplace was the small town of Nashville in Barton County. Only 12 km away was Lamar, where President Truman had entered the world, one year before him. Harlow was one of non-identical twins, his brother's name being Horace. They were called after their grandfathers. He had an older sister, Lilian, and a younger brother, John. His father, Willis Shapley, was a farmer and hay dealer who had 'gone west' as a teenager in the 1870s. The Shapleys were an 'old' American family whose ancestors had lived in Connecticut during the Revolutionary War and later in the Hampton, New York area. Land in Missouri had been granted to Harlow's grandfather for his service in the Civil War between the Northern and Southern states.

When Shapley was an old man, he wrote, or gave interviews for, an autobiography called *Ad astra per aspera; Through Rugged Ways to the Stars*, (Shapley 1969) which, according to his protegé Bart Bok 'is not the very best of autobiographies, but it does show the true Harlow Shapley with all his wonderful ideals, his vanity, his compassion, and his greatness'. Shapley was fond of giving public lectures and had developed a style for amusing his audiences by means of humorous asides. This habit carried over into his book, in which he over-emphasised the humbleness of his origins and tried to present himself as a 'simple country boy', writing of his farm as 'a pretty average place'. But he also let slip that his uncle Lloyd Shapley was a 'navy man', who was at one time governor of Guam, a United States possession in the Marianas Islands, east of the Phillipines. This uncle, on one of his visits, taught the boys to shoot with rifles and shotguns even though Shapley described his family as 'non-militaristic'.

The picture Shapley painted of his home life was of a rustic household where there was a daily newspaper, the St Louis *Globe-Democrat*, but only a few books. One of these was *Three Men in a Boat* by Jerome K. Jerome. There were also some books of poetry and an organ. The twins were persuaded by their mother, Sarah Stowell Shapley, to attend a Presbyterian Sunday school in a neighbour's house. She was nominally a 'hard-shell' Baptist—a strictly Calvinistic group, though Shapley claimed later that this was because her father was one; she was not a fundamentalist herself. 'Our father was without religion; the whole family has been without formal religion and has got along pretty well that way'. His autobiography, read between the lines, gradually reveals an ambitious family—his grandfather ran for State Senator more than once, though never successfully. He was a rabid Republican in a thoroughly Democratic area. His father Willis was 'something of a leader in the community', though not given to praising his son's achievements and abilities—and Shapley was a person who needed praise. Democrats were described as 'tobacco-spittin' people'. How his forbears would have reacted to Harlow's later political activities is hard to imagine.

At first the twins attended a country schoolhouse near their farm, travelling on horseback. One of the teachers was his own sister, who told him later that he was the best student she ever had. For a few weeks, they went to a proper school in Warrensburg, Missouri, but they had to leave when the family ran out of money. According to his autobiography, Shapley narrowly escaped death at one point when a team of mules ran away with him, dragging him along the ground. His father came to the rescue and cut the straps to set him free.

At the age of 10, Harlow and his brother were sent to stay with their paternal grandfather Horace Stowell in Hamilton, upstate New York. They lived in an old stone farmhouse and went to another country school, now called in his memory the 'Shapley School'. This grandfather was a strong abolitionist—somebody who felt very deeply that slavery was wrong—and had helped escaped slaves from the South to reach Canada. An outlying building of his property had been a station of the 'underground railway', a network of safe houses that escapees could make use of on their journey. Perhaps it was this unconventional grandfather who gave Harlow his later interest in helping victims of the Nazis and in fighting political rightists.

After a year it was back to Missouri and the farm. He spoke of going at the age of 15 to a 'sort of business school' in Pittsburg, Kansas, for a few months. He then became a crime reporter on the *Daily Sun* of Chanute, Kansas, a rough oil town. Here at 16 he witnessed a shoot-out at close quarters during an election. His paper adopted a strong political line and he took pleasure in reporting the opponent's rough language, complete with four-letter words! Though the right candidate won, he was threatened for his pains by the loser. Chanute was followed by Joplin, Missouri, an even less inspiring mining town with even worse tough characters. Altogether, he spent two years as a reporter. This experience was useful in that it forced him to learn shorthand and also taught him how to handle the press in an effective manner.

His aim at this time was to raise enough money to go to college. Chanute had a Carnegie Public Library, one of many founded worldwide by the philanthropist Andrew Carnegie in his old age, and Shapley took good advantage of it to broaden his outlook. He also began to buy serious novels such as the works of Tolstoy and Dostoievsky.

When the family thought they had discovered zinc and lead on the property their mother, always ambitious for her sons, who, she hoped 'should amount to something, get somewhere, go to school' [university] expressed the hope that there would soon be enough money to send them to college. Though the promise of instant riches did not materialise, years later Shapley was happy to think that his mother had lived long enough to witness his academic success.

When they had scraped together a little money, Harlow, then seventeen or eighteen, and his younger brother John set off for the high school at Carthage, about 20 miles (32 km) away, in the hope of entering during the second semester. They were refused as 'not qualified'. Fifty-seven years later, the town of Carthage celebrated 'Shapley Day', complete with fourteen marching bands and thirty floats. On that occasion, the school that had rejected him gave Shapley an honorary high school diploma and the State Legislature passed a resolution of commendation.

In spite of their setback, the brothers pressed on. Next they tried the Carthage Collegiate Institute, a less prestigious school of Presbyterian origin a few blocks away which turned out to be willing to accept them for the little money they could afford. An extraordinary source of funding was a cheap railway excursion between terms to New Orleans. They experienced two 'mild train wrecks and a fire', which the railway paid them to keep quiet about. He also took up reporting again on the Joplin *Times* for a short while. He later claimed that he had learned Latin and geometry while working on his father's hay wagon. He found that he could flatter his teachers and get high marks in his examinations. After completing his final school project on 'Romanticism in English Literature (Hoagland 1964), he graduated in 1907 having done six years work in just two semesters. He was the valedictorian (top pupil) of a class of three! The way to university was now open.

8.1 University of Missouri

Shapley went to the University of Missouri, in Columbia, Missouri, with $200 in his pocket and the idea of studying journalism. To his initial dismay, its new School of Journalism was not due to open for another year. In a story he loved to repeat, but which sounds highly improbable, he claims that he opened the University Catalog to look for another subject to study. Unable to pronounce 'Archaeology', he went on to 'Astronomy'. Probably, his first contact really occurred when he was looking for a job on campus and was offered one as an assistant for 35 cents an hour at the Laws Observatory, a small but fairly well-equipped installation, typical of its time, with a meridian instrument for finding the positions of stars and a $7\frac{1}{2}$-in. (19 cm) general-purpose telescope.

Here he was extremely lucky: Frederick H. Seares, who was later to become an important staff member at Hale's Mount Wilson Observatory, was in charge of the subject and rapidly took a liking to him. Seares had written a practical textbook and was according to Shapley a good teacher, 'prim and neat'. He occasionally praised his protegé, who had needed such encouragement, by saying 'That's well done'. In many little ways he directed Shapley's talent. Missouri also had two good mathematicians, Profs Oliver Kellogg and Earl Hedrick, who saw that Shapley was mathematically talented. Nevertheless, he very nearly failed physics at one point—partly due to a teaching assistant who had marked a test paper incorrectly. Two years later he graduated with 'high honors' in mathematics and physics.

Shapley's undergraduate years at the University of Missouri were a period of excitement and intellectual awakening for him. He recounts how he enjoyed learning English literature, French, and Latin from teachers who he found outstanding and inspiring. He thought afterwards that, were it not for the job he got at the Laws Observatory, he might have become a classicist. He particularly enjoyed the *De Rerum Natura*[1] of Lucretius. He even wrote a paper for *Popular Astronomy* on *Astronomy in Horace*[2] (Shapley 1909).

His account shows that he led a quiet and studious life at the University of Missouri, taking little part in student activities—even avoiding the student newspaper in spite of his journalistic experience. He played tennis until he injured his elbow. He stayed with a roommate in a University dormitory after his second year. He described how after dinner 'they all went upstairs and danced for a while. That was a sort of social affair, but it was mostly men dancing with men'. The summer school was livelier and more enjoyable.

It was at Missouri, in his third year, that he met his future wife, Martha Betz, of Kansas City, Missouri, in a mathematics class. 'She was a clever lady in those days. She took five full courses and got the top mark in all five. Her field was philology and German literature and such things. Eventually, it became astronomy'. One wonders what exactly he meant by 'in those days'. He made her learn the Gregg shorthand system that he had acquired as a reporter and they used this 'language' to correspond.

Seares left Missouri for the better equipment and clearer skies of Mount Wilson while Shapley was still there. His place was taken by Eli S. Haynes, who seems to have been a rather crude character. Shapley related how, while listening to a Brahms quartet some years later at Princeton, his mind wandered and he burst out laughing, very inappropriately:

> I got to thinking dreamily of the past ... when Eli Haynes came into the observatory office and instead of putting his hat down on the typewriter and using the spittoon, did it the other way around. He spit on his typewriter and ... I suddenly remembered how Haynes' eyes bugged, I burst out, "Ha, Ha, Ha!" Then I nearly died of humiliation. (Shapley 1969, p. 34)[3]

In 1910 he received his bachelor's degree (A.B.) and he stayed on an extra year to get a master's (A.M.). He already had two year's experience of teaching elementary astronomy by that time. Professor Kellogg encouraged Shapley to apply for the Thaw Fellowship in astronomy at Princeton so that he could go

---

[1] Of the Nature of Things, by Titus Lucretius Carus, (99–55 BC) Roman poet and Epicurean philosopher.

[2] Quintus Horatius Flaccus 65–8 BC, another great Roman poet and a follower of Epicurus.

[3] Reprinted with the permission of Scribner, an imprint of Simon & Schuster Adult Publishing Group, from THROUGH RUGGED WAYS TO THE STARS by Harlow Shapley. Copyright ©1969 by Harlow Shapley.

on for a doctorate. With good recommendations from people like Seares, now at Mount Wilson, and Kellogg, he was successful. His family were very happy and his father even boasted a little about his clever son. 'My mother of course thought that I was on my way to the presidency of the United States'.

In contrast, his twin brother Horace did not follow the academic route. He was eventually to take over the family farm and free it from debt. In his eighties he decided it was time to acquire a little learning and he became an extension student at the University of Missouri.

8.2   Princeton and Henry Norris Russell

The Thaw Fellowship was worth $1000, the best-paid that Princeton offered. Shapley received free tuition and subsidised accommodation. For the first time he found that he was comfortably off.

Once again fate looked after him. The head of the department of astronomy was Henry Norris Russell, the son of a strict Presbyterian minister and the recipient of a substantial private income who was used to being taken care of by a doting family and several servants (DeVorkin 2000). Russell, unusually for American astronomers of the time, was at the very forefront of theoretical astronomy. The sizes, masses, and luminosities of typical stars were just becoming known. He and the Danish astronomer Ejnar Hertzsprung had recently shown that stars could be enormously different in luminosity, even if their spectra appeared to be almost the same. They had reached their conclusions completely independently of one another and graphed their results similarly. Their plot is now known as the H–R or 'Hertzsprung–Russell' diagram (see Fig 7.2).

Meeting the conservative Russell on his arrival in Princeton in the autumn of 1911 was a strange experience for the still unsophisticated Shapley:

> Russell was a very shy man, and he had to meet this unknown graduate student that Missourians spoke well of. The first day Russell asked me to come with him and Robert Williams Wood of Johns Hopkins [optical physicist]—the famous Robert Wood—to look through the 23-inch telescope for sulphur deposits on the moon. Russell was a high-class Long Island clergyman's son and very high hat.
>
> 'Mr. Shapley, perhaps you would open the door for us?' he would say. Mr. Shapley did open the door. He didn't like it! I don't quite know what was expected. Perhaps he thought I would kowtow to him, which I suppose I did to some extent; he intimated that I was, after all, a wild Missourian of whom no one should expect much.
>
> But that attitude lasted only a little while. In a few days I came to him with an orbit I had solved by a method that he had used. That woke him up, and pretty soon we were the chummiest of creatures. Students were interested when Shapley, the Missourian, and Russell, swinging a cane, would stroll across the campus. If students got in the way, Russell would just brush them off with the cane. (Shapley 1969, pp. 31–32)

At the time, Russell's main interest was in the orbits of binary stars and Shapley joined him in his investigations. Binaries are stars in close orbit around

FIG. 8.1. Harlow Shapley in 1923, photographed by Bachrach (O. Gingerich).

each other but they are usually too far away from the earth to be seen separately. However, in certain cases where we happen to observe them from the right direction, they eclipse each other. Thus, from the study of brightness changes amongst the stars binaries can be found. By graphing brightness against time a period can be extracted. Using Newton's laws, Shapley and Russell found that it was possible to estimate their sizes and luminosities and hence their distances from their light curves and their velocities as obtained from their spectra. He and Russell developed approximate mathematical methods for analysing their behaviour. Shapley's practical energy was in stark contrast to the somewhat mollycoddled and more theoretically inclined Russell but they complemented each other very well.

One of Russell's colleague at Princeton,

> Raymond Smith Dugan, used to say that if there was a piece of work which would have taken him (Dugan) half an hour to accomplish, Russell would spend half a day thinking how to do it more easily; and should he succeed in doing it in a way 'which does not take any time', he would regard this as an intellectual triumph. (Kopal 1986, pp. 181–182)

Shapley learned much from the well-informed Russell and his colleagues and also from the colloquia offered by the department's many visitors. In this way, he came to know what was happening at the frontiers of research and gradually found he was able to make significant contributions of his own. Russell always regarded him as his best graduate student. He was encouraged to expand his mind in other, non-astronomical, directions: Dugan was an expert in the wild flowers of the area and Shapley acquired this interest from him. He also attended lectures in paleontology and physiology out of curiosity. He describes how Princeton graduate students were encouraged to 'get culture' by taking advantage of the proximity of New York to attend concerts and other events.

8.3 'Standard candles' and the distances of the stars

In 1911, the distances of only the nearest stars were known and the true size of the Universe could not even be imagined.

To find the distance of a star, one uses the earth's orbit as a survey baseline and observes by what small angle the object of interest seems to move against the background of more distant stars as the earth moves around the sun. This effect is known among astronomers as 'parallax'. A triangle is formed by the lines joining three points—two places on the earth's orbit, six months apart in time, and the star itself. From the angles and the baseline, one can calculate the distance by trigonometry. The problem is that this method is only suitable for the very closest stars—the angles become too small to measure accurately enough for more distant ones.

One method for measuring greater distances is to find a group of stars of known luminosity or 'wattage', with characteristics that make them easy to recognise among the myriads of others. Such well-behaved stars are known as 'standard candles', because a lot of early scientific work on measuring brightness was done in the days before electric lights and the standard of luminosity was then a particular kind of candle. The apparent brightness of any light source like a light bulb or a star is reduced the further away it is—according to an inverse square law—rather like the force of gravity. Thus, if a star can be recognised as being of a standard type and its brightness can be measured, one has a means of finding its distance. The search for astronomical 'standard candles' goes on to the present day and is critical to our understanding of the size of the Universe. It was Shapley's hope that the eclipsing binaries could be used in this way: his thesis project with Russell was on this subject.

As a former reporter, Shapley realised very quickly the virtue of publicity if he was to make his way in astronomy. He presented his first results before the Pittsburg meeting of the American Astronomical Society in 1912 and took care that his lantern slides were professionally produced, to make a good impression. As he began to see that he could work independently of Russell, his confidence grew. Soon he was publishing about six research papers per year and was becoming known to astronomers elsewhere. His PhD thesis of 1913 on *The orbits of eighty-seven eclipsing binaries* was published in summary in the *Astrophysical Journal* (Shapley 1913).

After finishing his thesis, he and his younger brother John scraped together enough money to make a trip to Europe. 'We went steerage, and lived cheaply, as students do and should'. John's interest lay in linguistics, archaeology and art and he afterwards had a successful career at New York University and the University of Chicago. While John looked at churches and art galleries, Harlow visited observatories, spending just a little time on botanising. He travelled Europe from end to end. In England he attended a meeting of the Royal Astronomical Society, where he met 'everybody of stature'. At the Astronomische Gesellschaft (German Astronomical Association) in Bonn he met Hertzsprung, the co-inventor of the Hertzsprung–Russell diagram, and the famous German astronomer Karl Schwarzschild.

While in Paris he received the tragic news that his father had been struck by lightning and killed. He described how he wandered the streets for a long time in shock—a greater shock than he experienced at his mother's and his sister's deaths.

Returning to Princeton, in the autumn of 1913, Shapley held a Proctor (post-doctoral) Fellowship. He remarks in his autobiography (Shapley 1969, p. 43) how easy it had been to get a PhD in the days when the thesis counted for everything—'very loose management compared to today'. In fact, the thesis is still almost the only requirement for a PhD degree in many countries. 'But at Princeton the thesis was [then] the thing, and mine was considered rather outstanding'. At his oral examination he was asked by Dugal 'How would you adjust a photographic telescope?'—a question he could not answer, but his degree was awarded *magna cum laude* in spite of it.

Nearing the end of his PhD studies, Shapley had begun to think of his future. He had written to his undergraduate mentor Seares at Mount Wilson Observatory to ask about a job. As a result he was invited to meet George Ellery Hale, who was about to visit New York.

> It was a remarkable meeting. I went from Princeton to New York in time to stay overnight restfully, and I saw two operas. One was *Pagliacci*; the other was *Versiegelt* by Leo Blech. The next morning I had breakfast with Seares and Hale, who were on their way to Europe. Hale asked me what I had done the night before, and I told him about the operas ... We talked at length. Then, unexpectedly, he said, 'Well, I must be going.'
>
> Not one word had been said about astronomy or about my going to

Mount Wilson, or anything like that. He had met me, and we had chatted about something neither of us knew much about. Then he went away.

I didn't know what this meant. Had my table manners slipped up? (Once my uncle gave me a gun for my good table manners!) I said to Seares 'I don't quite understand this. Have I done something wrong?'

'Oh no, no, don't worry about that', Seares said. I think he was sadistically enjoying the whole affair. He later explained that Mr. Hale knew that I would know astronomy, perhaps more astronomy than he did, and that I had the Princeton touch. When he found that I seemed to be a decent guy, that was all he wanted to know ...

Then, not long afterward, I got a letter. It said: 'Please come to Mount Wilson'. (Shapley 1969, pp. 44–5)

8.4   First visit to Harvard

With his immediate future assured, Shapley made a number of trips to get himself better known in other places. He visited Yale and Brown Universities, receiving a job offer at the latter. His first visit to Harvard was the highlight of this period and he picked up several ideas there that eventually led to his most important discovery. He met the leading members of the Harvard College Observatory. He had dinner with 'the famous and jolly' Annie Jump Cannon. She teased him by saying 'Young man, I know what you're going to do. You're going to be the director of the Harvard Observatory.' and laughed. But she remembered her words ten years later when he *did* become director.

He met E.C. Pickering, the then director, who was an expert in photometry— measuring the brightness of stars. He was a large, impressive man. After dining at his house, Pickering played the flute for his benefit.

Solon I. Bailey, the second-in-command, 'so New England that it made you ache', suggested to Shapley that he should use the Mt Wilson 60-in. for measuring stars in globular clusters. Globular clusters are dense groups of up to billions of stars, so-called because they are globular in overall shape. They are not very numerous; only about 150 are known in our Milky Way galaxy. Bailey had made a detailed study of globulars himself and had found that they contain many interesting variable stars, some with periods less than about one day, called RR Lyraes.[4] Shapley was later to follow up Bailey's suggestion with dramatic results.

8.5   On to Mount Wilson

Shapley headed for California by train in April 1914 for what was to become a seven-year stay. He stopped in Kansas City, Missouri, to get married to his fiancée, Martha Betz. She had in the meantime been a graduate student at Bryn Mawr, a women's college of high standing near Philadelphia. They spent their honeymoon on the train, calculating the orbits of eclipsing binaries! On arrival in

---

[4]Types of variable stars are usually named after the first one found of each kind. In this case, RR Lyrae was the first one found outside a globular cluster, in the constellation of the Lyre.

Pasadena, they found a place to live near the Santa Barbara Street headquarters of the Mount Wilson Observatory. Hale was already mentally ill by then, with periods of remission. Shapley had to get to see him because he was being paid less than he had expected—an error which was rapidly corrected. His salary went from $90 per month to $135. Later, as a famous observatory director, he liked to point out to new employees that 'When I was your size they paid me $90 a month and I was already internationally famous'. In fact, they had been expecting him to live in the observatory hostel—the famous 'Monastery' on the mountain—but had not realised he would arrive with a wife.

His arrival at Mount Wilson occurred at the beginning of a great period in its history—a history that he helped to make. With the best climate of any observatory then in existence, for both good clear weather and steady atmospheric conditions, as well as possessing the largest professional telescope—the 60-in. reflector—it was to become the astronomical Mecca. Besides Shapley, Hale around the same time engaged other young astronomers such as Adriaan van Maanen and Roscoe Sanford, who were to have illustrious careers based on the observations they obtained on Mount Wilson. Already present was his former professor, Seares, to whom he was assigned as an observing assistant, with the task of measuring the colours and brightnesses of stars. He found the conditions on the mountaintop somewhat rugged. In particular, access to the observatory was on foot or by mule up a winding dirt road to an altitude of over 5800 ft (1770 m). Strangely for an astronomer, Shapley admitted later on that he did not like telescope work and that he found the night-time stints very hard to take. He froze during the cold winter nights and was not able to get enough sleep in the daytime.

8.6 The Cepheid and RR Lyrae standard candles

Just after reaching Mount Wilson, Shapley published his first important paper. The Cepheid variables, which change their brightnesses periodically with a cycle time of a few days to a few tens of days, were believed until then to be eclipsing binaries. From his own work he realised that they were nothing of the sort: instead he suggested that they are single stars that pulsate by getting bigger and smaller in a 'rythmic' way. His paper stimulated Eddington to his investigations of the conditions inside stars, in the course of which his own model was confirmed.

Then, following up the suggestion made to him earlier by Bailey, he turned his attention to studying the globular clusters. As mentioned, Bailey had found that the clusters contain many RR Lyrae variables with periods of less than one day. Bailey had also found some somewhat longer-period variables (now called 'Type II Cepheids' or W Vir stars) that had very similar periods to the ordinary Cepheids and were for many years mistaken for them.

Before his work, an interesting property of the Cepheids had been found by a rather formidable New Englander at Harvard College Observatory, Miss Henrietta Swan Leavitt (1868–1921). Her concern had been the discovery of variable stars. She was descended from the early puritan settlers of New England:

Miss Leavitt inherited in a somewhat chastened form the stern virtues of her puritan ancestors. She took life seriously. Her sense of duty, justice and loyalty was strong. For light amusements she appeared to care little. (Bailey 1922)

She had found nearly two thousand variable stars in the Magellanic Clouds and had determined the periods—the time from one maximum to the next—of several in the smaller of the two Clouds. The Magellanic Clouds are two large nebulous objects in the Southern Hemisphere sky that look like detached parts of the Milky Way.[5] The great advantage of studying stars in one of the two Magellanic Clouds is that they are all at about the same distance from us, although that only became evident as a by-product of Miss Leavitt's work. In 1908 she published a paper in which she showed that there is a relationship between average brightness and period obeyed by some of her newly discovered variables. In another important publication dated 1912, she even suggested that they were similar to certain nearby stars for which parallaxes might be measureable (Leavitt 1912).

Hertzsprung saw Miss Leavitt's paper and realised that her variables must be Cepheids. He attempted to perform a calibration for stars of this type in the Milky Way by means of a relatively crude method called 'statistical parallax' (for example, by using the Sun's movement through the other stars, taken over a few years, as a trigonometric baseline). His paper, published in 1913, was spotted in turn by Russell, who repeated his calibration work and probably discussed it with Shapley.

Later, following an intensive observing programme with the 60-in telescope and with the aid of many leaps of faith, Shapley re-calibrated the Cepheids using much the same method that Hertzsprung and Russell had done before. He smoothed the data in a way that would now be considered outrageous—a process politely referred to later by Heber D. Curtis as Shapley's 'elaborate system of weighting' (Gingerich and Welther 1985).

8.7   The distances of the globular clusters

Shapley guessed that the RR Lyrae variables that Bailey had found in the globular clusters might also be standard candles. Thus, if he measured their average apparent brightnesses, he could get the *relative* distance of the clusters, but not their *actual* distances.

Fortunately one of the globular clusters, known as ω (omega) Cen, contained both RR Lyrae variables and (type II) Cepheids. Since Shapley knew how bright Cepheids were, he proceeded to calibrate his RR Lyraes using the ω Cen Type II Cepheids as if they were ordinary Cepheids, since all the stars within the cluster could be assumed to be the same distance away from us. Strangely enough, his calibration was, by a series of flukes, not very different from later values (Baade

[5]The Magellanic Clouds are now known to be separate 'dwarf galaxies' that are members of the local group of galaxies that includes the Milky Way.

1963) even though he was wrong to have treated the type II Cepheids and the ordinary Cepheids as though they were the same.

He was now in a position to measure the actual distances of the globular clusters. He just had to extract the apparent brightnesses of their variable stars. This was, however, a major task, and required many photographic plates to be taken of each cluster. The sensitivity of each batch of plates had first to be found using stars that did not vary. Then, the brightness of each variable star image had to be measured. The periods of the variables had to be extracted using hand computations and their median magnitudes determined. Only then could his calibrations be used to find the distance of the clusters. By this time, around 1917, he was working extremely hard and publishing at a furious rate, sometimes writing as many as twenty research papers in a year.

8.8   The centre of the Milky Way

Shapley's greatest discovery came in 1917. Combining the directions of the clusters on a map of the sky with the distances he had obtained, he was able to plot their positions in three dimensions and conclude that they formed a swarm around a point located 55,000[6] light years away in the direction of the constellation Sagittarius (See Fig 8.2 for modern versions of his diagrams). He looked again at the questions raised by Hinks and Eddington: why are most globular clusters in or near the constellation Sagittarius and what is so special about that direction? The answers had to be that the centre of their distribution must be the centre of the Galaxy. In 1919 he published a famous paper called *Remarks on the Arrangement of the Sidereal Universe* (Shapley 1919), where he presented his results.

Copernicus had suggested that the Earth was not the centre of the Universe. He had thought that honour was due to the sun. Now Shapley had shown that the sun was just an ordinary star at a great distance from the centre of the Milky Way! With Seares, he calculated that our sun was just one out of a hundred thousand million stars or so in the Galaxy. He delighted in pointing out

> the solar system is off centre and consequently man is too, which is a rather nice idea because it means that man is not such a big chicken. He is incidental—my favourite term is 'peripheral'. (Shapley 1969, pp. 59–60)

His general picture of the shape of the Galaxy and the position of the sun within it were accepted by most astronomers rather quickly, though Kapteyn, the discoverer of the 'star drifts', who died in 1922, never took to it. Also, his radical assertion that the actual diameter of our galaxy is 300,000 light years was not believed by all.

---

[6]The present-day figure is about 24,000 light years.

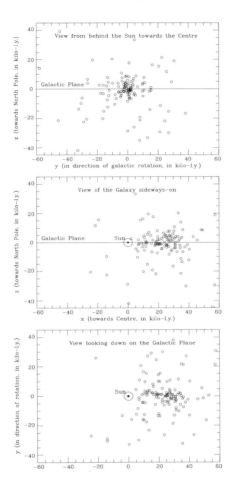

FIG. 8.2. The distribution of globular clusters in space, derived from modern
data (Harris 1996). The positions of the sun ⊙ and the Galactic Centre × are
marked in the bottom and middle panels. Both are in the centre of the top
panel. Using diagrams of this kind, Shapley guessed that the centre of the
globular cluster system is the centre of the Milky Way galaxy. Each diagram
is about 120,000 × 90,000 light years. The middle panel can be compared
with the Herschel (Fig. 4.4) and Eddington (Fig. 7.3) sun-centered diagrams.

8.9   A near miss

Very strangely, in late 1917, after a spate of discoveries of novae in spirals such
as the Andromeda Nebula (M31), Shapley (1917) had written a paper in which
he took very seriously the possibility that they were located well beyond the
Milky Way. If M31 were indeed an 'island universe' (i.e. what we would now

call a galaxy), from the fact that its individual stars were undetectable it had to be at least of order a million light years away.[7] The brightness of its recently observed novae, when compared with those that occur in the Milky Way, was consistent with this figure. Further, the globular clusters of the Milky Way, if placed at the estimated distance for M31, would resemble some of its observed features. But Shapley immediately abandoned his estimate of the distance of spirals! Measurements made by a Mount Wilson colleague, Adriaan van Maanen, in whose work he placed too much faith, ('After all, he was my friend') seemed to show that the spirals were rotating rather fast. If the observations were correct, they would have to be quite close by. Van Maanen was a trusted observer of the proper motions of nearby stars, but the position of a fuzzy object such as a spiral galaxy cannot be measured very precisely; this is what led to his erroneous observation. As an individual, he was pleasant and socially polished. Shapley had taken to him: 'He was rather an alert-minded person and I liked his nonsense'. Unfortunately, it took many years before his result could be disproved in a definite manner and indeed the whole question later became the subject of an acrimonious disagreement between van Maanen and Hubble (see Chapter 9).

8.10 The 'Great Debate'

One man, Heber D. Curtis of Lick Observatory, took on the mantle of 'leader of the opposition' in relation to Shapley's view of the universe, opposing him on several points.

Hale was excited by Shapley's discovery and was well aware of its sensational, not to mention its philosophical, implications. He decided that a debate about the new picture of the universe should be staged between Shapley and Curtis under the auspices of the National Academy of Sciences in Washington, where it would have the greatest scientific impact. Funds were available from the William Ellery Hale Foundation, named after his father. The event was scheduled for 26 April 1920. He handed Shapley a cheque for $250 to cover his expenses. Shapley recollected 'I would have gone anyway; we had already resigned ourselves to poverty'. As it happened, he and Curtis found themselves on the same train headed east for Washington. They studiously avoided the subject of the debate, talking about 'flowers and classical subjects'.

Shapley believed:

1. (more-or-less correctly) that the universe is large and that the galaxy is of order 300,000 light years across.
2. (wrongly) that the spiral nebulae, which we now know to be galaxies, were part of our own Milky Way.

while Curtis maintained

1. (correctly) that the spiral galaxies are beyond the Milky Way.
2. (wrongly) that the universe is small.

[7]The modern value is about twice this.

For Shapley, the Great Debate had personal significance. The Director of Harvard College Observatory, Edward Charles Pickering, had died suddenly in February 1919 and Shapley was interested in becoming his successor. 'Should I or should I not? Should I curb my ambition?'. Not much given to modesty, he soon concluded 'All right, I'll take a shot at it'. After all, Harvard Observatory enjoyed immense prestige and the large telescopes of their West Coast rivals were only just beginning to show their immense potential. He was met in Washington by Harvard representatives—George R. Agassiz, a member of the 'Visiting Committee' of the astronomy department and Theodore Lyman, Chairman of the Physics Department, both friends of the President of Harvard University, Abbot Lawrence Lowell (brother of the eccentric astronomer Percival Lowell who believed that the surface of Mars was criss-crossed by canals.). They had been asked to 'look him over' to see if he might be a suitable candidate for the job. Even the deportment of Shapley's wife, Martha, had been checked, at Harvard's request, by Prof John C. Duncan of Wellesley College to see if she would fit in. 'Harvard took pains about that kind of thing'.

Russell advised Shapley against taking such a position and William Pickering and others felt that he would be foolish to give up his astronomical career 'just to be director of that observatory'. However, his reputation was already high. Even Kapteyn, a critic of his discoveries, mentioned him as a possible director, Russell being his other choice. Hale thought Shapley the best candidate, but was worried about his youth 'He is about 30, very brilliant, but [I am] not sure how well he could manage men'. A Harvard official, sent to supervise the Harvard team at the annual Rose Bowl football tournament in Pasadena, gave tickets for the game to the Shapleys and used the occasion to observe his behaviour. '[I] consider him a man who would work well with others and others with him. Agreeable and serious personality. Attitude of his fellow-workers toward him good ...'. Hale, however, prompted by Edwin Frost, the Director of Yerkes Observatory, started to hedge, and even suggested Seares as a suitable person. Fortunately, Lowell asked him to explain his change of mind:

> As I said ... I really believe he would prove a great success at Harvard.
>
> On the other hand, I clearly remember saying that in view of Shapley's daring and comparative youth, you would necessarily take some chances in offering him the place. My own regard for his exceptional qualities is so much in advance of that of the majority of American astronomers that I felt bound to point out this aspect of the case. He is much more venturesome than other members of our staff and more willing to base far-reaching conclusions on rather slender data. Nevertheless, the best judges among them, after critical analysis of his work, are in accord with his views.[8]

This was the situation just before the Great Debate. The audience of a couple of hundred people was a distinguished one, mostly consisting of non-astronomers.

[8]Hale to Lowell, 29 March 1920; quoted by Gingerich (1988).

Shapley was very nervous and misjudged his hearers by giving a rather poor presentation at a relatively unsophisticated level.

In later life Shapley claimed to be surprised that the debate had come to be regarded as astronomical history. He said that he had forgotten about the whole thing. However, it has since then been the subject of numerous articles; probably because it encapsulated the opinions about the universe as seen at the time. In retrospect, both men are seen to have been partly right and partly wrong. As additional evidence accumulated, for example the discovery by the Swede Bertil Lindblad and the Dutchman Jan Oort that the stars of the galaxy are rotating around its Centre, Shapley's intuitive discovery came to be accepted more and more. However, up to around 1930, there were still those who believed in a sun-centered Universe (Gingerich 1999).

Agassiz and Lyman were not impressed by Shapley's showing at the debate. They reported to the President of Harvard that Shapley 'does not give the impression of being a big enough personality for the position' (Gingerich 1990). It was felt that Russell was perhaps the best candidate after all. He was approached, but got cold feet and turned the appointment down. A period of negotiation with Shapley then commenced. He was in effect appointed as interim director in April 1921, on probation. By the end of October, Lowell had been sufficiently impressed that Shapley was made permanent director of the Observatory.

After the 'Great Debate', Shapley became for several years the champion of those who did not believe in external galaxies or 'Island Universes'. Apart from van Maanen's erroneous conclusion that the spirals were close by, Shapley had previously estimated from his standard candles that the Milky Way galaxy was really huge—about three times the currently accepted value, and that the external galaxies were small and close.

There was a curious episode, related many years later by Milton Humason to Allan Sandage, that happened just before Shapley returned to the East. He showed Shapley a plate of M31, the Great Nebula in Andromeda, now known to be the nearest spiral galaxy, on which he had found some new images that seemed to differ in brightness from previous ones in the manner expected of variable stars. He had marked the plate on the back to show where they were located. Shapley, fresh from the Great Debate, told him that he could not be right and coolly wiped his marks off the plate! If this story is true, Shapley had again missed making the great discovery that later fell to Hubble's lot, that the spirals lie beyond the Milky Way![9]

8.11 Director of Harvard College Observatory

Harvard College Observatory had an immense reputation for the photographic work it had been doing with small telescopes, mainly on the spectral classification of large numbers of stars. Shapley's predecessor as director, Pickering, maintained a number of very capable lady astronomers on pitiful wages (though perhaps not

[9]See also Trimble 1996.

FIG. 8.3. Shapley sitting beside Ejnar Hertzsprung at the hundredth anniver-
    sary meeting of the Royal Astronomical Society in 1922, the year after his
    appointment as Director of the Harvard College Observatory (SAAO).

unusual for the time) to analyse the plates, not allowing them much scope for
interpretative work. Of these, Annie J. Cannon and Henrietta Leavitt were the
most famous, each making significant contributions to the progress of astronomy.
There was also the photometrist, E.S. King, and Bailey. Photometry, or the
measurement of the brightnesses of stars, is central to understanding the distance
scale of the universe, but is laborious work that lacks the glamour of many other
areas of astrophysics. The Harvard 'A, B, C ...' regions are still the backbone of
photometric calibrations, though much better measured today through the use
of photo-electric techniques.

Bailey had been Acting Director during the interregnum and he tactfully
headed off to Harvard's field station in Arequipa, Peru, to get out of Shapley's
way. We should recollect that Shapley owed to him the hint to get into globular
cluster work, the area in which he made his name.

Shapley lost no opportunity to make himself into a public figure and gave many popular lectures, some of which his New England audience clearly felt to be 'over the top'. He was reproved by Charles William Elliot, a former President of Harvard, after one of these affairs, with the words 'You're a very young man and you have a great subject. But you don't need to emphasise it'. He was by no means slowed down and continued on his publicity-seeking path.

> As a popular lecturer and a public speaker he had few rivals. He could hold an audience spellbound, using all the arts of the actor to enliven the most arid subjects. His timing was flawless ... A photograph of Messier 8 was on the screen. He pointed out the patches of obscuration. 'Some people might say they are the fingerprints of God; but perhaps they are only the fingerprints of the careless devil who made the plate'. (Payne-Gaposchkin 1984, pp. 157–158).

> The young director was everywhere, running upstairs two steps at a time, pushing his soft sandy hair off his forehead, greeting everyone with the same casual cheerfulness. He knew exactly what each member of the staff was doing. He made a regular stop at each desk, and with a few well-chosen words ... made each of us feel important ... Women still carried out the subordinate work, and very well they did it ... Dr Shapley cynically measured his projects in 'girl-hours'. Part of the compensation [pay-off] was in the form of personal encouragement, which conveyed the sense that everyone's work was of real importance. It was the secret of his early success. More than once I heard him say: 'I think *I* could do this, so I'm sure *you* can'. (Payne-Gaposchkin 1984, p. 154)

The kinds of programmes carried out at Harvard College Observatory were quite unlike those at other observatories. Mount Wilson represented the trend in the subject—the pursuit of limited programmes, especially about individual objects. Harvard, on the other hand, were concerned with getting a general picture of the contents of the Heavens. When photography had become a standard technique in astronomy in the 1880s, they had concentrated on work with small, wide-angle, telescopes, determining the brightness and spectra of vast numbers of stars. In this way, they obtained a systematic knowledge of the variable stars, obtained through thousands of exposures covering all parts of the sky. Miss Leavitt's great discovery of the period-apparent brightness relation for stars in the Small Magellanic Cloud was but one example of the usefulness of this work.

Harvard College Observatory, located in Cambridge, Massachusetts, suffered from weather that was rarely completely clear. Nevertheless, the photographic techniques that they used needed absolute clarity only when making comparisons from one region of the sky to another. In a normal exposure, the stars that did not vary could be used as standards against which to measure those that did.

Mount Wilson, on the other hand, represented the new approach to astronomy. Before the growth of Los Angeles in the valley below, with its smog and city lights, it was the ideal site on which to place large telescopes. It and the other great Californian Observatory, Lick, on Mount Hamilton, soon led the way in spectroscopic investigations and in the examination of the faintest objects in the sky, which contained the clues to the size of the Universe.

Harvard had already established a 'station' on a mountain rather ominously called 'El Misti' in Arequipa, Peru, where studies could be made of the Southern Hemisphere's stars. However, it had only small telescopes and could not compete with the Californian institutions for ease and certainty of access.

8.12   The Harvard Graduate School

Shapley sought an area of astrophysics at which Harvard could excel in spite of the poor Massachusetts climate. His idea, and one of his greatest contributions as director, was to establish a Graduate School of Astronomy. If they did not have the best skies, they could concentrate on theoretical work. It was a good decision—physics was making enormous strides based on the revolutionary quantum theory, introduced by Max Planck in 1900, and the theory of relativity introduced by Albert Einstein in 1905. Niels Bohr had produced his model of the atom in 1913 and the early years of Shapley's directorship were to see the development of quantum mechanics under Schrödinger and Heisenberg.

The first graduate student to obtain a PhD under Shapley was the English girl, Cecilia Payne. She had heard a lecture by Shapley at the Royal Astronomical Society in London and she decided she would like to study with him. Taking her on was a brilliant choice. It was she who discovered that most stars have similar chemical compositions, the variations in their spectra being mainly the result of the physical conditions, especially the temperature, in their atmospheres. She also found that hydrogen and helium are extremely abundant in the sun and stars, even though they are relatively rare on the earth. Convincing the old-timers such as Russell and Eddington took several years. According to her, 'Russell had been Shapley's teacher and mentor, and his word was law. If a piece of work received his imprimatur, it could be published . . . His word could make or break a young scientist'. To obtain his approval of her thesis, she had to tone down her conclusions. Only in 1929 did Russell, at the end of a comprehensive paper on the subject, acknowledge that she was right after all (DeVorkin 2000; Gingerich 2001). In 1925 she received her PhD from Radcliffe, the sister college of a Harvard which at the time did not have women students itself, let alone award degrees to them (Payne-Gaposchkin 1984, p. 177).

Payne was attracted to Shapley's lively personality but could see his weaker points:

> He was the most wonderful talker I have ever known. A discussion with him was like a rousing game of ping-pong, ideas flashing back and forth, careening off at unexpected angles and often coming to earth in a breathless finish . . . In spite of this vigorous scientific companionship, Dr Shapley kept his distance. He never forgot, or let me forget, that he was the Director of the Observatory. I knew him for more than 50 years, and never once did he call me by my first name . . . I would gladly have died for him, I think.

But this did not prevent me from taking a critical view of my idol. He liked to be flattered, and I certainly flattered him. That was one of the reasons why he enjoyed my company ... With his susceptibility to flattery went a less endearing trait. He never forgot or forgave a slight. He was vain and vindictive. A generous supporter, a stimulating companion, he could also be an implacable enemy. In his published recollections he says that he does not remember disliking anyone. I think this was an exaggeration. (Payne-Gaposchkin 1984, pp. 155–156)

Not wishing that the observatory should continue as a one-man band like that of Princeton under Russell, he hired Henry H. Plaskett to help with his graduate department. Plaskett specialised in theoretical investigations of spectra, a highly fruitful area of work between the two World Wars. He was particularly interested in the application of physics to the formation of spectral lines in 'gaseous nebulae', the sort of nebulae originally identified by Huggins. Bart Bok, who had studied for his PhD in Groningen, was one of many Dutch astronomers attracted to work in the United States. He joined Shapley's department in 1929 and completed his PhD work there. Students flocked to Harvard from all over the world. Donald Menzel, who had studied with Russell at Princeton and had worked at Lick Observatory, also joined the staff. He and his students, such as Lawrence Aller, in a brilliant research campaign, came to understand the physics of gaseous nebulae. Shapley himself reckoned that 1930 was about the high point of the Observatory's influence and its 'joy of life'. The Harvard Observatory, in spite of its situation in a poor astronomical climate, prospered at a time when the California astronomers had become somewhat set in their ways. It was also the epoch of the Great Depression, and the relatively well-endowed Harvard College Observatory could use its funds to attract many excellent young researchers.

Shapley created an enthusiastic intellectual atmosphere at Harvard Observatory. One of his ideas was to conduct get-togethers called the 'Hollow Square Conferences' in which students and staff sat around a big square table with the centre empty. People gave short reports on their own work and exciting developments they had heard of. No subject could continue for more than 10 minutes.

Being director of the largest observatory and with more and more of his former students making original contributions to the journals, Shapley became the acknowledged leader of astronomy on the East Coast. He communicated frequently with Russell and they made use of their power to influence a whole generation of astronomical appointments. They were often consulted when positions became vacant. Woe betide anyone who had crossed these two powerful figures (Fig 8.4). Shapley's interests in scientific societies increased and he found himself a very busy person. As if to emphasise the many activities he was involved in, he had in his office a huge desk, inherited from Pickering, that could be rotated to bring each of his preoccupations in front of him in turn. Any small child that penetrated to his sanctum could expect a ride on it!

FIG. 8.4. Shapley and Russell at the meeting of the International Astronomical Union in Cambridge, Massachusetts, in 1932. These two figures dominated the astronomical life of the East coast of the USA for many years, controlling many of the appointments that were made (SAAO).

8.13 'Shapley's Universe'

Because he was unwilling to accept that the Milky Way was an ordinary galaxy like many others, Shapley developed a model of it as a kind of 'super-galaxy', in whose outer parts were situated the spirals. However, even at the time of the time of the 'Great Debate', there were many, such as Eddington, who believed that our galaxy was nothing special and that van Maanen's results had to be wrong (Gingerich 1990).

Payne was present at a dramatic moment in Shapley's life, when in February 1924 he opened a letter from Hubble, who had been his colleague at Mount Wilson in the period 1919–21. Though they were both Missourians, there was no love lost between them. Hubble's letter contained the announcement that he had discovered two faint long-period Cepheid variables in the Andromeda galaxy (M31; see also section 9.8). Based on Shapley's own calibration of these stars as standard candles, the new result clearly showed that M31 lies far beyond the bounds of the Milky Way. 'Here is is the letter that has destroyed my universe' he said as he handed it over to her.

His reply a few days later showed that, while impressed, he was not quite convinced that Hubble's case was watertight. 'Your letter telling of the ... two variable stars in the direction of the Andromeda nebula is the most entertaining piece of literature I have seen for a long time'. His choice of the words 'in the direction of', rather than 'within', of course suggested that the directional coincidence could be a chance one. He went on: 'My experience with Cepheid variables that have periods of more than twenty days is that they are generally not dependable—they are likely to fall off the period luminosity curve'. He also warned that underlying nebulosity could, thanks to the non-linear response of photographic plates, lead to spurious variability. Though, whether or not they truly belonged to M31, 'The distance of your variables from the nucleus and the lovely number of plates you have now on hand of course assures you of genuine variability for these stars'.[10]

In August, Hubble wrote again to announce the discovery of more Cepheids. Shapley replied: 'Your very exciting letter [was] received here at Woods Hole [where he was on vacation]. What tremendous luck you are having with Messier 31.' He still insisted that the presence of variables could be coincidental and suggested that other fields close to M31 should be checked to make sure they were not common foreground stars. But the evidence was, in reality, overwhelming.

'I do not know whether I am sorry or glad to see this break in the nebular problem. Perhaps both. Sorry because of the significance for the measured angular rotation [and the reputation of van Maanen!], and glad to have something definite and interesting come to hand'.[11]

He continued to believe in the super-galaxy concept for some years further, after making modifications to try to take care of Hubble's discoveries. However, by the mid-1920s, Lindblad and Oort had begun to find other evidence for the Milky Way's true size from arguments concerning the motion of the stars, which they showed were revolving around the centre of the galaxy. What finally killed the Shapley super-galaxy concept was Robert Trumpler's discovery that the light of distant stars was dimmed by the dust between them and us and not only because of their great distance. This was what had led Shapley to believe they were further away than they really are. The galaxy was therefore much smaller than he had thought. Having his pet idea exploded was a serious psychological blow to him. What was worse, at the time of Trumpler's discovery he was finishing a book on *Star Clusters*, which had assumed that interstellar absorption was negligible (Gingerich 1990)!

As late as 1930, Shapley wrote *Harvard College Observatory Circular* No. 350 entitled *The Supergalaxy Hypothesis*, which repeated his outdated ideas. Payne was officially the editor of these publications, but Shapley deliberately did not show it to her.

---

[10]Shapley to Hubble, 27 February 1924, Harvard University Archives, Harvard College Observatory, Shapley correspondence, 1921–1930, Box 9, folder 71.

[11]Shapley to Hubble, 5 September 1924, *ibid.*

When I read the paper I was appalled: I was sure that it was all wrong—
that I could prove from my own knowledge that he was mistaken. He
must have sensed that, and deliberately kept it from a critic. To this day
I cannot imagine what inspired him to write it, if not some impish urge to
fly in the face of the contemporary trend of thought. (Payne-Gaposchkin
1984, p. 174)

8.14   Exploring the southern sky

Knowing that he was unable to match Mount Wilson's superior telescopic fire-
power, Shapley decided that on the observational side he should capitalise on
Harvard's Southern Hemisphere astronomical experience. Because the (southerly)
Magellanic Clouds are so much nearer to us than the Andromeda Galaxy (M31),
they are much easier to study with small telescopes. The other Southern Hemisphere observatories were still traditional position-measuring institutions and
showed little interest in astrophysics: Harvard was to have a near-monopoly in
Southern Hemisphere astrophysics until the move of the Radcliffe Observatory
from Oxford to South Africa, effectively completed in 1948. Shapley was able to
state proudly in 1952 that 'I was known as "Mr. Magellanic Clouds"' for three
decades'.

To make Southern Hemisphere observing easier, he moved the Harvard Southern Station from Peru to near Bloemfontein in South Africa around 1927. South
Africa seemed to offer better climatic conditions and was a lot easier for access
and convenient living. In those days, there was no air travel and observers had to
spend long times away from home; today a North American astronomer can fly
quickly to Hawaii or northern Chile, the main locations of the world's largest telescopes. He spent a lot of time raising funds for this distant operation. He also set
up a field station in Harvard, Massachusetts, beyond the city lights of Boston. In
spite of the erection of moderately large instruments at both new observatories,
their best work continued to be photography with the smaller telescopes, which
they called 'cameras' because of their wide-angle coverage. These could be used
for making surveys of the skies to get a general picture of their contents, leaving
detailed studies of individual objects to the larger instruments on Mount Wilson.
The staff of the Harvard College Observatory numbered about 100 at this time,
which made it one of the largest in the world. It became in part an astronomical factory for mass-production of data, with a large number of women working
on large-scale plate-measuring and classification projects. The atmosphere was
quite different from that at Mount Wilson where the astronomers worked mainly
on their own.

8.15   The distribution of galaxies in space

Harvard's small telescopes were soon put to use on a survey of bright galaxies
by Shapley and his graduate student Adelaide Ames. In 1932 they produced
an important catalogue of galaxies, *A Survey of the External Galaxies Brighter
than the Thirteenth Magnitude*, better known as the *Shapley-Ames Catalog*. Ames

was drowned in a canoe accident in the same year, to the great distress of the Observatory community.

Shapley and others, such as his colleague Bart Bok, also took advantage of the wide angular coverage of the small Harvard 'cameras' to study the Milky Way in detail.

One of the first discoveries made at Bloemfontein was what is now called the 'Shapley Supercluster' of galaxies. A second survey was intended to go to much fainter limits. The surveys showed that the distribution of galaxies in space was far from uniform, in contradiction to Hubble's often-expressed (see Section 9.17) belief that they were evenly spread. In 1931 and 1932, Shapley emphasised this conclusion in correspondence with Hubble, who kept insisting that the data became smoother if taken to fainter limits. Hubble was so annoyed by Shapley's work that once, when Bart Bok was visiting him, he dramatically threw a copy of the *Shapley-Ames Catalog* into the waste-paper basket (Levy 1993). In the next few years Shapley increased the numbers of galaxies he had identified from 100,000 to ultimately about a million. Modern work by de Lapparent, Geller, and Huchra (1986) have vindicated Shapley by showing that the distribution of galaxies is full of voids separated by 'thin' walls, resembling a large-scale sponge.

Hubble's work (see Chapter 9) led him towards denying the expansion of the universe since his data would then imply it was closed, small and 'suspiciously young'. All three of these possibilities went against his prejudice in favour of an infinite universe without an origin in time. In a letter dated 12 January 1937 to Howard P. Robertson, a cosmologist at Princeton, Shapley put his finger on a potential problem with Hubble's work, namely that his distance scales might be wrong. Robertson, however, was inclined to believe Hubble. Many years later, Shapley was shown to be right, though he had not published his suspicions at the time (Gingerich 1990).

Shapley raised enough money from the Rockefeller Foundation to build a 60-in. telescope at his new station in South Africa, perhaps in the hope of determining the distances to the galaxies himself. The gift was conditional on his securing matching funds, which he got from the Harvard Corporation. His selling point was that a southern 60-in. telescope looking at the Magellanic Clouds would be as powerful as a 200-in. looking at the Andromeda Galaxy (M31). Unfortunately, he compromised its effectiveness by using a rather thin primary mirror made around 1890 by the English amateur A.A. Common with the result that the instrument required continuous adjustment during exposures. There was also nobody on the spot with a deep knowledge and enthusiasm for astrophysics who could have driven the observing programmes more successfully.

Bart Bok, in his obituary of Shapley, says that 'The 1920s and 1930s were the best years of the Directorship of Harlow Shapley at Harvard'. His achievements were well-recognised; in 1934 he received the Gold Medal of the Royal Astronomical Society and was invited to give the George Darwin Lecture on his view of the Milky Way Galaxy.

8.16   Diversions from astronomy

As the years wore on Shapley had to admit that more and more of his time was being taken up by popular lectures and fund raising. 'They said that people generally crossed to the other side of the street when they saw me coming'. He trebled the endowment of the Observatory in a short time. 'I think I was able to convince my colleagues that we should not work for money—we should work for the glory of the contribution'. He became concerned with various humanitarian issues, which diverted his attention from astronomy. In particular, he started to help the victims of totalitarian regimes, particular the Hitler one in Germany, to make their escape. The organisation he set up to aid them was called the 'National Research Associates'. Many of the refugees that Shapley supported were left-wingers, and this caused the FBI to take an interest in his activities. As war approached, although he was a loyal citizen, Shapley, like Eddington before him, could not bring himself to hate individual members of enemy nations. Consequently, he was not amused when one of his rescued scientists said 'I feel rather obliged to kill someone, but who?'

According to Hoffleit (1992), he played a major role in setting up the 'Westing-house-Science Service Science Talent Search', the idea of which was to select and sponsor high school students with a special aptitude for science. He also enthusiastically promoted amateur astronomical groups, such as the local Bond Astronomical Club, the Amateur Telescope Makers of Boston and the American Association of Variable Star Observers. Almost every year he reported on the highlights of the past year's astronomy at the annual meeting of the latter group.

8.17   The Sculptor and Fornax dwarf galaxies

The Sculptor and Fornax 'dwarf galaxies', named after the constellations in which they lie, were found at Harvard in 1938 on wide-angle plates taken in South Africa. This was perhaps the last important discovery associated with Shapley. The dwarf galaxies are minor members of the 'local group' of galaxies that are bound together by gravity. They contain relatively small numbers of stars and are quite spread out on the sky, so that they are hard to distinguish from the stars of our own galaxy. The other members of the local group, apart from the Milky Way galaxy, include M31 (the Great Nebula in Andromeda), NGC6822 and the two Magellanic Clouds. These discoveries were immediately followed up by Hubble and Walter Baade, working at Mount Wilson. They used the large Mount Wilson telescopes, which could fortunately reach far enough south, to study the kinds of stars they contained. They soon became aware that there were no young luminous stars in either but there were plenty of (old) RR Lyrae variables in Sculptor. The conclusions of their joint paper contain a hint of Baade's later discovery of 'Populations' I and II, but also a dour statement, obviously due to Hubble, that 'discussion of the data now available would be largely speculative, and hence of little permanent value' (Osterbrock 2001).

Among the foreign astronomers who ended up at Harvard around this time was the young Czech Zdenek Kopal. He found Shapley was in his most communicative mood between dinner and midnight, after days spent mostly on administrative matters. This was when he had time for astronomical activities. Kopal related that Shapley was still smarting from the discovery of absorption by interstellar dust nearly a decade earlier. Allowing for this effect, his estimated distance from the sun to the Galactic Centre had to be reduced to about a half of its previous value. Kopal was warned by Bok 'not to mention interstellar absorption more than necessary in Shapley's presence if I wanted to stay on the right side of the Director' (Kopal 1986, p. 159). This kind of emotional attachment to one's ideas is not unusual amongst astronomers. In fact, Shapley's great discovery had been adjusted in its details rather than overthrown.

8.18   The sociable Director

Shapley and his wife, Martha Betz Shapley, had five children, Willis, Alan, Mildred, Lloyd, and Carl, the first three born in California and the last two in Cambridge. Their upbringing during the twenties and thirties took up most of the Martha's time, but not all of it. Taking advantage of the large house that he occupied as director, Shapley enjoyed giving big parties. Musicales and plays were organised. Cecilia Payne-Gaposchkin (1984) remembered 'I can still see Shapley impersonating Napoleon (he prided himself on the resemblance), and Henry Norris Russell, swathed in a sheet, acting out the death of Archimedes'. Bart Bok reckoned that he owed Shapley's appreciation of him to his performance in the 'Harvard Observatory Pinafore', a parody of life at Harvard Observatory written in the time of Pickering, based on the Gilbert and Sullivan opera. This was presented at the 1929–1930 meeting of the American Astronomical Society.
    Kopal remembered these years with fondness:

> at Harvard Mrs Shapley's career as a warm and gracious hostess blossomed out to the full, entertaining with equal grace the rich and the poor, famous and humble (sometimes, as with Albert Einstein or Igor Stravinsky, both in one person), the geniuses and amateurs, or scared undergraduates. The Shapleys loved to entertain; and any occasion for a party was always welcomed. On each Friday nearest the full moon, when astronomical photography (and most astronomical work in those days was photographic) was curtailed by bright moonlight, Dr Shapley was host to a Full-Moon Club at his home, for graduate students as well as staff.
>
>   The highlight of the season used always to be the Christmas parties at the Residence, at which praises to the Heavens were raised by carols sung by many astronomers of all ages. One particular number which sticks vividly in my memory was the solo parts of 'We Three Kings', the role of Melchior being sung by the good Christian Bart J Bok (I sang Caspar), and that of Balthazar by Henry Norris Russell—a guest performance which would have penetrated the Gates of Heaven even more readily if that distinguished amateur had been better able to carry the tune! Miss Sibyl Chubb used to act as honorary conductor of most such performances, with Dr Shapley as a genial master of ceremonies, seeing

to it that the festive spirit prevailed at all times; though it was his wife who, inconspicuously, but with unfailing charm, made everyone feel truly a member of the same family. (Kopal 1986, p. 167)

Einstein stayed with the Shapleys on the occasion of receiving an honorary degree from Harvard. He was asked to bring his violin and he played Bach and Brahms with the assembled company for several hours.

8.19   Other activities

The Second World War saw many of the Observatory staff involved in the war effort, some as optical workers. Martha Betz Shapley, as a competent mathematician, helped calculate the trajectories of shells for the Navy.

Harlow Shapley received a great many honors and sat on the boards of many organisations such as the Woods Hole Oceanographic Institution, the Massachusetts Institute of Technology (MIT) and the Worcester Foundation for Experimental Biology. He served as President of the American Academy of Arts and Sciences from 1939 to 1944.

Shapley's output of research papers dropped considerably during the War years and thereafter. Popular astronomy and non-astronomical subjects began to supplant serious astronomy in his publications (list in Bok 1978).

8.20   International affairs

Bok (1978), in his obituary of Shapley, states:

Following World War II, Shapley turned increasingly away from astronomical endeavours; national and international affairs claimed him. Looking back at it all, it seems a pity that about 1946 he did not resign his post as Director of Harvard Observatory ...

After World War II, Harvard Observatory began to lose the leading position that it had reached in astronomy during the 1930s. This was in part due to the fact that so much of Shapley's time was dedicated to national and international affairs. These, Shapley felt, had higher priority than research and teaching.

Very soon after the end of the war, Shapley was in Moscow on the occasion of the 220th anniversary of the (then Soviet) Academy of Sciences and found himself at a dinner attended by the dictator, Joseph Stalin. A hobby from his Mount Wilson days had been the study of ants. He had noted that they obey a kind of physical law, that they moved faster the higher the temperature, and had even written a papers on their 'thermokinetics'. However, while at that dinner an ant crawled out of a basket of fruit. He caught it and placed it in a little vial that he always carried with him for such occasions and poured a little vodka in with it as a preservative. This, of course, attracted the attention of Stalin's bodyguards. Some discreet explanations had to take place. Shapley later said that Stalin would never have noticed him but for this incident!

In 1946 he made a trip to India and got to know the future Prime minister Jawaharlal Nehru and his sister, Mrs Pandit (later a President of the United

Nations General Assembly). In 1950, when Nehru visited the United States, he went to the Observatory and participated in one of the 'Hollow Square' meetings.

These trips, and other activities such as the presidencies of the American Astronomical Society, of the American Society for the Advancement of Science and of the National Society of Sigma Xi (a scientific honour society), were of more interest to the post-war Shapley than his astronomical work, and his productivity as a researcher inevitably suffered.

8.21   Losing the initiative

The lead in Magellanic Cloud investigations started to slip away from Harvard when the Radcliffe Observatory's 74-in. (1.9 m) telescope was completed in South Africa in 1948 and became the largest telescope in the Southern Hemisphere.

It is now known that Shapley's 'cluster variables' are not the same as Cepheids, though this error only became obvious around 1950. It was Walter Baade, a German astronomer working at Mount Wilson, who found that the RR Lyrae variables and 'Type II Cepheids' are not related to ordinary Cepheids. In fact, during the Second World War he had made the highly important discovery that stars could be separated into two distinct 'populations' called I and II, according to whether they were young or old. The RR Lyraes *do* vary by pulsating in a somewhat similar way to Cepheids, but their periods are shorter. They and the Type II Cepheids occur only among groups of stars formed when the universe was younger and had fewer atoms of the heavy elements. The ordinary Cepheids are comparatively young stars in which the heavy elements are more abundant. Shapley and everybody else had mixed up the two kinds, which have different luminosities (wattages). They can still be used as standard candles, but must not be confused with one another.

Because the cluster variables seemed very faint, Shapley thought that the globular clusters were quite a bit further away than they are now known to be. The solution to the problem came with the aid of a more powerful telescope, the 1.9 m Radcliffe instrument in South Africa. Its new director, David Thackeray, had written to Baade in 1949, asking for advice on things to do with the relatively simple equipment at first available to him. Baade encouraged him to search for RR Lyrae variables in the Magellanic Clouds and went on:

> Both Hubble and I hope that Shapley's tendency to consider the Magel-
> lanic Clouds as his personal property will not deter you from from attack-
> ing this problem [of searching for RR lyrae variables]. He has monopolised
> the Clouds all too long and it is high time that the barbed wire fences
> and the warning signs 'Keep out. This means you!' are taken down. Mo-
> nopolies in science are intolerable and should never be respected ... The
> whole situation has become intolerable and a good fresh breeze is most
> desirable ... [12]

[12]Baade to Thackeray, 20 March 1949, quoted by Feast (2000).

FIG. 8.5. Harlow and Martha Betz Shapley at the International Astronomical Union in Rome in 1952. At this meeting his calibration of the Cepheid distance scale was shown to be wrong by a factor of two (SAAO).

Baade announced his conjecture as to what was wrong with the distances at the meeting of the International Astronomical Union in Rome in 1952 (Fig 8.5). Immediately afterwards, Thackeray made the dramatic and unexpected announcement that he had actually *found* RR Lyraes in the Small Magellanic Cloud and could thus confirm Baade's suggestions. This had the consequence that the true Cepheids had to be *more* luminous than Shapley had thought and that therefore he had underestimated their distances. One consequence was that the universe now had to be twice as large as previously believed!

This is probably what Shapley was referring to when he said in 1952 that he had ceased to be 'Mr. Magellanic Clouds'. Either the Harvard plates did not go faint enough to detect these variables or Shapley had not searched them to faint enough limits. Worse, in his own mind, he had become too attached to his view of the distance scale to believe that it would ever have to be revised. Perhaps his proprietary attitude towards the Clouds, even asserted within Harvard College Observatory (Payne-Gaposhkin 1984, p. 213) had inhibited fresh ideas.

In what can only be described as a moral lapse, Shapley soon afterwards attempted at a meeting of the American Astronomical Society to claim much of the credit for the revised distance scale for himself. His speech was reported in newspapers such as the *New York Times* and in the news magazines. Baade, who never minced words, was particularly incensed and described Shapley's assertions

as 'simply shameless', accusing him of plagiarising the remarks he had made at
the meeting in Rome. In fact, even Bok, who usually had the highest opinion of
Shapley, had to admit that he 'did not like it' and seems to have been the only
person who told him to his face that he should have referred to Baade's work
properly. A concerted effort was made by leading American astronomers to make
sure that the true facts became known (Gingerich 1990; Osterbrock 2001).

8.22   Post-war social concerns

One of Shapley's efforts post-war was to help restore Eastern European observatories that had been devastated during the fighting. These included Pulkova,
outside St Petersburg, and Toruń, the birthplace of Copernicus, in Poland. The
latter observatory was given one of the small telescopes from Harvard to help it
get going again.

Shapley saw that private donors could no longer be counted on to provide
the funds for astronomical research and he became involved in getting the US
Congress to make the National Science Foundation into a funding agency for science. He also helped to organise US support for the foundation of UNESCO—the
United Nations Educational Scientific and Cultural Organisation. In fact, it has
been stated that he was largely responsible for putting the S in UNESCO (Gingerich 1990). By this time he was involved in so many liberal and international
causes that he came to be regarded as politically suspect by right-wingers.

> I knew him well enough to realize that he was no radical: he was a conser-
> vative at heart, and ran his own observatory like a benevolent dictator.
> But he was an implacable foe of injustice and fraud. The hostility that
> he incurred during the internationally turbulent years cast a shadow that
> saddened and weakened his final years as director of the Observatory.
> (Payne-Gaposchkin 1984, p. 210)

8.23   The 'Communist in the State Department'

In one of the many bizarre incidents of the 'cold war' era, Shapley had a notorious
clash with a cabal of right-wing politicians—the House Un-American Activities
Committee, under the chairmanship at the time of John Rankin, a Congressman
from Mississippi. They summoned him to appear before one of their hearings in
Washington. Realising his danger, he brought with him a lawyer recommended
by a friend in the Harvard Law School. Shapley found the committee atmosphere
to be like that of the 'Court of Star Chamber', a draconian 'law court' set up by
the English King Henry VII to enforce his will in a quasi-legal manner. Rankin
first told Shapley's lawyer to 'get out of here', which he protested against but
eventually complied with. His secretary, who was brought along to take notes
of the proceedings, was shouted at by Rankin 'Throw her out', and she was
intimidated enough to comply.

> I said, 'Very well, if she can't be here, I'll be my own recorder'. I have
> mentioned that I write shorthand. I started to write down all that was

being said. That made him even madder. He came crawling over the intervening table and grabbed the notes out of my hand. I rose in my great dignity and said, 'This is a case of assault'.

But that did not bother Rankin. He said, 'You are going to be cited for contempt of Congress.'

I get compliments; sometimes I fish for them. Here was an opportunity. I came out into the anteroom—flash, flash, flash, flash: the newspaper people were there. And out came Congressman Rankin. He came out, pointing his scornful finger at me, and shouted, 'This man has shown more contempt for our committee than anybody who ever appeared before it.' That was the greatest compliment that I ever had. It was flattering to be held in contempt of that committee. (Shapley 1969, p. 153)

Despite his flippant account of what had happened, Shapley was deeply upset. On his return to Harvard, he told Bok, with his voice breaking, 'That man had the nerve to tell me that I am un-American'.

Fortunately, he was supported by the President of Harvard, James Conant, against right-wing contributors of funds who tried to use their influence to get him dismissed. Some academics at other institutions, also accused, were not so fortunate. Senator Joseph McCarthy, another right-wing demagogue, was later to state in a speech that Shapley was 'one of half a dozen communists in the State Department [the department of foreign affairs of the United States]'. Of course, Shapley was not a communist nor was he a member of the State Department. He told reporters that 'the Senator succeeded in telling six lies in four sentences, which is probably the indoor record for mendacity'.

8.24 Retirement

In 1952 Shapley retired. He admitted to having become progressively more and more out of touch with contemporary research—the papers in the *Astrophysical Journal* were being 'written by authors whose names are completely unknown to me'. The atmosphere at the Observatory was by then rather poisonous. Shapley's political opinions and those of the successor he would have preferred, Bok, were considered a serious liability by many Harvard administrators, and Donald Menzel was appointed to succeed him instead. Because private donors were no longer easy to find and the Observatory was forced to seek government money, it had to toe the current U.S. administration's political line. Kopal (1986, pp. 170–171) wrote 'what cannot be excused by any reason known to this writer was the mean and shabby way in which the new incumbent set out after 1954 to push the Director Emeritus (and his wife) out of the observatory at which they had served science so faithfully and so long, for no apparent reason ... '.

He moved from Harvard Observatory to a farmhouse just across the Massachusetts border, in Sharon, New Hampshire, where he was to enjoy a long and active retirement. Here he continued to offer hospitality to his friends as he had done at Harvard.

Although a declared agnostic, Shapley was deeply interested in religion and was a genuinely 'religious' person from a philosophical point of view. 'I never go to church', he told Cecilia Payne-Gaposchkin, 'I am too religious for that'. He took a great interest in the 'Institute for Religion in an Age of Science', whose meetings he attended annually. He even edited a volume of its proceedings, called *Science Ponders Religion.*

Near the end of his life he wrote the autobiography from which several passages have been quoted. He died on 20 October 1972, in Boulder, Colorado, and is buried in Sharon.

In his touching obituary of Shapley, Bart Bok (1978) wrote:

> Historians of the future should not only take note of Harlow Shapley's great scientific achievements, but I hope they will also remember him as a fine human being, as an independent, bold human spirit with a healthy distrust of all authority. Harlow Shapley loved to push beyond frontiers, scientific and human.

References

Baade, W., 1963. *Evolution of Stars and Galaxies*, ed. Payne-Gaposchkin, C., Harvard Univesity Press, Cambridge, MA.

Bailey, Solon I., 1922. Henrietta Swan Leavitt, *Popular Astronomy*, **30**, 197–199 & plate xvii.

Bok, B.J., 1978. Harlow Shapley November 2, 1885 – October 20, 1972, *Biographical Memoirs of the National Academy of Sciences*, **49**, 241–291. See also Bok, Bart J., 1974. Harlow Shapley, *Quart. J. Roy. Astr. Soc.*, **15**, 51–55; Bok, B.J., 1972. Harlow Shapley—Cosmographer and Humanitarian, *Sky and Telescope*, **44**, 354–357.

de Lapparent, V., Geller, M.J., and Huchra, J.P., 1986. A Slice of the Universe, *Astrophys. J.*, **302**, L1–L5.

DeVorkin, David H., 2000. *Henry Norris Russell, Dean of American Astronomers*, Princeton University Press, Princeton, NJ.

Feast, M.W., 2000. Stellar Populations and the Distance Scale: The Baade-Thackeray Correspondence, *J. Hist. Astr*, **31**, 29–36.

Gingerich, O., 1988. How Shapley Came to Harvard; or, Snatching the Prize from the Jaws of Debate. *J. Hist. Astr.*, **19**, 201–207.

Gingerich, O., 1990. Through Rugged Ways to the Galaxies, *J. Hist. Astr.*, **21**, 77–88.

Gingerich, O., 1999. A Brief History of Our View of the Universe, *Pub. Astr. Soc. Pacific*, **111**, 254–257.

Gingerich, O., 2001. The Most Brilliant PhD Thesis Ever Written in Astronomy, in *The Starry Universe: The Cecilia Payne-Gaposchkin Centenary*, ed. Davis Philip, A.G. and Koopman, R.A., L. Davis Press, Schenectady, NY.

Gingerich, O. and Welther, B. 1985. Harlow Shapley and the Cepheids, *Sky and Telescope*, **70**, 540–542.

Harris, W.E., 1996. A Catalog of Parameters for Globular Clusters in the Milky Way, *Astron. J.*, **112**, 1487.

Hoagland, Hudson, 1964. Harlow Shapley—Some Recollections, *Pub. Astr. Soc. Pacific*, **77**, 422–430 .

Hoffleit, Dorrit, 1992. The Selector of Highlights: A Brief Biographical Sketch of Harlow Shapley, *J. Amer. Soc. Variable Star Observers*, **21**, 151–156.

Kopal, Z., 1986. *Of Stars and Men*, Adam Hilger, Bristol.

Leavitt, H., 1912. Periods of 25 variable stars in the Small Magellanic Cloud, *Harvard Coll. Obs. Circ.*, **173**, 3 March 1912.

Levy, D.H., 1993. *The Man Who Sold the Milky Way; a Biography of Bart Bok*, University of Arizona Press, Tucson.

Osterbrock, D.E., 2001. *Walter Baade, a Life in Astrophysics*, Princeton University Press, Princeton, NJ.

Payne-Gaposchkin, C., 1984. *An Autobiography and Other Recollections*, ed. Haramundanis, K., Cambridge University Press, Cambridge.

Shapley, H., 1909. Astronomy in Horace, *Popular Astronomy*, **17**, 397–401.

Shapley, H., 1913. The orbits of eighty-seven eclipsing binaries—a summary, *Astrophys. J.*, **38**, 158–174.

Shapley, H., 1917. Note on the Magnitudes of Novae in Spiral Nebulae, *Publ. Astr. Soc. Pacific*, **29**, 213–217.

Shapley, H., 1919. Studies based on the colours and magnitudes in stellar systems: Remarks on the arrangement of the sidereal universe, *Astrophys. J.*, **49**, 311–336.

Shapley, H., 1969. *Ad astra per aspera; Through Rugged Ways to the Stars*, Charles Scribner's Sons, New York.

Shapley, H. and Ames, A., 1932. A Survey of the External Galaxies Brighter than the Thirteenth Magnitude, *Ann. Harv. Coll. Obs.*, **88**, (2), 43–75.

Trimble, V., 1996. H_0: The Incredible Shrinking Constant 1925–1975, *Publ. Astr. Soc. Pacific*, **108**, 1073–1082.

9

EDWIN HUBBLE: JOURNEYING TO THE EDGE

Thus the explorations of space end on a note of uncertainty. And necessarily so. We are, by definition, in the very center of the observable region. We know our immediate neighbourhood rather intimately. With increasing distance, our knowledge fades, and fades rapidly. Eventually we reach the dim boundary—the utmost limits of our telescopes. There, we measure shadows, and we search among ghostly errors of measurement for landmarks that are scarcely more substantial.

The search will continue. Not until the empirical resources are exhausted, need we pass on to the dreamy realms of speculation. (Hubble (1936), concluding lines of *The Realm of the Nebulae*)

9.1 Birth and early years

Like Shapley, Edwin Powell Hubble was born in Missouri, but there the resemblance ends. The place of his birth was the small town of Marshfield and the date 20 November 1889. After his death, his wife, Grace, who survived him by many years, wrote a manuscript biography, treating him as an old-fashioned hero. The following passage about his ancestry, though basically factual, is typical of her romantic account:

> His ancestors came from England, Ireland and Wales, with no strains of foreign blood [!] The first of them to come to America, in the middle of the 17th Century, was an officer of the Royalist Army. In the Revolutionary War and the Civil War they were soldiers, in times of peace, pioneers, living on the land. Tall, well-made, strong, their bodily inheritance had come down to him, even to the clear, smooth skin that tanned in the sun, and the brown hair with a tint of reddish gold. They had handed down their traditions as well, integrity, loyalty as citizens, loyalty to their families ... and a sturdy reliance on their own efforts.[13]

Grace's biography is highly selective about the life of her hero. Her version, with its embellishments of the facts, was generally accepted as correct by his obituarists (Adams 1954; Humason 1954; Mayall 1954; Robertson 1954). Researching original sources and records relating to Hubble's early career, Osterbrock et al. (1990) were able to paint a much more accurate picture, which in no way detracts from his science but does bring out some of the stranger aspects of his character. The biography by Gale E. Christianson (1995), *Edwin Hubble, Mariner of*

---

[13]Hubble, G.B., to Mayall, ca. 1 January 1957, M.L. Shane Archives, Lick Observatory, quoted by Osterbrock et al. (1990).

the Nebulae, whose outline is followed here, is a deeper and more comprehensive study still.

Hubble's father, John Powell Hubble, grew up in Springfield Missouri. He attempted to become a lawyer but his studies were not at first successful and he became an insurance salesman instead. The family had a farm in Marshfield, about 30 km away, where John Hubble met Edwin's mother, Virginia Lee James (Jenny), the daughter of the local doctor. John Hubble eventually did become a member of the Missouri Bar, but insurance was to remain his main source of income. His family lived comfortably, usually in large rented houses, and had the services of a maid, although the children were expected to help with household chores. His work as a travelling salesman took him away from home for long stretches. He was unpleasantly stern and Calvinistic in outlook, ruling his children with a rod of iron, to the extent that they later wondered how their mother could have tolerated him. His son later remarked that his father had blighted his life. Because of his absences, the Hubble children were to spend a lot of time with their grandparents.

Edwin Powell Hubble was the third and brightest of the seven children that survived. His middle name, which he later avoided using, had been his father's also, and originated as his Hubble grandmother's maiden name. He learned to read and count even before his formal education started. Tall for his age, he wished that he could have entered school with his older brother Henry and sister Lucy. When in 1895 he eventually did go to elementary school in Marshfield, the fact that he knew the work already made him rather a nuisance to his teachers. To most of his classmates he seemed older than he actually was.

Hubble's grandfather William James had built himself a telescope and gave Edwin his first look through it at the age of 8. This was the probably the trigger for his lifelong passion. How much time he spent on astronomy as a child we do not know; it is said he sat up all night two years later to observe an eclipse. To counter boredom, he read a good deal. As a country boy, he spent a lot of time outdoors, learning to swim and shoot and to become aware of the nature around him.

In 1900, his father's work took him to Chicago. The family moved with him, eventually settling in Wheaton, a commuter town to the west. There he went to the Central grade and high schools, located together in a building called the 'Old Red Castle'. One of his teachers was Harriet Grote, whose son, Grote Reber, was to become one of the pioneers of radio astronomy. Edwin was a bright pupil. He obtained high marks in mathematics, physics, the sciences and modern languages, although his spelling remained erratic throughout his life. As a teenager, he must have gone through a rebel phase, because his marks for 'deportment' declined sharply at one point. By 1906, when he completed high school, his behaviour had stabilised and he once again was earning high marks. In his bearing he was somewhat aloof from his classmates, giving the impression that he felt himself superior to them. He was about two years younger than the average in his class, but performed well, finishing school at the age of 16.

He was by no means completely devoted to academic work; he also showed prowess at sports. As a tall boy, he was well-suited to baseball and was one of the stars of his high school team. He wanted to play football, but was forbidden to do so by his parents, who were worried that he would be injured. He competed successfully in many athletic activities such as jumping, discus, and relay, and even set a state record for the high jump on one occasion.

The young Edwin earned pocket money in various ways typical of teenagers in those days, such as by caddying at golf courses and by delivering ice. The latter activity, now a largely forgotten one, consisted of bringing blocks of ice, cut from frozen ponds in winter and stored in insulated warehouses, to customers who used them in summer to cool primitive refrigerators. In his last summer before leaving school, he got a job assisting surveyors in the wilds of northern Wisconsin, an experience which his sister, Helen, thought had turned him into an adult. His vacations were otherwise spent with his grandparents on their farm in Marshfield.

At his high school graduation ceremony in 1906, the Superintendent called him forward when awarding the class honours. Then came what he remembered as an 'awful moment'. 'Edwin Hubble, I have watched you for four years and I have never seen you study for ten minutes. Here is a scholarship to the University of Chicago'.[14]

9.2   University of Chicago

At this time (1906), the University of Chicago was only fourteen years old. It was a fortunate choice, because the new university had set out from the beginning to be a first-class research institution, modelled on the German universities of the time. Its first President, William Rainey Harper, who had appointed Hale to the faculty over a decade before, had just died. During his presidency he had persuaded many excellent academics to work there. On its physics staff were Robert A. Millikan, celebrated for his 'oil-drop' method for finding the charge on the electron, and Albert A. Michelson, who was and remains famous for his experiments on light.

Although his sister Betsy recollected that Hubble had always wanted to be an astronomer, their father had been none too keen on the idea, knowing how poorly academics were paid compared to solid citizens like lawyers. His son nevertheless took as many science courses as he reasonably could while nominally studying for entrance to law school. The university course was non-specialised in the first two years, after which the student received a degree called 'Associate in Science'. This structure, later abandoned, was still in force while Hubble was there. At first he studied mathematics, physics, chemistry, and astronomy as well as modern languages (French and German) and Latin. He evidently created a good impression and received a glowing recommendation from the astronomer

[14]Hubble, G.B., 'Family History', HUB 82(1), Box 7, 7, Huntington Library, quoted by Christianson (1995, p. 35).

Forest R. Moulton when he decided to take up astronomy more seriously a few years later. He obtained enough credits to acquire his Bachelor of Science degree a little early, in March 1910.

While at Chicago, Hubble joined the Kappa Sigma fraternity. This was one of a number of small, usually residential groups, common at American universities, and was particularly favoured by the university's star athletes. He was an obvious recruit, as an active sportsman, 6 ft 3 in. (1.91 m) tall, and good at running and basketball. He shared a room with another former Wheatonian. As the fraternity did not yet have a house of its own, their room was in 'Hitchcock Hall', a residence with an atmosphere somewhat like that of an Oxford college. Being a member of a fraternity, he was expected to be sociable and had to attend university dances, which, at seventeen, he seems not to have enjoyed. His fraternity living brought him in touch with the wealthier element among his fellow-students. Although the non-resident commuting students worked harder, they tended to be from more plebeian backgrounds than his fraternity brothers.

The Chicago basketball team—the Maroons—of which he was a member, won prestigious events such as the national intercollegiate championship and later the 'Big Ten' championship. He did well not only at basketball but also as a member of the university's athletic team. His coach was the famous Alonzo Stagg after whom Stagg Field was named—the stadium under which the first nuclear reactor was built in 1942. Stagg tried hard to persuade John Hubble to relax his order that his son was not to play football, but without success.

Hubble spent some of his university vacations also on survey work in the forests near the Great Lakes. His sisters were told stirring stories of his heroic behaviour during these expeditions, such as how a bear had caused a packhorse to bolt and how he, finding his way only by starlight, bruised and battered, had managed to lead the animal back to camp. His tent-mate was supposedly killed by a falling tree during a thunderstorm. He even had to fight off a pair of would-be robbers. According to another of his stories, he and a companion were accidentally abandoned in the forest at the end of the project and had a three-day walk back to civilisation on meagre supplies. The factual basis of all this is unknown.

9.3 Rhodes Scholar

Even at school, Hubble had been interested in becoming a Rhodes Scholar. This was a distinct possibility because he possessed the basic qualifications of a good academic record and sporting prowess. As a college junior (third-year student), he at last had the opportunity of taking the necessary examination.

The Rhodes Scholarships were founded by the English-born South African financier and politician, Cecil John Rhodes (1853–1902), in terms of his will and were intended to support British Empire, American, and German students at Oxford University. A fascinating character, Rhodes was an unscrupulous operator both as a politician and as a financier of gold and diamond mines. He nevertheless harboured mystical and idealistic beliefs about the British Empire and wished to encourage contacts between Englishmen, Americans, and Germans. Candidates for his scholarships were to be chosen for literary and scholastic attainments, success in outdoor sports, moral qualities, and qualities of leadership—in other words the essentials of a Victorian hero. Hubble was a bright and dynamic student and would easily have been able to convince the selectors that he met these criteria. He spent much of 1909 in learning Latin to prove that his had been a well-rounded education and was one of six Illinois candidates in 1909; two including him passed the formal examination and interview in front of the high-powered committee, consisting of four university presidents. Hubble was given a strong recommendation by Millikan, for whom he had worked as a laboratory assistant. He was reported to be a 'man of magnificent physique, admirable scholarship, and worthy and lovable character ... I have seldom known a man who seemed to be better qualified to meet the conditions imposed by the founder of the Rhodes scholarship than is Mr. Hubble'.[15] His application was, of course, successful, but he had to remain a further year in Chicago to finish his degree.

When he went to Queen's College, Oxford, Hubble was in his element. He adopted English clothing and habits and very rapidly became Anglicized. He took to wearing plus-fours (called knickers in the United States!) and a Norfolk jacket as well as to carrying a cane. His efforts at changing his accent caused some amusement to other American students. He developed numerous acquaintances among the wealthier students and enjoyed several visits to aristocratic country houses. Somewhat surprisingly, he placed his main emphasis on law studies, probably as the result of pressure from home. Work was for the mornings. As often with Oxford students, he relied more on his tutor, notes, and reading than on attending lectures. The afternoons were for sports. He carried on with high jumping, shot put, and running, becoming a member of the University team in shot put and hammer throw as well as in water polo. He earned a 'Half-Blue' in athletic events and even claimed to have boxed in an exhibition match with the French heavyweight champion, Georges Carpentier.

Nevertheless, he somehow met Herbert Hall Turner, Savilian Professor of Astronomy, who, in turn, knew everybody who was anybody in British astronomy and ran a famous gossip column, the *Oxford Notebook*, in *The Observatory* magazine. He completed his jurisprudence course after two years with second-class honours but in his third year turned rather surprisingly to learning Spanish. One of his American Rhodes scholar contemporaries felt that he had become 'very British' in manner. This affectation was never to leave him.

He took advantage of his vacations to make cycling tours of European countries with Oxford friends. He greatly enjoyed Germany, where he met some naval officers in Kiel and later told people that he had fought a duel with one of them for allegedly having paid too much attention to his wife!

Although his father's letters encouraged him to keep up his religion, it appears

---

[15]Millikan to James, 8 January 1910, HUB, Box 18, f. 840, quoted by Osterbrock et al. (1990).

that he drifted away from the Baptists during his stay in England, though he tried to re-assure his parents of his continuing interest. After three years and the expiry of his scholarship it was time to return home, with the intention of practising law. However, his English experience had changed his behaviour for life. Photographs after this time often show him in plus-fours and with a pipe in his mouth. His pose was always serious and dignified. He was clearly aware of his male good looks.

9.4 Back to the United States

Family circumstances had changed dramatically for the worse by the time of Hubble's return to the United States. His father had moved to Louisville, Kentucky, in 1910, but had died aged 52 of kidney failure in January 1913, some months before Edwin got back home. The family was left fairly short of money, even though his brother Henry was now at work in the insurance field.

Hubble later on had something of a complex about this period of his life and avoided talking about it. He implied that he had passed the Kentucky Bar examination and had had a successful legal practice. This was believed by his close associates Milton Humason (1954) and Howard P. Robertson (1954), who mention it in their obituaries of him. There is in reality no record that he passed the Kentucky Bar examination, informal though it was, or that he ever practised law. The facts suggest that he never became a lawyer at all. During his first, unsettled year back in the United States, he first obtained temporary summer work as a translator for an import company in Louisville and then, in the autumn, became a high school teacher in New Albany, Indiana. Here he taught physics, mathematics, and Spanish. He took the schoolchildren on day-hikes but his main success was as a basketball coach. He was a popular figure with his pupils, but his mind was elsewhere. Why he should have kept silent later about this year of teaching is not known. He may have felt that it represented a false step in an otherwise clearly directed career (Osterbrock et al 1990).

9.5 Graduate student at Yerkes

Somehow, towards the end of his teaching year, his astronomical ambitions returned and Hubble pulled himself together. He decided to become a graduate student in astronomy at the Yerkes Observatory of the University of Chicago. He wrote to his former undergraduate professor, F.R. Moulton, for advice on entering graduate school and obtaining financial assistance. Moulton contacted Edwin B. Frost, then Director of Yerkes, encouraging him to take on Hubble with the recommendation:

> Personally, [Hubble] is a man of the finest type. Physically he is a splendid specimen. In his work here, altogether, and especially in science, he showed exceptional ability. I feel sure you would find him just the sort of man you would wish to have.[16]

[16]F.R. Moulton to E.B. Frost, 27 May 1914, YOA, quoted by Osterbrock et al. (1990).

Frost decided to offer Hubble the opportunity to study further at Yerkes, starting in October 1914, with tuition fees paid and $30 per month for support. In the meantime, he was invited to attend the annual meeting of the American Astronomical Society, due to take place in the Chicago area that August. This was the occasion at which Vesto M. Slipher of Lowell Observatory made the startling announcement that a number of nebulae[17] were approaching or receding from the sun with much higher velocities than any stars. He had been able to do this work because he had a specially efficient spectrograph with an ultra-fast camera that could register the spectra of very faint objects. In particular, he found that the 'Great Nebula in Andromeda', known to astronomers as Messier 31 or M31, was approaching the solar system at 300 km/sec. He received a standing ovation at the meeting for this discovery, but was unwilling to make an outright claim that his nebulae were outside the Milky Way and therefore much more distant than the stars—the issue was then still considered an open one. However, the young Danish astronomer, Ejnar Hertzsprung, on hearing of his work was sure that Slipher had clinched the argument that they were.

The Yerkes Observatory where Hubble studied was no longer the dynamic institution that Hale had founded some twenty years previously. The glory had departed when Hale left for the clearer skies of California and took most of the more active researchers with him. Frost himself was gradually going blind and was unable to use the telescopes. The best-known astronomer remaining was E.E. Barnard, a straightforward observer with little theoretical knowledge.

Hubble very soon specialised in photographic observations of nebulae, using the 24-in. reflector at Yerkes that had been built by the optician George Ritchey while working with Hale before they left for Mount Wilson. He made a particular study of a cometary-shaped object called NGC 2261 (from the *New General Catalogue of Nebulae and Clusters*, a nineteenth-century compilation based on the surveys of nebulae made by William Herschel and his son, John, that is still referred to). He was able to compare his own photographs with one taken in 1900 by a British amateur, Isaac Roberts. To his surprise, quite considerable changes had occurred in the interval. This showed that NGC2261, now called 'Hubble's Variable Nebula', although of appreciable angular size, must be less than a few light-years across and therefore relatively nearby. The following academic year he was offered a more valuable scholarship.

At this point in his life, Hubble began to distance himself from his rather impecunious family. His brother Bill effectively assumed responsibility for his mother and sisters. However, he did occasionally see the latter, who visited him in the beautiful surroundings of Yerkes. His heroic tendencies apparently found another outlet when he rescued the wife of a middle-aged professor who had

---

[17]The word 'nebula' then referred to any object in the heavens that looked like a cloud. When Hubble proved somewhat later that some of the nebulae are systems of stars beyond the Milky Way, another term had to be found for them. Hubble's choice of 'extragalactic nebulae' was, by the end of his life, old-fashioned. He was reluctant to call them galaxies, evidently because the word had been suggested by Shapley.

fallen into the nearby Lake Geneva. He stated afterwards that the professor had shown no emotion when thanking him for rescuing his wife 'nor did he seem particularly glad to see her again'.[18]

The last part of Hubble's education consisted of courses in mathematics and celestial mechanics that he took at the University campus in Chicago. For the half-year that he spent there he was graduate resident head of Snell Hall, a particularly rambunctious student residence. His reputation as a tough athlete was sufficient to exact good behaviour.

As his student days drew to a close, Hubble started the inevitable search for a job and was fortunate enough to attract the attention of the Mount Wilson astronomer Walter S. Adams, one of Hale's former Yerkes group who had been resident in California since 1904. Hale was given a good report on Hubble by his friend Henry Gale of the University of Chicago Physics Department. The consequence was an offer of a position at $1200 per year on completion of his thesis.

9.6   Major Hubble

In April 1917, the United States Congress joined the First World War by declaring war on Germany and Hubble's patriotic reaction was to enlist. He applied for a commission in the Officer Reserve Corps and finished off his thesis in a hasty manner. Its title was 'Photographic Investigations of Faint Nebulae', and even he had to admit that it was rather skimpy, notwithstanding which it was published a few years later in *Publications of the Yerkes Observatory*. He realised that there was good work to be done in classifying the nebulae on a better basis than had been attempted before that time although he made temporary use of a scheme devised by Max Wolf of Heidelberg. Clearly he favoured the point of view that they were extra-galactic (i.e. outside the Milky Way galaxy); his thesis contains the prophetic sentence, probably echoing Eddington's words of 1914: 'Suppose them to be extra-sidereal and perhaps we see clusters of galaxies; suppose them within our system, their nature becomes a mystery' (see Section 7.6). If the spiral nebulae were like the Milky Way, whose size had been estimated by Eddington at about a tenth of its currently accepted value, they had to be millions of light years away.

Hubble wrote to Hale to ask him to keep his position at Mount Wilson open until after the war, which he, as a strong supporter of the war effort, was quite willing to do. In May 1917, Hubble reported for duty with other graduates at Fort Sheridan, near Evanston, Illinois. After his basic training, he found himself helping to prepare a new Division. He took charge of a battalion with 25 officers and 600 men and got them ready for the trench warfare they were expected to face in Europe. He showed himself to be an enthusiastic officer and an accomplished pistol shooter.

[18]G.B. Hubble, 'E.P. Hubble, University of Chicago 1914-1917', HUB 82(2), 10-11, quoted by Christianson (1995, p. 98).

Only in September 1918, after fourteen months in the army, was Hubble finally sent to Europe. He had been promoted to Major in January of that year. While his unit did reach France, they never saw action, although Hubble later implied he had: 'The hardest thing was to see wounded men fall, and go forward without stopping to help them'.[19] He spoke of being in an observation balloon and having been knocked senseless by a bursting shell. The latter episode seems to have been genuine, as his right elbow could afterwards not be straightened. However, his extant military record does not reflect his presence at any actual engagement.

Following the war, he stayed with the occupation forces for a few months and reputedly dealt with reparations claims and was a judge-advocate at courts martial. In 1919 he found himself in Cambridge, England; he later stated that he had been in charge of American army students at British universities. He was associated with Trinity College and attended lectures by Eddington on spherical astronomy. He also got to know Hugh F. Newall, the wealthy English astronomer and friend of Hale who lived next to the Cambridge Observatories in his house, Madingley Rise. He was proposed as a Fellow of the Royal Astronomical Society at their May 1919 meeting. Later, in July, he attended a special meeting of the Society with a group of American astronomers and was present at a dinner given in their honour. In August he sailed for New York. He stopped in Chicago for a day with his mother and sisters and then headed west to an impatient Hale.

9.7 Mount Wilson

Major Edwin Hubble joined what was then the Mount Wilson Solar Observatory on 3 September 1919. The 100-in. telescope had only recently come into regular use. He was given an office in the Observatory's headquarters building on Santa Barbara Street, Pasadena. For living quarters he shared a house with two others. All bachelors, they did not cook for themselves but preferred to eat out at a nearby restaurant.

This was a particularly exciting time to have had access to the largest telescopes in the world because it was the start of the decade in which the shape of our own galaxy was finally settled and also when many of the nebulous objects in the sky were realised to be galaxies, a process in which Hubble himself took part. Admittedly, Slipher had already discovered their high velocities, but the significance of his work had not yet become clear. The idea that some of the nebulae might be galaxies had been floating around for a long time, but no conclusive proof had yet appeared. As far back as the eighteenth century, an obscure English instrument maker, Thomas Wright, had put forward the hypothesis that they might be 'island universes'. His speculations were afterwards elaborated by the philosopher Immanuel Kant.

Adams, who had originally recruited Hubble, was, in fact, the effective head of the Mount Wilson Observatory at the time of his arrival. Others on the relatively

---

[19]Hubble, G.B.: 'E.P. Hubble: World War I', p. 13, HUB 82(4), Box 7, 1, quoted by Christianson (1995, p. 108).

young staff included Adriaan van Maanen, Paul Merrill, Harold Babcock, and Harlow Shapley, all of whom were to carve out important niches in astronomical research. The homespun Shapley, who had kept his Missourian accent, found Hubble pretentious and felt that he had an unjustifiable attitude of superiority:

> Hubble and I did not visit very much. He was a Rhodes scholar and he didn't live it down. He spoke with a thick Oxford accent. He was born in Missouri not far from where I was born and probably knew the Missourian tongue. But he spoke 'Oxford'. He would use such phrases as 'to come a cropper'. The ladies he associated with enjoyed that Oxford touch very much. 'Bah Jove!' he would say, and other such expressions. He was quite picturesque.[20]

A few weeks later, on his first observing run with the 60-in. telescope, the young night assistant Milton Humason, who was later to become a famous astronomer in his own right, watched Hubble closely. Later he wrote:

> My own first meeting with Hubble occurred when he was just beginning observations on Mount Wilson. I received a vivid impression of the man that night that has remained with me over the years. He was photographing at the Newtonian focus of the 60-in. standing while he did his guiding. His tall, vigorous figure, pipe in mouth, was clearly outlined against the sky. A brisk wind whipped his military trench coat around his body and occasionally blew sparks from his pipe into the darkness of the dome. 'Seeing' [the steadiness of the atmosphere] that night was rated extremely poor on our Mount Wilson scale, but when Hubble came back from developing his plate in the dark room he was jubilant. 'If this is an example of poor seeing conditions', he said, 'I shall always be able to get usable photographs with the Mount Wilson instruments'. The confidence and enthusiasm which he showed on that night were typical of the way he approached all his problems. He was sure of himself—of what he wanted to do, and of how to do it. (Humason 1954)

Hubble immediately concentrated on photography of nebulae (Fig 9.1), one of the first to be observed being his favourite, NGC 2261. Using a small telescope capable of photographing large areas of the sky, he made a survey to discover and classify more nebulae. He was able to divide them into several classes. Some were diffuse and lay near the plane of the Milky Way. Using photographs of individual bright examples, he could show that they shone either by reflecting the light of associated stars or by re-processing that light in a way that was not yet understood. Others, lying in directions well away from the Milky Way, he called 'non-galactic nebulae', though it was still unclear whether they lay beyond the Milky Way in actual distance. Having the use of the world's largest telescope he was able to take better and more detailed photographs and could classify the nebulae better than anybody else. He rapidly became an accepted authority on the subject. The 'non-galactic nebulae' were not only spirals, but

[20]Shapley (1969, p. 57).

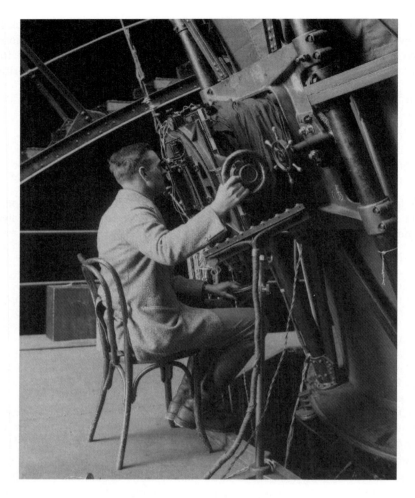

Fig. 9.1. Hubble at the Newtonian focus of the Hooker 100-in. telescope. He is
supported by a movable platform, far above the floor of the dome (Copyright
unknown).

could be divided into spirals, spindles, 'ovates', globulars, and irregulars. Their
classification was to become a lifelong interest. He published a paper *A General
Study of Diffuse Galactic Nebulae* in the *Astrophysical Journal* (Hubble 1922)
in which he speculated that his 'non-galactic nebulae' lay outside the Milky
Way but admitted that the evidence for his view was sparse. As a member of
the Commission of the International Astronomical Union for Nebulae and Star
Clusters, he began to promote a scheme for classifying nebulae according to
shape, but he met with little support.

Not long before Hubble's arrival at Mount Wilson, Shapley had presented his outline of the Milky Way galaxy and the question of the nature of the spiral nebulae was coming to a head. The 'Great Debate' (see Section 8.10) between Shapley, who believed that they were part of the Milky Way, and Heber D. Curtis who thought they were extragalactic (beyond the Milky Way), had taken place a few months later. Shapley's belief that they were relatively close had arisen from an erroneous claim by van Maanen that rotation could be detected by comparing plates taken at different times. The measurements involved were extremely difficult both to make and to check. As a friend and admirer of van Maanen, Shapley preferred to accept his views rather than those of the extra-galactic faction.

9.8   The distances of the galaxies

Van Maanen's results were challenged in print by Knut Emil Lundmark, a Swedish astronomer who spent 1921–1923 in the United States and worked at Lick and Mount Wilson. Using deep spectroscopic observations of M33, Lund-mark came to the conclusion that some bright points that showed up in photographs of this object were actually very distant stars and he correctly took this as proof that there had to be something wrong with van Maanen's work. At this point, however, Shapley still had faith in his friend.

By 1923, Hubble had embarked on a new programme of photography of the spiral nebulae (Fig 9.1), hoping to detect nova outbursts [a nova is a type of stellar outburst, less dramatic than a supernova], which he thought would turn out to be useful as standard candles or distance indicators. In 1917, Ritchey had found a number of these events in spirals and his work had been noticed by Curtis, who realised that their faintness implied they were very distant. To him, this was clear evidence that the spirals were 'island universes', or galaxies as we call them today. Shapley, on the other hand, wished to believe that they were local and suggested that Ritchey's alleged novae were not of the familiar type, but were related to the intermittent obscuration of stars by fast-moving nebulosity. This disagreement had been an important issue in the 'Great Debate' already referred to. Hubble's own search achieved success in October 1923, when he found his first nova in the Andromeda Nebula (M31), the nearest spiral galaxy and one which somewhat resembles the Milky Way. However, there were two other star-like variable objects on his plates. While searching older plates for previous evidence of these 'novae', he discovered that one of them was varying in a cyclic manner— in fact it was a not a nova but a Cepheid variable, clearly resembling those that Shapley had already used so successfully as standard candles! Its period was 31.4 days and, using Shapley's calibration, Hubble found that the Andromeda nebula had to be around a million light years from the earth, or several times more distant than the furthest parts of the Milky Way.

Hubble had proved definitively that the spirals were outside the Milky Way system! Further, he had shown that they are composed of stars similar to those in our own galactic neighbourhood.

He wasted no time in writing to Shapley to announce his discovery. This was the missive of February 1924 that was shown to Cecilia Payne (Payne-Gaposchkin 1984) with the words 'Here is the letter that has destroyed my universe' (see Section 8.13).

> You will be interested to hear that I have found a Cepheid variable in the Andromeda Nebula (M31). I have followed the nebula this season as closely as the weather permitted and in the last five months have netted nine novae and two variables . . . I have a feeling that more variables will be found by careful examination of long exposures. Altogether next season should be a merry one and will be met with due form and ceremony.[21]

The news of the discovery spread quickly. James Jeans, the famous British theorist and rival of Eddington, heard it while visiting Mount Wilson as a Research Associate. He in turn told Henry Norris Russell of Princeton, the spider at the centre of the web of American East Coast astronomy. As a result Hubble was asked to send a paper about his work, with the possibility of winning a $1000 prize, to the American Association for the Advancement of Science, which was due to meet in Washington, DC. His paper, 'Cepheids in Spiral Nebulae', arrived at the eleventh hour and was read to the Association by Russell, who pointed out its significance:

> it has already expanded one hundred fold the known volume of the material universe and has apparently settled the long–mooted question of the nature of the spirals, showing them to be gigantic agglomerations of stars almost comparable in extent with our own galaxy.[22]

Hubble's contribution tied for first place with that of a microbiologist from Johns Hopkins University, with the result that he received $500 instead of the full amount!

His first publication on Cepheid-based distances, *N.G.C. 6822, A Remote Stellar System* (Hubble 1925), again showed incontrovertibly that galaxies are remote and outside the Milky Way. The Great Debate was over and decades of speculation about the existence of 'island universes' had been brought to an end (Fig 9.2).

9.9 Marriage

In June 1920, William H. Wright, an astronomer from Lick Observatory, had been invited to Mount Wilson to make some observations there. His wife Elna had come with him and she brought her sister-in-law, Grace Leib, along. Grace's husband was still alive at the time and the meeting might have had no further significance. The ladies saw Hubble at a window, studying an astronomical plate. They were duly introduced to 'The Major', as Hubble was called by the Lick astronomers. Grace, writing long afterwards, was impressed:

[21]Hubble to Shapley, 19 February 1924, HUB 611.

[22]Stebbins, J., Memoir 'Edwin Hubble and the A.A.S. Prize', M.L. Shane Archives, Lick, quoted by Christianson (1995, p. 161).

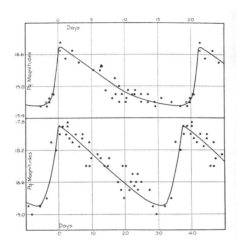

FIG. 9.2. The light curves of two Cepheid variables in the galaxy NGC 6822, from Hubble (1925). The upper star has a period of about 21 days and the lower about 37 days. From their average stellar magnitudes (brightnesses) and Shapley's calibration work, Hubble could derive the distance of NGC 6822, showing that it was far beyond the confines of our own galaxy, the Milky Way.

> This should not have seemed unusual, an astronomer examining a plate against the light. But if the astronomer looked an Olympian, tall, strong, and beautiful, with the shoulders of the Hermes of Praxiteles, and the benign serenity, it became unusual. There was a sense of power ... [23]

As it happened, one year later Grace's husband, Earl, a mining engineer, was asphyxiated by poisonous gas in an accident. They had lived their married life of nine years in the home of her parents, the Burkes, a family of Irish descent. Her father had worked his way up to become a Vice-President of the First National Bank of Los Angeles and was a millionaire. Grace had had a private education at a private school in Los Angeles and had later studied English at Stanford University.

Within a year of Earl's death, Grace and Edwin were dating. He would visit the Burke house after observing runs on Mount Wilson and read to her. It was apparently at this time that he began to gild his past with tales of legal expertise, heroism and adventure, even though the reality was impressive enough. He offered to give up astronomy and practise law in order to keep her in the style to which she was accustomed, but she would not hear of it. They were married privately in a Catholic ceremony on 26 February 1924 by the family priest of the Burkes' and started their honeymoon with a short stay at the Burke cottage

---

[23]Grace Hubble, EPH the Astronomer, quoted by Christianson (1995, p. 133).

in Pebble Beach, a private enclave near Monterey, California. Very soon they headed for New York and Boston, where they had a meeting with Shapley. Then they crossed the Atlantic to Liverpool, encountering heavy seas which made Edwin very seasick, but which Grace took in her stride. They went by train to London, where they took in the conventional sights, and moved on to visit old friends at Oxford and Cambridge. While staying with Newall in his large house, which incidentally did not contain baths, they met Eddington at lunch (he was regarded as too radical to invite to dinner!). Later, Hubble spoke at the Royal Astronomical Society on his discoveries. This English interlude was followed by a continental tour through France and Italy. They were so impressed by the interior of the Palazzo Vecchio in Florence that they resolved to design the rooms of their own house according to the same taste. They also visited the tomb of Galileo and his haunts around Arcetri.

They arrived back in California in May and Hubble went immediately to Mount Wilson for an observing run. On coming down the mountain he was able to move straight into their first home, a small apartment that Grace had found near Caltech.

9.10 The 'Tuning Fork' diagram

Living in a small space did not inhibit Grace's taste for entertaining. Among their first guests was Eddington, whom she took swimming at a local country club and found silent and far away in his thoughts. The latter's great rival in theoretical astronomy, Jeans, had been appointed a Research Associate at Mount Wilson in 1923 and had returned two years later. The normally shy Jeans also became a friend at this time and encouraged Hubble in his scheme for classifying nebulae.

Hubble tried to persuade Slipher, Curtis, and other members of International Astronomical Union's Commission on Nebulae and Clusters, to have his scheme adopted officially. The method of classification which he now settled on for describing galaxies, probably after discussions with Jeans, is shown in Fig 9.4. The spiral nebulae were placed in an order that depended on how tightly they were wound, but were also divided into two separate series according to whether they had a 'bar' in them or not. Hubble and Jeans believed that the ellipticals were younger than the spirals and envisaged the diagram as an evolutionary sequence. He called the ellipticals 'early type', thinking that they evolved to become spirals of 'late type'. This nomenclature is still used, although it is no longer regarded as the actual order of evolution. Hubble's evolutionary scheme, although first published in 1926, was illustrated for the first time in his famous book, *The Realm of the Nebulae*, published in 1936. It became known as his 'Tuning Fork' diagram (Fig 9.4).

Unfortunately for Hubble, Slipher did nothing about his classification scheme and the other Commission members were not convinced of the need for a change. Another problem was that Shapley was now on the Commission and he objected to the clumsy name of 'non-galactic nebulae' that Hubble like to use. He sug-

FIG. 9.3. Undated photograph of Edwin Hubble, published in Journal of the British Astronomical Association in 1940, but probably taken many years before. (Copyright unknown.)

gested 'galactic nebulae' or preferably 'galaxies', the modern term. Hubble could not bring himself to accept the word advocated by his rival, and opted for the phrase 'extra-galactic nebulae', which he continued to use for the rest of his life, even though most astronomers came to prefer Shapley's version. At its next meeting, in 1925, the IAU again declined to adopt the Hubble scheme.

Meanwhile, in 1926, Hubble was scooped in the area of galaxy classification by Lundmark, who published a very similar scheme. He became quite furious and wrote to Lundmark as follows:

> This is a very mild expression of my personal opinion of your conduct and unless you can explain in some unexpected manner, I shall take considerable pleasure in calling constant and emphatic attention, whenever occasion is given, to your curious idea of ethics. Can you suppose that colleagues will welcome your presence when they realize that it is necessary to publish before they discuss their work?[24]

[24]Hubble to Lundmark, 26 August 1926, quoted by Smith (1982, p. 152).

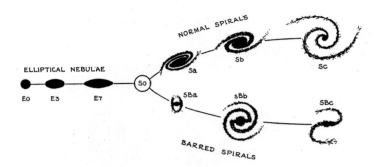

Fig. 9.4. The Hubble 'Tuning Fork' diagram of galaxies, from his book of 1936, *The Realm of the Nebulae*.

This curious letter appears rather ridiculous today. Although Hubble had undoubtedly thought a great deal about galaxy classification, neither he nor Lundmark were the only ones who had put forward schemes, and there was no reason why he should have expected to have a monopoly.

Hubble finally published his own classification scheme in a paper in the *Astrophysical Journal* called *Extragalactic Nebulae* (Hubble 1926), a few months later than Lundmark. In a footnote, he accused Lundmark of having learned about his scheme at the Cambridge IAU meeting and having passed it off as his own, without acknowledgements or references. Lundmark, in his next paper, *Studies of Anagalactic Nebulae*, as he called them, replied in kind:

> In his paper, Extragalactic nebulae, ... E.P. Hubble makes an attack on me which is written in such a tone that I hesitate to give any answer at all ...I did not have any access whatsoever to the memorandum [submitted to the IAU] or to other writings of E.P. Hubble ... Hubble's statement that my classification except for nomenclature is practically identical with the one submitted by him *is not correct* ... As to Hubble's way of acknowledging his predecessors I have no reason to enter upon this question here. (Lundmark 1927)

Although he mentioned Jeans's idea of a possible evolution of galaxies from left to right in the tuning fork diagram, Hubble emphasised that the scheme was purely a descriptive one. It gained widespread acceptance and is still almost exclusively used, with modifications in detail, and has turned out to be adequate for almost all cases.

9.11   The Hubbles at home

Hubble was soon promoted from Assistant to full Astronomer at Mount Wilson, in spite of his habit of taking very long vacations, one that went against the puritanical New England grain of Adams, who had finally taken over the Directorship in 1923. The increased salary eased his financial position. In addition, Grace received the proceeds of a life insurance policy on her first husband. Just before their marriage, Hubble had found an ideal housing site in the suburb of San Marino, next to Pasadena. This was an area greatly favoured by the wealthy. Grace's family, the Burkes, paid for the construction of their house as a wedding gift. It was a small but elegant creation, with a study for Edwin that showed off his collection of antique books and featured a marble statue of Galileo. The living room contained a large fireplace and there was a dining room to match.

Hubble, though ready to help with technical problems at the telescope, would never fix anything at home. He took no photographs other than astronomical ones and showed no interest in laying out the garden. Nevertheless, he remained fond of physical exercise and some of his leisure time was spent hiking in the nearby mountains. The climate of Los Angeles area was ideal for outdoor activities and in the nearby San Gabriel range there were numerous trails where a person could walk for hours on end. At the end of a strenuous day he could be sure of finding one of the private camps that offered meals and a bunk for the night.

Sad to relate, the couple's only attempt at parenthood, Grace then being thirty-seven years old, resulted in a stillbirth, at a moment when Hubble was away from home, working on the mountain. In a gesture of extraordinary stoicism, rather than interrupt his observing, Grace avoided telling him of this tragedy until after he had returned. Edwin's scientific career and reputation seems to have dominated the lives of both partners.

9.12   The recession of the nebulae

By 1928, Hubble had become quite famous and was often photographed in the role of a 'deep thinker', clad in tweed and with a pipe in his mouth. By the time of the triennial meeting of the IAU Commission on Nebulae and Clusters of that year in Leiden, Netherlands, his reputation had increased to the extent that he was elected chairman. At about the same time, probably at the instigation of Willem de Sitter, he conceived the idea of making a more systematic study of the recession of the galaxies that had been discovered by Slipher fourteen years before. The latter, working with the small telescopes of Lowell Observatory, was effectively out-gunned. Hubble had a further advantage: by finding and measuring Cepheid variables, he could also determine distances. His privileged access to the largest telescope in the world gave him, in effect, a monopoly.

De Sitter, a decade before, had taken Einstein's equations for treating the universe according to general relativity and had found another solution besides Einstein's own. Neither one turned out in the end to be wholly satisfactory, for as Eddington (1932) remarked, 'Einstein's universe contains matter but no motion; deSitter's universe contains motion but no matter'. Nevertheless, de Sitter thought that his solution could be a reasonable approximation to the truth since matter was known to be spread very thinly in the universe and he took a close interest in the redshift question. The opportunity seemed to be open for Hubble to decide which of the two solutions was the correct one.

Working with Hubble was Milton Humason, the assistant who had started his involvement with Mount Wilson as a pack-train driver and ultimately became a formidable observer (Mayall 1973). Hubble divided their task into two parts. Humason was to photograph the spectra and find out by how much each galaxy was redshifted. Hubble would take direct photographs to look for variable stars, determining their brightnesses and their periods. With Hale's support, a new camera based on an inverted microscope lens was built for the spectrograph. He now had the most sensitive possible system for measuring spectra, even though the images were tiny, making the redshift information more difficult to extract. They soon confirmed the existing results and extended them to fainter galaxies. They could observe galaxies more than four times as distant as Slipher could.

In 1929, Hubble came to the conclusion that the galaxies were receding from us at about 500 km/s for every megaparsec of distance (the megaparsec is a million parsecs; a parsec is an old astronomical unit of distance equal to about three light years). Thus, for every 3,000,000 light years of distance from us, the galaxies retreat by 500 km/s (Fig 9.5). This number Hubble called K, but it is now known as H, for the 'Hubble constant'. He published his result as *A Relation between Distance and Radial Velocity among Extra-Galactic Nebulae* (Hubble 1929) in *Proceedings of the National Academy of Sciences*.

The final paragraph of this paper began with the words:

> The outstanding feature, however, is the possibility that the velocity–distance relation may represent the de Sitter effect, and hence that numerical data may be introduced into discussions of the general curvature of space.

Duerbeck and Seitter (2000) cynically remark that 'Hubble was the *last* of the early cosmologists who believed that his result confirmed the de Sitter model'. While this is superficially true, Hubble's work was by far the most systematic, especially as to finding distances, so he was not just the *last* person to point out the result, but he had said the *last word* on the matter. Other investigators, such as de Sitter himself, Wirtz, Lundmark and Silberstein had been hinting that the high velocities of the galaxies were of cosmological origin. Hetherington (1986) suggests that Hubble's pre-eminence was a result of his legal training that made him adept at advocating his point of view in a highly convincing manner.

By this time, Einstein and de Sitter were not the only ones to have found solutions to the equations of general relativity as applied to the universe. Alexander Alexandrovich Friedmann, a Russian mathematician, published general solutions for both expanding and non-expanding universes in 1922 in the German journal *Zeitschrift für Physik*. Unfortunately, he died relatively young and his work went almost unnoticed. Similar results were found independently, though somewhat later (1927), by the Abbé Georges Lemaître (Fig 9.6), a Belgian mathematician, who had worked for a year (1924–1925) in Cambridge with Eddington and had recently been appointed Professor of Astronomy in Louvain. He published

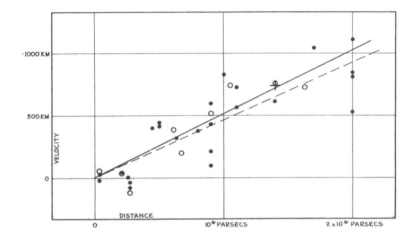

FIG. 9.5. Hubble's first diagram of the velocity–distance relation for galaxies, from his paper of 1929. The horizontal axis is the distance of the galaxy in parsecs (1pc $\sim$ 3 light years) and the vertical axis is the velocity in km/s. The black dots are individual galaxies; the open circles are the same data averaged into nine groups. Hubble gave two solutions—the two lines—one for each set of data, viz 465±50 and 513±60 km/s per megaparsec (1 megaparsec $\sim$ 3,000,000 light years). The cross is an estimated mean value for twenty-two additional galaxies without measured distances.

in French in a Belgian journal but it was one that few astronomers read, even though it was also circulated to observatories worldwide as a Louvain Observatory publication. At the time, he sent a copy of his fundamental paper to his former supervisor, Eddington, who unfortunately laid it aside. In this paper Lemaître actually published a value for the 'Hubble constant' before Hubble did, using Slipher's radial velocities, but with no proper data on the distances of the galaxies. He had had to assume that they were all of equal intrinsic luminosity and he took an average value from Hubble's work. The number that he found (625 in the same units as Hubble's) was about 25% higher than Hubble's own first estimate, just referred to, that only came out two years later. A year after that, when Eddington had set essentially the same problem to a later PhD student, G.C. McVittie, Lemaître happened to hear about it and guessed that his original contribution had never been read. He sent another copy. This time Eddington realised its importance and reported on it to the Royal Astronomical Society in May 1930. Later he organised for an English translation (which unaccountably omitted Lemaître's value of the Hubble constant) to be published in the Society's *Monthly Notices*, a journal with a much wider circulation and where the work was much more likely to have an impact. It soon became known

FIG. 9.6. The Abbé Lemaître, the first person to publish a value for the Hubble
constant, at the Rome meeting of the International Astronomical Union in
1952 (SAAO).

to the Caltech theoreticians that Hubble associated with.

In 1930, de Sitter published an analysis of the redshift-distance relation and
openly admitted that neither his nor Einstein's solution of the equations was
able to explain the observations. In fact, he generously pointed out that Lemaître
had found the correct answer. In spite of de Sitter's long and open interest in
the question, Hubble's blood boiled over this article, particularly because he
mentioned the earlier interest of astronomers other than himself in the velocity–
distance relation. He wrote to de Sitter:

> I consider the velocity–distance relation, its formulation, testing and con-
> firmation, as a Mount Wilson contribution and I am deeply concerned in
> its recognition as such.

As in his complaint to Lundmark, he went on

> We have always assumed that, where a preliminary result is published
> and a program is announced for testing the result in new regions, the
> first discussion of the new data is reserved as a matter of courtesy to
> those who do the actual work. Are we to infer that you do not subscribe
> to this ethic; that we must hoard our observations in secret?[25].

Hubble and Humasons' definitive paper was published in 1931. The fact that
the relation is associated today with Hubble rather than any of the others just

---

[25]Hubble to de Sitter, 21 Aug 1931, HUB Box 15, f. 616, quoted by Christianson (1995, p.
230).

mentioned, is due to the high quality of his data, especially as to distances, his thoroughness, and his persistence in observing ever more distant galaxies.

9.13   Distance indicators

In extending his relation to more distant galaxies, Hubble asserted that the brightest stars in every galaxy are similar to each other and that they could also therefore be used as distance indicators. Since they are brighter than the Cepheid variables, they can be picked out at greater distances. Using observations of some relatively nearby galaxies where the Cepheids can also be seen, these stars could be calibrated and turned into standard candles.

As a further step, to distances so great that even the brightest stars can no longer be picked out, he assumed that the brightest galaxy in a cluster of galaxies could be regarded as yet another standard candle. To make the brightest cluster galaxies useful as distance indicators, they had in turn to be calibrated using examples whose distances were already known. Fortunately, in a few clusters, some bright stars could be discerned and used as standard candles. This method could be refined by taking not the brightest member of the cluster (which might be anomalous), but say the tenth or twentieth brightest, as the most representative.

Obviously, errors could creep in at any stage of this step-by-step process and they could be cumulative. As a matter of record, the present-day estimate for the Hubble constant, the result of an enormous international effort with the largest telescopes available,[26] is about a factor of 7 smaller than Hubble's one, i.e. the galaxies are that much older and further away. The change is purely due to revised distance estimates. The first major error in the chain of calibrators was discovered by Walter Baade, David Thackeray, and Adriaan Wesselink (see Section 8.21). These authors showed that Shapley's calibration of the Cepheids was wrong. The Hubble constant had to be approximately halved, from around 526 km/s per Megaparsec to about 280. Subsequently, Humason, Mayall, and Hubble's student Sandage (1956) showed up errors in the calibration of the brightest stars in galaxies, further reducing the constant to 180. Sandage (1958) soon afterwards found that some of them were not single stars at all but were small, dense, clusters.

Hubble's values for the recession velocities were correct and he had established the recession as an indubitable fact. He simply got the distances wrong. His methodology is still essentially in use, although it has undergone considerable refinement in the meantime. Modern detectors, such as Charge-Coupled Devices (CCDs), introduced in the 1970s, now enable calibrators to be studied in much more distant galaxies. The current (2005) (one hardly dares to say accepted!) value of the Hubble constant is around 75 km/s per Megaparsec.

---

[26]Including the Hubble Space Telescope, named after Edwin Hubble.

9.14 Inner doubts

Hubble's aggressive defence of what he saw as his scientific territory was probably symptomatic of some inner lack of confidence. The single-mindedness of his pursuit of the redshift–distance relation shows that he was fully aware of the fundamental nature of his work. He was well-read in philosophy and in the history of science and understood clearly that he had reached conclusions that were leading to a permanent change in our picture of space and time. Yet, in print, he was quite reluctant to interpret his data in theoretical terms, preferring to describe what he had found as a simple observational relation between redshift and distance.

He often seems to have felt insecure and perhaps was worried that his fame might be ephemeral. The arguments he had early on with van Maanen are even said to have given him intestinal pains. An instance of how petty he could get in the matter of personal prestige occurred on Mount Wilson. It was the tradition in the Monastery dining room that the 100-in. observer sat at the head of the table and the 60-in. observer sat on his right side. One dinnertime, van Maanen happened to be the 100-in. observer. Hubble was spotted moving his napkin-ring and replacing it with his own. Van Maanen, a smaller person than the intimidating Hubble, accepted this insult in silence.

9.15 The Hubble and van Maanen problem

The disagreement between Hubble and van Maanen, concerning the latter's (erroneous) claims that the galaxies could be seen to rotate, ultimately led to a serious problem for Adams, the Director of Mount Wilson. The photographic plates on which van Maanen claimed to have seen the evidence were examined by Hubble and, at Adams's request, by two other staff members of Mount Wilson, Seth B. Nicholson, and Walter Baade. Of course, the effect, if it had existed, would have been a very small one and difficult to be certain about, but the balance of probabilities was that van Maanen was wrong. Adams's confidential report on the matter to the President of the Carnegie Foundation continues the story:

> Hubble then wrote a long statement for publication which both Seares [who acted as editor of Mount Wilson publications] and I felt could not be published in the form in which it was written. Its language was intemperate in many places and the attitude of animosity was marked. He objected to any material change in the wording and a deadlock seemed to be indicated ... Seares undertook a long and careful analysis of the measures and prepared an accurate statement of the results. This was to be published under the names of van Maanen, Hubble, Nicholson and Baade. The paper was submitted to those concerned and all agreed to its publication with minor changes with the exception of Hubble who opposed it violently. I do not feel that Hubble's attitude in this matter was in any way justified ...
>
> The solution of the problem represented a compromise which was not altogether satisfactory but which at least presented the results without any violently controversial features ... Hubble, who had much the better

of the general weight of evidence, showed a distinctly ungenerous and almost vindictive spirit. This is not the first case in which Hubble has seriously injured himself in the opinion of scientific men by the intemperate and intolerant way in which he has expressed himself.[27]

In fact, Hubble's opinion in the matter was that he had given van Maanen ten years too many, during which he could have retracted his claims. The last straw was when a speaker at the Royal Astronomical Society suggested that his results could be accepted but for van Maanen's work. It was this that brought matters to a head. Their public disagreement only ended when van Maanen was induced to publish a statement admitting that his observed rotation of the galaxies should be viewed with reserve (Hetherington 1986).

9.16 Celebrity status

The Hubbles basked in Edwin's increasing fame. In 1928 he was asked to give one of the prestigious lectures offered by the Carnegie Institution of Washington, the foundation that owned Mount Wilson, to audiences of invited (and hopefully generous) guests. His talk on that occasion, *The Exploration of Space*, was published afterwards in *Harper's Magazine*, reaching a wide audience and paying its author very well. This was followed by many other invitations from the most prestigious universities to give public lectures, sometimes resulting in published books. He was offered several positions that carried higher pay than his job at Mount Wilson, one among them being the Presidency of the Carnegie Institute of Technology in Pittsburg. He turned them all down, preferring to stay a researcher with access to the unique facilities of Mount Wilson.

When Einstein visited Mount Wilson early in 1931, he was naturally interested in meeting Hubble, who had had so much to say on cosmological matters. They were photographed together for the newspapers. Grace Hubble was happy to have the Einsteins as dinner guests and they in turn were flattered to meet some of the Hollywood 'celebrities' whose acquaintance the Hubbles enjoyed cultivating. The Hubbles also met, for example, William Randolph Hearst, the notorious newspaper proprietor, at his ornate castle 'San Simeon', north of Los Angeles. By this time, Hubble had cut off all contact with his own family. Grace never met any of them. Their closest friends seem to have been their neighbours, Homer and Ida Crotty, the former being a retired lawyer. Aldous Huxley, the author of *Brave New World* and grandson of 'Darwin's Bulldog', T.H. Huxley, was a frequent guest at their house.

In 1934, Edwin went to Oxford to deliver the annual 'Halley Lecture' and there received an honorary degree. While in England, he was the guest of wealthy friends. His prolonged absences while being paid continued to cause raised eyebrows at the Carnegie Institution, but he managed to avoid any serious repercussions.

---

[27]W.S. Adams, 'Mount Wilson Observatory Confidential Statement for President Merriam', Adams Papers, Box 43, Mount Wilson Observatory Archives, Huntington Library, August 1935, quoted by Hetherington and Brashear (1992).

9.17   Counting the galaxies

If galaxies are uniformly distributed throughout space, and space itself behaves as everybody prior to Einstein thought it should, the numbers of galaxies at each brightness level should just depend on their intrinsic luminosities and how far away they are. Just as Herschel attempted to gauge the distribution of stars in space by counting, Hubble did the same for galaxies.

By 1934, he had counted 44,000 galaxies in areas totalling about $1\frac{1}{2}\%$ of the sky. He made plots of the numbers of galaxies brighter than a certain level and was surprised to find that there were fewer faint ones than were expected from classical theory. Even applying relativistic corrections to the data according to the best available advice, he found that the numbers dropped off too fast at large distances.

From this result, and a personal belief that the galaxies *should* be uniformly distributed in space, he came to feel that general relativity was probably not valid and that the redshifts had some other explanation, not yet understood. In fact, the belief that galaxies are uniformly distributed in space runs as a *leitmotif* through nearly all Hubble's later work.

Although this conclusion was wrong, the reasons for his error lie in the lack of knowledge at the time as to the proper way to correct the data and even of how to correct distances according to general relativity (Sandage 1997).

9.18   'The Realm of the Nebulae'

By the mid-1930s, Hubble and Humason had reached the limit of the Mount Wilson 100-in. telescope's capabilities so far as distance was concerned. The photographic plates of the time were, in retrospect, very inefficient, recording only 1% or 2% of the light quanta that reached them. Furthermore, even on nights without a bright Moon, the airglow or background radiation of the sky swamped the faint light of the most distant galaxies. Modern electronic detectors are almost fifty times more efficient than the plates that Hubble used and allow, within certain limits, the subtraction of the sky background.

In the autumn of 1935, Hubble delivered the Silliman lectures at Yale University. This was a famous series, previous speakers having included Ernest Rutherford and Niels Bohr, two of the founding fathers of nuclear physics. They had been established by one Augustus Silliman in memory of his wife Hepsa and were supposed to confirm 'the presence and wisdom of God as manifested in the Natural and Moral World', without recourse to 'Polemical or Dogmatical theology'.[28] The lectures were written up as a book called *The Realm of the Nebulae* (Hubble 1936).

The Realm of the Nebulae is a scientific classic, written in an authoritative style. It starts off with the statement 'Science is the one human activity that is truly progressive.' This is followed by a quotation from the scientific historian George Sarton that seems to have reflected Hubble's personal philosophy:

[28]HUB Box 20, f. 1027, quoted by Christianson (1995, p. 246).

> The saints of today are not necessarily more saintly than those of a thousand years ago; our artists are not necessarily greater than those of early Greece; they are likely to be inferior; and, of course, our men of science are not necessarily more intelligent than those of old; yet one thing is certain, their knowledge is at once more extensive and more accurate. The acquisition and systemisation of positive knowledge is the only human activity that is truly cumulative and progressive. (Sarton 1927)

The book gives a fair history of the development of the subject, dealing in the main with Hubble's own investigations. He acknowledges the contributions of Humason and Baade, his colleagues at Mount Wilson. He mentions Friedmann but not Lemaître, preferring to quote the Caltech relativist Richard Tolman. He hedges a little on accepting the redshifts as real velocities of recession: but he takes seriously an unorthodox theory put forward by the Caltech professor Fritz Zwicky, who contended that light could somehow be altered on its way from distant galaxies to the observer.

In 1936, Hubble delivered the Rhodes Memorial Lectures at Oxford on *The Observational Approach to Cosmology* (Hubble 1937). To an even greater extent, he played down the cosmological nature of redshifts and seemed to favour a non-cosmological origin. To him the cosmological explanation implied a finite universe which went against his personal preference for an infinite one. He concluded the lecture series by saying:

> Two pictures of the universe are now sharply drawn. Observations, at the moment, seem to favour one picture, but they do not rule out the other. We seem to face, as once before in the days of Copernicus, a choice between a small, finite universe, and a universe indefinitely large plus a new principle of nature.

Eddington found Hubble's reluctance to accept cosmological redshifts hard to understand (and his letter on the subject to Miss Douglas is quoted again):

> I just don't understand this eagerness to find some other theory than the expanding universe. It arose out of difficulties ... in Einstein's theory. If you do away with it, you throw back relativity theory into the infantile diseases of 25 years ago. And why the fact that the solution then found has received remarkable confirmation by observation should lead people to seek desperately for ways to avoid it, I cannot imagine. They do not seem to have the same urge to find some explanation of light which avoids identifying it with electromagnetic waves.[29]

The next step in the exploration of space–time, it was anticipated, would require the power of the 200-in. telescope, by then under construction. However, it was not an easy task to jump from the technology of the 100-in. telescope to that needed for a telescope of double its size. The honeycomb Pyrex blank required for the primary mirror was successfully cast only after the partial failure

[29]Eddington to Douglas, 6 December 1943, quoted by Douglas (1956, p. 113).

of the first effort and the grinding and polishing work required many years, interrupted by the Second World War, to accomplish.

9.19   Second World War

The coming of the Second World War in 1939 rekindled Hubble's patriotism. Though a conservative who vehemently disliked President Franklin D. Roosevelt's Democratic administration, he was not an isolationist.[30] He became a strong advocate of going to war on Britain's side and of helping her in every way possible. He made speeches in favour of the pro-war candidate in the Republican primaries and opposed the 'America First' isolationists on every possible occasion. He and Grace received threatening phone calls as a result but brushed them aside.

Hubble at first turned down several offers of posts through which he could help the war effort but finally accepted the position of 'Head of Exterior Ballistics' at the Aberdeen Proving Grounds in Maryland, on the East Coast. He later declined an invitation from Vannevar Bush, the Director of the Office of Scientific Research and the most important scientific administrator of the War, to work at Los Alamos. In fact, it is claimed that he did not know much about the technical side of his war work (Osterbrock 2001), which involved the testing and evaluation of all kinds of guns, bombs, shells, and bullets and the preparation of firing tables. Grace was eventually able to join him in a small cottage, almost a shack, on the base, that he had had fixed up for them. He was only one of several famous scientists who worked at Aberdeen. John von Neumann, the great numerical analyst and S. Chandrasekhar also served there.

After the War, Hubble was awarded the 'Medal for Merit', with a citation signed by the President, Harry S. Truman.

Grace recorded some of the 'heroic' stories that seemed to accumulate around her husband and were never denied. One senior officer told her with a straight face that a German submarine had been captured in Chesapeake Bay, near the Aberdeen Grounds, and that in it were found written orders from Hitler himself for the destruction of the Grounds and 'Dr Hubble'. He was still trying to impress people. According to Christianson (1995), he had a habit, if a visitor arrived while he was sorting out his mail, of throwing the letters into the wastebasket one by one. Later on, when alone, he would quietly retrieve them. He was a notorious procrastinator in the matter of answering letters anyway. He would also read about some obscure person in an encyclopaedia and then casually introduce this subject into the mess-table conversation just to show how knowledgeable he was. Of course, his companions realised what was happening and were not particularly impressed.

---

[30]One who believed that the United States should isolate itself from international affairs.

9.20   The post-war period

Hubble returned to California at the end of 1945 (Fig 9.7). Like many astronomers who had been involved in war work, he had not kept up with astronomical developments and people like Adams, still the Director of Mount Wilson, regarded him as out of date in his ideas. During the period of hostilities, Baade, who was a German national and neither wanted to nor could participate in the war effort, had had the Mount Wilson 100-in. almost to himself. He had made a series of brilliant discoveries concerning the age and chemical composition of stars and was now at the centre of astronomical progress. Hubble did not know of these developments and at a talk he gave in 1946 had asserted that astronomy had stood still during the war.

By the time that the observatory had returned to a fully functioning state, Adams was well over retirement age and a new director had to be found. Vannevar Bush, in addition to his other positions, was also the President of the Carnegie Institution and he consulted privately with Adams on suitable candidates. The Palomar 200-in. would soon come into operation, but under the aegis of the California Institute of Technology (Caltech) and the Rockefeller Foundation rather than Mount Wilson Observatory and the Carnegie Foundation. Whoever took over would need to forge an agreement between these two rival bodies. Hubble was top of the list of internal candidates for the Directorship, but had not impressed Bush by the way he had handled his wartime job. Clearly, he had no liking or aptitude for administration and lacked subtlety in human relations. Adams for his part felt that Hubble was too keen on his personal fame and was not loyal enough to the Carnegie Institution. Shapley was an outside candidate who had many good points and was considered seriously, but was also regarded as a publicity hound and was liked by neither of them. The President of the Rockefeller Foundation was another who thought Hubble too arrogant and self-serving to be appointed. Hubble's collaborator, Richard Tolman of Caltech, questioned how he would react if not appointed, but eventually decided it was better to risk the possibility that he might resign than to appoint him. They were certain, however, that Hubble was too dedicated to his life's work to abandon his access to the world's largest telescopes. In the end, the Caltech physicist Ira S. Bowen became the prime candidate. Although not formally an astronomer, he had solved the problem of the nature of 'nebulium' that had been around since the time of Huggins. Furthermore, he was regarded as more honest and straightforward than the other candidates.

Just after witnessing the first atomic explosion, the test firing at Alamagordo, New Mexico, Bush visited Pasadena and offered the job to Bowen who promptly accepted. To sweeten the pill for Hubble, they appointed him chairman of a 'Scientific Advisory Committee' which would work with the Director. They broke the news to him as diplomatically as possible but Hubble protested strongly to Bush (and Bowen!) against 'the appointment of a physicist as director of the astronomical center of the world' (Osterbrock 1992). Bowen mollified Hubble by reassuring him that he would continue as the leader of extragalactic research. In

FIG. 9.7. Hubble examining a photographic plate in 1946 (Carnegie Institution
of Washington).

addition, Bush raised his salary to $10,000, the same as that of the Director.

The normally conservative Hubble took a very strong anti-nuclear stand and
made many speeches (even to such groups as the Daughters of the American
Revolution) along the lines of 'The war that must not happen', in which he
clearly foresaw the ruin that would befall the world after an all-out nuclear war.
He was in favour of a world government backed by a powerful international police
system.

When the Palomar 200-in. telescope was inaugurated on 1 July 1948, the
initial idea had been to hold a short symposium of invited lectures on galaxies.
The speakers were to have been Hubble, Mayall, and Chandrasekhar (by then

involved in the dynamics of galaxies from a theoretical point of view). However, the organisers feared that Hubble would steal the show and decided that the telescope's builders should be honoured instead, with a single astronomical lecture by Baade.

Baade started by paying tribute to Hubble's work and discussing how it should be extended but he also laid emphasis on his own discoveries concerning the ages of the stars and how they were related to their chemical compositions (Osterbrock 2001). Hubble kept a low profile on the occasion, placing himself inconspicuously at the back of the gathering.

On his first observing run with the new telescope he photographed his old favourite, the variable nebula NGC 2261 (Fig 9.8).

Unfortunately, the 200-in. mirror and its mount were found to need some improvements and the instrument was taken out of service the following May. This was a great disappointment because, not only did the 200-in. have four times the light-gathering power of the 100-in., but Palomar was a much darker site than Mount Wilson, where the city lights of the Los Angeles basin were making faint nebulae harder to measure. Hubble was itching to use the new telescope on ever more distant galaxies for a 'third phase' of his investigations.

Hubble's hubris was such that he hoped his career would be rounded off by the award of a Nobel prize. He had become the best-known astronomer in the country, partly through the efforts of a publicity agent that he employed! His face appeared in many newspapers and magazines. In 1948, his picture was on the cover of *Time*, confirming his 'celebrity' status. However, the supreme scientific accolade, the Nobel prize, was not awarded to astronomers in those days, though his name was indeed put forward at about the time of his death. His most original work was well in the past, when he had discovered the Cepheid standard candles in the nearby galaxies. The study of these stars in ever more distant galaxies is still the Holy Grail for many astronomers.

9.21 Last years

In May 1949, Hubble telephoned Jesse Greenstein, the mainspring of the Caltech Astronomy Department, looking for a good student to help him with a new project. Greenstein suggested Allan Sandage, then a graduate student who had been at Caltech for a year. Sandage had started college in Miami University, Oxford, Ohio, had seen some service in the Navy, and had completed his degree in physics at the University of Illinois. He duly went to meet Hubble at his Santa Barbara Street office. He found him to be overly formal in manner, 'more an actor than a natural patrician' (Christianson 1995, p. 326) even when with Grace, to whom he was introduced later. It turned out his job would be to make galaxy counts from plates that had been taken at the Palomar Schmidt telescope, a new wide-angle instrument that was the best of its kind for making sky surveys.

Very soon afterwards, while on his annual holiday at Rio Blanco Ranch in July 1949, Hubble suffered a near-fatal heart attack. At first his life was despaired of, but he made a good recovery over the next few months. Grace tried to keep the

FIG. 9.8. Hubble posing in the 'prime focus cage' at the top of the 200-in. telescope. When used at prime focus, the telescope is at its most efficient. The light from a faint object is reflected only once, directly from the primary mirror, far below (Carnegie Institution of Washington).

seriousness of his condition secret. It was October before his colleagues saw him again at the office. Even then he looked 'awful' to Baade and was easily depressed (Osterbrock 2001). He was not allowed to risk working at the telescopes or to visit Rio Blanco on account of their altitude. By mid-1950, he was well enough to undertake a long summer trip to the East Coast, where he and Grace had an evening with their friend Aldous Huxley. Then they went on to France and England. The highlight of this trip was a stay with Sir Richard Fairey, the founder of Fairey Aviation, at his country house, Bossington, in the valley of the

Test, the favourite river of fly-fishermen.

On his return to Pasadena, Hubble asked Sandage to examine a set of plates of M31 and M33 that he had made over the previous thirty years. He was to measure a set of blue irregular variables that were among the most luminous stars of these galaxies and then extract their light curves. This he did to Hubble's satisfaction. These objects are among the most luminous stars known and were afterwards called the 'Hubble–Sandage Variables'. Their nature is even now not well understood. Sandage soon became Hubble's assistant, accompanying him whenever he went observing. Hubble described Sandage in a letter as 'a young astronomer of great promise and ability, who is working with me' (quoted by Christianson 1995, p. 342). Sandage was destined to become Hubble's successor and has devoted a good part of his subsequent career to continuing and extending his mentor's programmes.

At last, in October 1950, Hubble was allowed to make observations again. He seemed to be suffering no ill effects and characteristically complained about excessive informality at the Monastery dinner-table.

When Baade's work on the revision of the calibration of the Cepheid standard candles came out (1952), Hubble felt that the whole issue of the distances of the galaxies had to be investigated over again in the light of the new knowledge. He put in a bid for half of the time of the Palomar 200-in. telescope to be devoted to this question and the determination of redshifts. The decision of the time allocation committee to grant considerably less than this had to be put very gently to him at a meeting held in his own home.

Soon Hubble was fishing again, but at a somewhat lower altitude than his favourite Rio Blanco. In May 1953, he travelled abroad for the last time, to deliver the George Darwin Lecture on *The Law of Redshifts* at the Royal Astronomical Society (Hubble 1953). He also gave a lecture at the Royal Institution.

The Darwin lecture was given at an interesting time in the history of the subject. He was certainly aware of the concept of 'stellar populations' that Baade had introduced and of the revisions to the distance scale that had been announced the previous year at the Rome meeting of the International Astronomical Union. Further, the measurement of the brightness of stars and nebulosities could now be made more accurately through the introduction of a new technique—photoelectric photometry.

> As for the future, it is possible to penetrate still deeper into space—to follow the red-shifts still farther back in time—but we are already in the region of diminishing returns; instruments will be increasingly expensive, and progress increasingly slow. The most promising programmes for the immediate future accept the observable region as presently defined, hope for only modest extensions in space, but concentrate on increased precision and reliability in the recorded description.

His Darwin lecture turned out to be his swan song. In September 1953, he observed again at Palomar, for what was to be the last time. He had seemed to be well again and was given a clean bill of health by his doctor. Following his return from the telescope, he spent the morning of the 28th at his office but, while being fetched for lunch by Grace, suffered a massive stroke in his car, from which he died almost immediately.

He had long since ceased to attend any church and did not want any funeral service. His ashes were buried in a secret location.

The histrionic conclusion of his Darwin lecture is perhaps a fitting way to end this account of his life:

> For I can end as I began. From our home on the Earth, we look out into the distances and strive to imagine the sort of world into which we are born. Today we have reached far out into space. Our immediate neighbourhood we know rather intimately. But with increasing distance our knowledge fades, and fades rapidly, until at last the dim horizon we search among ghostly errors of observations for landmarks that are scarcely more substantial. The search will continue. The urge is older than history. It is not satisfied and it will not be suppressed. (Hubble 1953)

References

Adams, W.S., 1954. Obituary, Dr. Edwin P. Hubble, *The Observatory*, **74**, 32–35.

Christianson, G.E., 1995. *Edwin Hubble, Mariner of the Nebulae*, University of Chicago Press, Chicago, IL.

Douglas, A. Vibert, 1956. *The Life of Arthur Stanley Eddington*, Thomas Nelson and Sons Ltd., London.

Duerbeck, H.W. and Seitter, W.C., 2000. In Hubble's Shadow, in *Homage to Miklós Konkoly Thege*, eds Sterken, C. and Hearnshaw, J.B., C. Sterken, Vrije Universiteit, Brussel, pp. 231–254.

Eddington, A.S., 1932. The Expanding Universe, *Proc. Phys. Soc.*, **44**, 1–16.

Hetherington, N., 1986. Edwin Hubble: legal eagle, *Nature*, **319**, 189–190.

Hetherington N.S. and Brashear R.S., 1992. Walter S. Adams and the Imposed Settlement between Edwin Hubble and Adriaan van Maanen. *J. Hist. Astr.*, **23**, 53–56.

Hubble, E., 1922. A General Study of Diffuse Galactic Nebulae, *Astrophys. J.*, **56**, 162–199.

Hubble, E., 1925. N.G.C. 6822, A Remote Stellar System, *Astrophys. J.*, **62**, 409–433.

Hubble, E., 1926. Extra-galactic Nebulae, *Astrophys. J.*, **64**, 321–369.

Hubble, E., 1929. A Relation between Distance and Radial Velocity among Extra-Galactic Nebulae, *Proceedings of the National Academy of Sciences*, **15**, 168–173.

Hubble, E., 1936. *The Realm of the Nebulae*, Yale University Press, New Haven (Reprinted by Dover Publications, New York, 1958).

Hubble, E., 1937. *The Observational Approach to Cosmology*, Clarendon Press, Oxford.

Hubble, E., 1953. The Law of Redshifts, George Darwin Lecture, *Mon. Not. Roy. Astr. Soc.*, **113**, 658–666.

Hubble, E. and Humason, M.L., 1931. The Velocity–Distance Relation among Extra-Galactic Nebulae, *Astrophys. J.*, **74**, 43–80.

Humason, M.L., 1954. Edwin Hubble, (obituary), *Mon. Not. R. Astr. Soc.*, **114**, 291–295.

Humason, M.L., Mayall, N.U., and Sandage, A.R., 1956. Redshifts and Magnitudes of Extragalactic Nebulae, *Astr. J.*, **61**, 97-162.

Lemaître, G.E., 1927. Un univers homogène de mass constante et de rayon croissant, rendant compte de la vitesse radiale des nébuleuses extra-galactiques, *Ann. Soc. Sci. de Bruxelles*. **47**, Ser A, p. 49; also *Publications du Laboratoire d'Astronomie et de Géodésie de L'Université de Louvain*, IV, 101–111.

Lundmark, K., 1927. Studies of Anagalactic Nebulae, *Medd. Astr. Obs. Uppsala*, No. 30.

Mayall, N.U., 1954. Edwin Hubble, Observational Cosmologist, *Sky and Telescope*, **13**, 78–81 and 85.

Mayall, N.U., 1973. Milton L. Humason—Some Personal Recollections, *Mercury*, **2**, (1), 3–8.

Osterbrock, D.E., 1992. The Appointment of a Physicist as Director of the Astronomical Center of the World, *J. Hist. Astr.*, **23**, 155–165.

Osterbrock, D.E., 2001. *Walter Baade, a Life in Astrophysics*, Princeton University Press, Princeton, NJ.

Osterbrock, D.E., Brashear, R.S., and Gwinn, J.A., 1990. Self-made Cosmologist: the Education of Edwin Hubble, *Astr. Soc. Pacific Conference Ser. Vol 10, Evolution of the Universe of Galaxies; Edwin Hubble Memorial Symposium*, ed. Kron R., Astr. Soc. Pacific, San Francisco; reprinted (with illustrations) as Young Edwin Hubble, *Mercury*, January 1990, 2–14.

Payne-Gaposchkin, C., 1984. *An Autobiography and Other Recollections*, ed. Haramundanis, K., Cambridge University Press, Cambridge.

Robertson, H.P., 1954. Edwin Powell Hubble 1889–1953, *Publ. Astr. Soc. Pacific*, **66** 120–125.

Sandage, A., 1958. Current Problems of the Extragalactic Distance Scale, *Astrophys. J.*, **127**, 513–526.

Sandage, A., 1997, Beginnings of Observational Cosmology in Hubble's Time: Historical Overview, in *The Hubble Deep Field*, eds Livio, M., Fall, S.M., and Madau, P., Cambridge University Press, Cambridge, pp. 1–26.

Sarton, G. 1927. Introduction to the History of Science, **I**, 3, Carnegie Institution of Washington; Williams & Williams, Baltimore, MD.

Shapley, H., 1969. *Ad Astra per Aspera; Through Rugged Ways to the Stars.* Charles Scribner's Sons, New York.

Smith, R.W., 1982. *The Expanding Universe: Astronomy's "Great Debate 1900–1931"*, Cambridge University Press, Cambridge.

INDEX